FIRE SERVICE PERSONNEL MANAGEMENT

FIRE SERVICE PERSONNEL MANAGEMENT

3rd Edition

Steven T. Edwards

Pearson

Boston   Columbus   Indianapolis   New York   San Francisco   Upper Saddle River
Amsterdam   Cape Town   Dubai   London   Madrid   Milan   Munich   Paris   Montreal   Toronto
Delhi   Mexico City   Sao Paulo   Sydney   Hong Kong   Seoul   Singapore   Taipei   Tokyo

Library of Congress Cataloging-in-Publication Data

Edwards, Steven T.
  Fire service personnel management / Steven T. Edwards. — 3rd ed.
    p. cm.
  ISBN 0-13-512677-0
  1. Fire departments—Personnel management.   I. Title.
  TH9158.E3 2010
  363.370689 3—dc22

                      2009018722

Publisher: Julie Alexander
Editor in Chief: Marlene McHugh Pratt
Senior Acquisitions Editor: Stephen Smith
Associate Editor: Monica Moosang
Development Editor: Anne Marie Masters, iD8 Publishing Services
Editorial Assistant: Heather Luciano
Director of Marketing: Karen Allman
Executive Marketing Manager: Katrin Beacom
Marketing Specialist: Michael Sirinides
Marketing Assistant: Judy Noh
Managing Production Editor: Patrick Walsh
Project Liasion: Julie Li
Senior Media Editor: Amy Peltier
Media Project Manager: Lorena Cerisano
Production Editor: Lisa S. Garboski, bookworks publishing services
Manufacturing Manager: Ilene Sanford
Manufacturing Buyer: Pat Brown
Senior Designer Coordinator: Christopher Weigand
Manager, Visual Research: Beth Brenzel
Manager, Rights and Permissions: Zina Arabia
Image Permission Coordinator: Vickie Menanteaux
Manager, Cover Visual Research & Permissions: Karen Sanatar
Cover Designer: Bruce Kenselaar
Cover Photo: Ray Kemp/911 Imaging
Composition: Aptara Inc.
Printer/Binder: Courier Kendallville
Cover Printer: Lehigh Phoenix Color

Credits and acknowledgments borrowed from other sources and reproduced, with permission, in this textbook appear on appropriate page within text.

10 9 8 7 6 5 4 3 2

www.pearsonhighered.com

ISBN: 10: 0-13-512677-0
ISBN: 13: 978-0-13-512677-6

CONTENTS

Chapter 7 Training and Development 160

Chapter 8 Performance Appraisal 202

Chapter 12 Productivity and Performance 321

PREFACE

Not too long ago, I was sitting in my office working on a new personnel-related policy for our department. At the time, I was fire chief of the Prince George's County Fire Department, a large metropolitan fire department adjacent to Washington, DC. As I was thinking about the new policy, my executive assistant came in and informed me that there was a third alarm fire in the southern portion of the county. I could continue to work on the personnel matter or respond to the fire, even though I knew that my presence was not really needed there. In an instant, I was out the door to the emergency.

I returned several hours later and sat back at my desk to finish some departmental business. The new personnel policy was still there as I had left it. I began to think, Which was the more important function for me to perform? Going to the multiple alarm was fun; I got to see people I did not see on a regular basis and did a couple of press interviews about the incident. Our officers are highly qualified and can handle these types of incidents very well. My presence at this emergency was not really needed; it was a large fire, but relatively routine. I was thinking, Fire response is great; you make decisions with no committees and firefighters do what they are told with no questions asked. It is challenging, dynamic, and you get instant feedback on how well you are doing. Too bad personnel matters were not as easy.

I then looked down and thought about the new personnel policy. It occurred to me that this new policy would have substantial long-term value to the department. It would improve our management and our ability to work more closely with our most valuable resource—the members of the fire department. If fire officers put as much effort into personnel management as they did in response to emergencies, think how much better we would all be. I knew that the personnel policy would significantly enhance our capability of dealing with a host of complex and challenging issues. I felt guilty for going to the fire because I should have done what was more important. The new personnel policy was finished before I left work late that evening.

Much of the reason for my writing *Fire Service Personnel Management* was the lack of specific information available to fire departments on this topic. One can find bits and pieces about personnel management in fire service management texts, but not a concentration on this very important subject. I wanted to be able to contribute to what I have found to be the most important resource in any organization—people. It seemed logical that if we always say this, then we should be able to study personnel management as it specifically relates to the fire service environment. If we are successful in managing the people in our organizations better, that success will spread to other departmental functions and magnify their successes.

This book is for fire department officers and prospective officers—not management generalists. It uses established personnel management concepts and examines them as they directly relate to fire departments. Therefore, this book is dedicated to the improvement of personnel management of fire departments and, in the process, enriching the work environment of fire department employees. I hope that I have made a small contribution to this very important task.

Wherever you see the words *fire department* in this text, it means a fully functional department that provides an array of services to the public including fire suppression, emergency medical services, hazardous materials response, fire code, investigations, public education services, and others. *Fire department* in this context is not meant to be restricted to those that provide fire service only. The word *employees* in this book means any member of the fire department, be they career or volunteer, uniform or civilian.

There is a certain feeling of success and pride that one gets when he or she extinguishes a fire or makes a rescue. I hope that each of you has that same feeling when you successfully manage a difficult personnel situation. Good luck.

Please note that the Review Questions and Case Study at the end of every chapter are intended for class discussion and review. Answers may be found within the chapters. An Answer Key is not provided.

ACKNOWLEDGMENTS

My entire professional life has been devoted to the fire and rescue service, but it could never be as important as my family. Therefore, I dedicate this book to my children, Michelle, David, and Sara, with a special thanks to my wife, Angela Edwards.

Many experiences related in this book are from my career in the Prince George's County Fire Department and the Maryland Fire and Rescue Institute. Both of these excellent organizations and their people have educated and exposed me to fine examples of personnel management. To all of my coworkers, past and present, thanks for the opportunities and the cooperation.

List of Reviewers

Bernard W. Becker III
Fire Science Instructor
University of Cincinnati
Cincinnati, OH

Richard L. Bennett
Assistant Professor of Fire Protection
Department of Public Service Technology
University of Akron
Akron, OH

Craig Bryan
Fire Captain
City of Forest Park Fire Department
Forest Park, OH

Donald E. Bytner
Instructor
Department of Law Enforcement and Justice Administration
Western Illinois University
Macomb, IL

Bruce Evans, MPA
Fire Captain, Henderson Fire
Fire Programs Coordinator, Community College of Southern Nevada
Las Vegas, NV

Alejandro Garcia
Fire Lieutenant/Licensed Paramedic
Wichita Falls Fire Department
Wichita Falls, TX

Tommy Hill
Fire Science Faculty Leader
Alabama College System

Chattahoochee Valley Community College
Phenix City, AL

Charles H. Kime, Ph.D.
Professor of Practice, Fire Service Programs Coordinator
Arizona State University
Mesa, AZ

Judith Kuleta
Director
Bellevue Community College
Bellevue, WA

Jeffrey Lindsey, PhD, EMT-P
CFO Director Graduate Studies
George Washington University
Washington, D.C.

Richard A. Marinucci
Fire Chief
City of Farmington Hills
Farmington Hills, MI

James D. Piech, Ret. Bn Ch., M.S.
Instructor/Fire Science Instructional Chair
Milwaukee Area Technical College
Oak Creek, WI

Kenneth A. Teague, NR-EMT-P, BS
Educational Coordinator
Caldwell County Emergency Services
Lenoir, NC

LaRon Tolley
Distance Education Manager
Western Oregon University
Monmouth, OR

Karen Zucco-Gatlin
Human Resource Supervisor
Department of Human Resource Services
Washington State University
Pullman, WA

ABOUT THE AUTHOR

Steven T. Edwards is the director of the Maryland Fire and Rescue Institute of the University of Maryland. The Maryland Fire and Rescue Institute is the state's comprehensive training agency for emergency services, training more than 38,000 students each year. He is a former fire chief of the Prince George's County Fire Department in Maryland, where he served for 25 years in a variety of positions from high school cadet to fire chief. Edwards also served as past chair of the board of directors of the Safety Equipment Institute, as past chair of the Congressional Fire Service Institute National Advisory Committee, as well as in numerous local- and state-level appointments. In 1997, he was elected as the president of the North American Fire Training Directors.

Edwards is a graduate of the University of Maryland University College (UMUC) with a bachelor's degree in fire service management and a master's degree in general administration. Both degrees were achieved with summa cum laude honors. He has attended the Harvard University John F. Kennedy School of Government's Program for Senior Executives in State and Local Government and the National Fire Academy, and has presented at national conferences and seminars. Edwards is currently an Associate Professor of the adjunct faculty at UMUC, teaching its course in fire service personnel management.

During his career of more than 40 years in fire service, Edwards received numerous awards and honors, including the Prince George's County Fire Department "Gold Star of Valor" in 1979 for the rescue of two firefighters at a major fire and explosion. During his tenure as fire chief, the department received the IAFC Award for Excellence as well as twenty-eight National Association of Counties Awards for Excellence. While continuing his fire service career at the University of Maryland, the Maryland Fire and Rescue Institute was selected as the Congressional Fire Service Institute National Fire Service Organization of the Year for 1999. In addition, Edwards received the University of Maryland President's Distinguished Service Award in 2003 for exceptional performance, leadership, and service. At the 2004 International Fire Service Training Association (IFSTA) Spring meeting, Edwards was presented with the "Hudiberg" award, the highest honor within this association. In 2009, Edwards was awarded the CFSI Mason-Lankford National Fire Service Leadership Award for exceptional service throughout a career and for his improvements of the fire and emergency services.

The following grid outlines the course requirements of the Personnel Management for the Fire and Emergency Services course developed as part of the FESHE model curriculum. For your convenience, we have indicated specific chapters where these requirements are located in the text.

Course Requirement	1	2	3	4	5	6	7	8	9	10	11	12	13
Identify and explain contemporary personnel management issues.	X	X	X	X	X	X	X	X	X	X	X	X	X
Explain potential personnel management issues.	X	X	X	X	X	X	X	X	X	X	X	X	X
Classify the collective rules, procedures, laws, and policies that relate to personnel management issues.	X	X	X	X		X			X	X	X		
Analyze simple/complex personnel management issues from recruitment to retirement.		X			X	X		X	X	X	X	X	
Formulate recommendations and solutions to personnel management issues.	X	X	X	X	X	X	X	X	X	X	X	X	X
Explore organizational development and leadership styles and how they relate to personnel relationships.	X	X								X	X	X	X

New to This Edition

Students will benefit from a new four-color design and a variety of new content in this edition. Changes to this edition include:

- Correlation guide to FESHE curriculum learning objectives
- Updated color photos and illustrations throughout the text
- Learning objectives and key terms identified at the beginning of the chapters
- References to the NFPA 1021, Standard for Fire Officer Professional Qualifications, for each chapter
- Real life examples and experiences throughout the text and a Case Study for each chapter
- New information regarding managing diversity and compliance programs in the fire service in Chapter 2
- A new section on Volunteer Fire Department Recruitment/Retention in chapter 5
- Expanded discussion on selection techniques for hiring and promotion in Chapter 6
- Appraisal and feedback discussed under "Performance Assessment Techniques"
- Expanded information on training safety and the role of the instructor in chapter 7
- Expanded sections on the Candidate Physical Ability Test and Peer Fitness Training Certification under the IAFC/IAFF Joint Labor Wellness/Fitness Initiative topic in Chapter 10
- Additional information on Enlightened Labor Relations in Chapter 11, as well as grievances and arbitration proceeding
- A complete new chapter on "Professional Development"
- Additional student review material and instructor resources may be accessed via www.myfirekit.com. MyFireKit offers a wealth of tools to reinforce student learning and comprehension, including quizzes, audio glossaries, scenarios, links, videos and animations, and more. MyFireKit also includes a full suite of supplements just for instructors, including Instructor Resource Manuals, Test Banks, and PowerPoint slides.

FIRE SERVICE PERSONNEL MANAGEMENT

1

Overview of Personnel Management

MFRI Archives.

OBJECTIVES

After reading this chapter, you should be able to:

- Describe various elements of personnel management.
- Describe team-based organizations.
- Explain how values affect personnel management.
- Understand ethics and ethical situations.
- Explain how to establish an effective relationship with the personnel office.

Managing human resources is a critical task in any organization, large or small, regardless of the product or service. The more effectively an organization manages its human resources, the more successful the organization will be. The reality of this statement has been verified over the years by numerous studies across industry lines and organizational structures. Managing employees requires the coordination of many human resource activities. Every personnel activity implemented, whether it is a value, a policy, or a practice, sends a message to employees and thereby has consequences for the organization. The art of managing human resources is complicated further by the complex and changing environment in which it takes place.

Personnel or *human resource management* is the term that refers to the philosophy, policies, procedures, and practices related to the management of people within an organization. Fire service personnel management, or personnel functions as they specifically relate to fire service organizations, is the focus of this book.

Often, personnel management is not an easy task because there are undoubtedly many more exciting things to do in a fire department. Many in the fire service naturally gravitate toward the activities associated with the response to emergencies, rather than toward the personnel management of fire service organizations. Emergency service is the reason they joined the fire department and why they stay. However, as the reader will learn in this text, personnel management of the fire department that is conducted improperly can jeopardize response to emergency incidents and other departmental efforts.

The fire service uses a variety of technology, tools, and equipment; for example, pumpers capable of moving thousands of gallons of water per minute, aerial platforms capable of reaching great heights, and thermal imaging devices capable of finding victims in smoke-filled rooms. Behind each piece of equipment and each emergency response is a person who needs to be hired, trained, evaluated, and motivated to do the job well. Having individual firefighters ready and able to do their jobs is what fire service personnel management ensures. It involves creating an organizational environment that values people and what they can contribute to the success of the department. It necessitates fire service personnel managers who are capable of dealing with a host of complex, challenging, and at times controversial personnel issues.

Just who are these fire service personnel managers? In essence, they are any fire service officer or person of authority that has supervisory responsibilities within the fire department. Company level officers, such as lieutenants and captains, discharge numerous personnel management responsibilities every shift. In many circumstances these individuals have the greatest opportunity for success regarding the firefighters whom they supervise because they have the most direct impact as they implement the personnel policy of the organization of which they are a member. As one moves higher in the organization toward the chief officer level, the personnel management responsibilities increase and become more complex. At this level one may have the need to develop and/or work very closely with the **personnel department** to develop policies and procedures that affect the employees of the fire department in a substantial manner. It is clear that personnel management is not just the responsibility of the personnel department; it is also a very important responsibility of every fire department officer and should be viewed as such.

personnel department
■ an agency of government responsible for personnel management and includes at a minimum the activities of staffing, retention, development, adjustment, and managing change.

Behind successful and innovative fire departments, one will find successful and innovative people; behind the department and the people, one will discover purposeful and enlightened personnel management. It is in the personnel management of the fire department that all of the activities come together to guarantee the success that others admire. If the basic personnel functions are not done well, then frustration and mistrust prevail, setting the stage for poor performance and ineffective public service. The charge of a fire service personnel manager is to provide the leadership and motivation to prevent such a situation, which comes through education and being fully aware of the importance and the proper implementation of effective personnel management functions. Leadership in this area can make the professional lives of one's colleagues much more productive and enjoyable.

This book is dedicated to improving the personnel management of fire departments and, in the process, enriching the work environment for firefighters and fire officers. When an individual takes on the role of fire officer, that person assumes a solemn responsibility to fellow firefighters, the fire department, and the public. This responsibility extends as much to counseling employees on performance or other personnel functions as to circumstances such as when one is on a building fire and must be constantly looking for safety hazards. The fire officer has a tremendous responsibility to perform many functions effectively. The personnel assigned to the officer deserve no less than the officer's best effort. Personnel management should not be left to some other government agency to perform. Taking charge and assuming responsibility are qualities that will improve the personnel management within a department. Hopefully, this book provides some of the tools necessary to accomplish this.

Elements of Personnel Management

The large organization with its highly specialized personnel or human resources department is a relatively recent development. Until the time of the Industrial Revolution in the nineteenth century, organizations were small and operated without personnel specialists and departments. The functions of personnel management were performed by the owner of the business or perhaps by a few selected supervisors, which is still true for most small businesses today. During the twentieth century, a number of trends developed that placed increased emphasis and importance on the elements of personnel management.

One of the initial and most pronounced trends to affect personnel management was the growth in the size of organizations. As organizations grew to several thousand employees, there was a need for more formal personnel management policies and procedures to accommodate this increase. Many organizations grew from local and regional companies to national and multinational corporations during this period, a trend that continues today. It is difficult to pick up a newspaper without reading about an acquisition or merger of a major company to form a larger organization. Accompanying this growth in organization size was a rapid expansion of government regulation of organizations. For example, the Sixteenth Amendment to the U.S. Constitution, which gave the federal government the right to tax income and created the law that established the Social Security system, led to a host of personnel-related requirements. Major legislation affecting employees was enacted, such as the National Labor Relations Act of 1935, which gave employees the legal right to unionize and bargain collectively, and the Fair Labor Standards Act of 1938, which established regulations on work hours and other limits on employers. Substantial legislation and oversight affecting personnel management now exist at the federal, state, and local levels.

Another major trend that has affected personnel management is a result of the changes in the social environment. The changes in Western society in the past century have been remarkable. Research on the social needs of employees led to the human relations movement, and employers turned to human resource professionals to design more employee-oriented personnel policies. The result was the evolution of personnel policies regarding working conditions and employee benefits beyond even the most ambitious dreams of the union organizers of the 1920s. Recent increases in employee educational levels and expectations have prompted many organizations to become even more sophisticated in their approach to personnel management. Major changes in the social environment and the workplace can also be traced to the civil rights movement that led to the Civil Rights Act of 1964. The subsequent laws and regulations from this era generated many new protections in the employment arena for previously disadvantaged groups such as minorities, women, and people with disabilities. In addition to becoming increasingly racially and culturally diverse, the workforce also included a greater proportion of dual-worker families and single parents who have special personnel management needs.

Improvements in technology represent a major transition as we have entered the twenty-first century. The amount of information that average employees currently have at their fingertips is staggering. Employees can do more in a home office than entire departments of many companies could do as recently as the 1970s. This access to technology has led to new educational requirements and skills for employees and has granted them more freedom than in any other generation. Rapid technological change has created new jobs, while significantly altering some existing jobs and eliminating others. Technology has also increased the oversight of the workforce by allowing for improved quality control and focus on deficiencies. One of the foundations of the total quality movement inspired by W. Edwards Deming is the use of statistical quality controls for manufacturing processes. With advanced computer technology, genetic research, and virtual reality, personnel management will have to move quickly to stay abreast of new developments and maintain quality operations.

process
■ an identifiable flow of interrelated events progressing toward some goal or consequence.

system
■ a particular set of procedures designed to control a process in a predictable way.

Another way of viewing personnel management is through the use of process–systems terminology. A **process** is an identifiable flow of interrelated events progressing toward some goal or consequence. An example would be the promotional process of a fire department, which includes a flow of events involving recruitment, training, testing, and evaluation. A **system** is a particular set of procedures designed to control a process in a predictable way. The promotional system of a fire department might include policies on tenure, eligibility, application steps, and selection devices. Therefore, the term *process* refers to a combination of events that leads to some end result, and the term *system* identifies specific procedures and policies used to control those events.[1] The process–systems view takes into account the interdependence of all aspects of personnel management and recognizes the relationship between personnel management activities and organizational goals.

Personnel Management: Today and Tomorrow

As we move forward in society, personnel management is becoming more complex and its focus is changing. Fire service personnel managers must increasingly have better skills in order to deal with the future issues they will face. How people are managed makes a huge difference in the success of organizations. Motivated employees provided with the tools and training they need are making a big difference and will continue to do so in the future.

Technological change, globalization, diversity, and political instability will continue to change the way people are managed within an organization. Successful organizations of the future will be efficient, responsive, diverse, and flexible. Most people in the fire service today would not describe fire departments as possessing many of these qualities. Therefore, fire service personnel managers will have to contribute toward reversing the policies and practices that may detract from the potential of fire service organizations to achieve as much as they can in the future.

World events will continue to have a major impact on the management of fire service organizations. No one can dispute that the events of September 11, 2001, have changed the fire service. In a single day, fire departments went from planning and discussing terrorism response to actually implementing it, with disastrous consequences for the Fire Department of New York. Aside from the operational aspects of this situation, the personnel management issues will continue to evolve in terms of employee assistance programs, disability claims, training, and other personnel-related circumstances. The consequences for the people who responded and dealt with these events may be severe and prolonged. The long-term implications regarding the firefighters and fire officers who responded to these disasters must be studied and the information from these studies utilized to assist fire departments of the future, especially as it relates to safety and occupational health.

Team-Based Organizations

New personnel management methods will be necessary to deal with fire service personnel of the future. Fire departments are going to have to become more adept at reacting quickly to technological and societal changes. The successful organizations of the future will share many of the following traits in terms of how they manage people.[2]

New organizations stress cross-functional teams and interdepartmental communication. The fire service has an advantage here because its history and culture value and promote teamwork. However, in the future some of this may mean less emphasis on the "chain of command" to get decisions made.

EMPOWERED DECISION MAKING

Most jobs in the future will require constant learning, higher-order thinking, and increased employee commitment, resulting in more employee empowerment and a less rigid working environment. Managers in the future will increasingly focus on how best to ensure that the front-line employees can do their jobs. This is a change from the concept that managers are there to oversee and control the workers' actions. Organizations with rigid chains of command, such as fire departments, will be stressed in this environment. The goal will be to maintain the chain of command because it has very essential features, but at the same time increase flexibility and empowerment where appropriate.

FLATTER ORGANIZATIONAL STRUCTURES

Instead of pyramid-shaped organizations with seven or more layers of management, flat organizations with four or fewer layers will be more prevalent. The remaining managers will have more people to supervise and will be less able to monitor the work of employees closely. In my opinion, numerous fire departments in particular have too many layers of "chief" officers. These layers need to be reduced and more authority and latitude must be given to the battalion and company-level officers. However, this means these officers will have to be properly trained in order to accept increased responsibility.

NEW BASES OF MANAGEMENT POWER

Noted management theorist Rosabeth Moss Kanter indicates that leaders can no longer rely on their formal authority to get employees to follow them. Success depends on tapping into good ideas, figuring out whose collaboration is needed to act on those ideas, and then working with both to produce results. Managerial work will be based on very different ways of securing and using organizational power. Fire service managers will have to win the respect and commitment of their highly trained and empowered employees of the future. Managers will have to embrace and become "change agents" for their organization.

KNOWLEDGE-BASED ORGANIZATIONS

There is no question that organizations and work of the future will be more knowledge based. This has been a clear trend for some time now. Employees will have to be well trained in order to exercise individual judgment and decision making. Many fire chiefs who now think of themselves as the "boss" may have to rethink this approach and consider themselves instead as a "sponsor, team leader, or internal consultant." The old autocratic style of leadership will be gone, as will the tendency to hoard information and make decisions alone. Fire departments will seriously have to consider their "single-point" entry systems and allow more latitude for people of sufficient knowledge and education to join the organization at different levels.

EMPHASIS ON VISION AND VALUES

Formulating a clear vision and values to which employees can commit themselves is more important than ever. Fire service managers will have to communicate clear values as to what is important and what is not. For example, this may affect how emergency operations are conducted, especially when there is no life hazard present. If the fire service is going to improve its record regarding firefighter fatalities and injuries, some of these **values** must be reexamined.

values
■ serve as criteria to judge the actions of individuals and organizations.

LEADERSHIP IS THE KEY

In my opinion it has always been about leadership. Empowerment, knowledge-based workers, and getting firefighters to think like the fire chief will place a premium on the people aspect of managing. Good leadership ensures the loyalty of followers and a commitment to the organization that breeds success. Regardless of the level of resources, political circumstances, or external and internal challenges, strong and effective leadership has stood the test of time.

TECHNOLOGY

Technology is always advancing, and in the future, fire service organizations must become more adept at using more sophisticated types of technology. Just think about how differently we communicate within organizations today; for example, e-mail is a relatively recent development. Real-time reporting systems, virtual reality training systems, and information-centric command systems all will play major roles in the future of fire departments. The people that serve as firefighters must be able to deal effectively with increasingly higher levels of technology in the work environment.

Are fire service organizations up to the challenges of the future? They must be, because there is no other real choice. Fire departments are going to have to change their structures and how they deal with the people who make up their organizations. Different leadership styles and a rearrangement of the focus of the department will occur. The best course of action is to be proactive and learn about these situations, and then do a good job to bring about change where warranted.

Fire service personnel management is the systematic control of a network of interrelated processes affecting and involving all members of an organization. To control and improve these processes, personnel systems are planned, developed, and implemented through the combined efforts of fire department officers and the personnel experts within their organizations. The processes and systems at the department level are part of and further controlled by the processes and systems at the local or state government level. The identification and study of personnel processes as they relate to fire departments form the basis of this book. The processes or elements of personnel management that we will endeavor to study include a chapter on each of the following:

Workforce Issues of the Twenty-First Century. The demographics and challenges of the workforce of the future require an examination of the benefits of diversity and workforce issues specific to fire departments that may present complications to management.

Legal Issues. The law and personnel management are entwined in many ways. To prevent adverse legal judgments and be proactive, personnel policies and procedures must be in full compliance with the law. The legal basis for personnel-related decisions as well as topical issues such as sexual harassment will be reviewed.

Job Analysis and Design. Job analysis, as the basis for many personnel functions, is clearly one of the most important building blocks of personnel management and deserves a high priority in any organization.

Fire Service Recruitment. This chapter focuses on the process of recruitment with an emphasis on selective recruitment techniques. It involves attracting individuals on a timely basis, in sufficient numbers, and with the appropriate qualifications.

Selection for Employment and Promotion. Most managers admit that employee selection is one of their most complicated and important decisions. The criteria against which an applicant will be evaluated, or the particular standards used in the selection process, must be chosen with great care. Relevant legal and administrative guidelines must be followed in this process to ensure validity and reliability. The hiring and promotional processes used in fire departments will be examined.

Training and Development. Training, which is a planned effort by an organization to facilitate the learning of job-related knowledge and skills by its employees to improve performance and achieve organizational goals, is of particular importance in fire departments due to the dangers involved in the response to emergency incidents. Testing, evaluation, and certification systems will also be explored.

Performance Appraisal. The formal, systematic assessment of how well employees are performing their jobs in relation to established standards includes communication of that assessment to the employee and the organization. The goal of the performance appraisal process is to improve the quality of work and the individual employees involved in that work. Several techniques are examined that have been proven successful in the performance improvement process.

Discipline. Organizations need control and a way to address deficient performance in a fair and equitable manner. Progressive discipline, as well as methods to deal with the few employees who require this type of action within an organization, will be reviewed.

Health and Safety. Much more emphasis is rightfully placed on the health and safety of the workforce today. The role of the Occupational Safety and Health Administration (OSHA) and that of other legal and national standard-making organizations are studied in this chapter, with the focus on how to develop programs aimed at the prevention of accidents and injuries.

Labor Relations and Collective Bargaining. The career fire service has a higher percentage of union membership than do most occupations. Labor relations is an important aspect of personnel management and is necessary for the success of a manager in a fire department. A review of labor history and current techniques to develop a productive partnership with unions are examined, as well as the collective bargaining process.

Productivity and Performance. Measuring and managing performance are challenging tasks and keys to organizational effectiveness. These important issues as they relate to fire service organizations are explored.

Fire Service Professional Development. Individual and organizational opportunities are important to extend training and education to higher levels by developing personnel on a continuing basis. How to create and manage professional development for the long-term growth and success of the organization will be examined.

Discipline   Workforce Issues

Health & Safety   Legal Issues

Performance   Job Analysis

Labor Relations   Fire Department Personnel   Recruitment

Training   Selection

It is important to understand these elements of personnel management separately; but even more important is how to blend them together to create a motivated and productive fire department. Success as a fire service personnel manager depends on one's ability to combine these elements effectively in the current work environment.

Values and Personnel Management

Public personnel management, as a field of public administration, has seen major improvement in the last 50 years. Personnel professionals now have at their disposal a litany of techniques backed by research and analysis to use as they deem appropriate within their organizations. Personnel management in the public sector is different from that in the private sector. For example, in the public sector, the law, in many cases, is applied differently to government entities, the motivations and aspirations of employees often are different, and the values are different. These values affect the manner in which policy issues are debated and implemented within government agencies. The noted scholars Donald E. Klingner and John Nalbandian (1985) observed that "as a culture, we all share certain values that are particularly relevant to an understanding of the methods and constraints in public personnel systems." These values serve as criteria to judge the actions of government. The values shared by many public agencies are as follows:

Rights of Individuals. This highly important value is in many ways the defining right of our society. These rights are protected from arbitrary government actions. The judicial branch of government actively protects these rights in our society.

Administrative Efficiency. Services performed in the public sector should be as efficient and as cost effective as possible. Government agencies should make decisions based on merit and for the good of the public at large as opposed to specific individuals.

Political Responsiveness. In a democratic form of government the responsiveness of the government to the public is of paramount concern. Government is to represent the will of the people and be responsive to their desires.

Social Equity. The pursuit of this value ensures that opportunities are fairly distributed and that those who are deprived receive special accommodation if needed in the process of government.[3]

These values serve as benchmarks and provide the direction for the development of government policy, regulations, and laws. They are influenced by the impacts of the economic, political, social, and technological conditions during the time period in which they are reviewed. As one studies the history of personnel management in the United States, these values dominate at certain times depending on the conditions in society. For example, the Pendleton Act of 1883 was passed to prevent the widespread **political patronage** that existed within the federal government. Political patronage was rampant at the time, with government jobs being decided totally on the basis of political loyalty. Relevant to the fire service profession is that one of the largest political machines and controller of political patronage at the municipal level was the infamous Boss Tweed. He rose to power as a volunteer fire chief in New York City in the late 1800s. Today, with the use of the merit system for employment in government jobs, the use of political patronage is reserved for a minimum of positions.

Values within organizations can be in conflict depending on the conditions that exist in society. During the 1960s and beyond, the value of social equity became prominent with affirmative action and other efforts to provide equal opportunity for all citizens. Many felt that these programs conflicted with the value of individual rights, in that more attention was placed on race or culture at the expense of individual ability and on being judged solely by merit for a position. This issue has had a very tumultuous past and remains a topic of heated debate today in many political and judicial arenas. It is just one example of how these values conflict at times. Often the meaning of individual rights, administrative efficiency, political responsiveness, and social equity are judged from the eye of the beholder. Fire service personnel managers need to be aware of these issues because they affect a fire department in substantial ways.

Organizations that operate effectively over a long period of time, generally have shared values among the employees. Values in this sense are the conscious desires and wants of people that guide their behavior, both on and off the job. **Organizational values** define the acceptable standards that govern the behavior of people within the organization. Without such values, individuals may pursue behaviors that are more in line with their individual value systems as opposed to that of the organization. This results in conflict and inappropriate behaviors. An example would be the value of safety, which may be disregarded at the individual level by a firefighter and result in unnecessary death or injury. The organizational value of increased safety must be the standard.

For the organization's values to have meaning, they must be congruent with its purpose or mission. In this way the values provide a framework for all members

political patronage
■ a process whereby government jobs are decided on the basis of political loyalty.

organizational values
■ define the acceptable standards that govern the behavior of people within organizations.

of the organization that encourages decisions and behaviors that support the organizational goals and objectives. A mission or vision statement is just paper, unless the values espoused are the daily and accepted manner of accomplishing the work. The goal is to have the organizational values shape the culture of the organization. Here are four suggestions on how to accomplish this:

1. Communicate the values at every opportunity—Both internal and external communications should express the values of the organization. Frequently refer to the values in making decisions and enroll new employees into the value system.
2. Promote contemporary values—Revisit the values as appropriate and update as necessary using an inclusive process. If the values are well stated, then change should not be frequent, but it is a good process to go through.
3. Confront contradictory behavior—If employees are allowed to display contradictory values, then over time the stated organizational values will become negated. Appropriate feedback and response are necessary to maintain the values of the organization. An example would be a fire department that constantly espouses safety, but in reality allows unsafe acts to occur on a frequent basis.
4. Obtain feedback on values—This should occur both inside and outside of the department. Engage people about what their view is of the values of your organization from a number of different perspectives.[4]

Values are formed from a number of experiences and not all values are equally as strong. The stronger the values are, the more reluctant people are to change, and weaker values are more readily changed based upon new experiences. A leader's effectiveness is, in part, measured by his or her ability to influence the activities of subordinates. Value conflict can exist at the organizational or personal level. When values clash and the leader is not able to resolve the differences, his or her effectiveness as a leader is diminished. To that end, leaders are more likely to be effective in influencing behavior toward the organizational values when they themselves exemplify the organization's values.

It is important to understand and consider the values of people whom one supervises. Understanding their values, desires, and expectations will increase one's leadership ability, as well as the ability to motivate. The leader's goal is to understand these values as they will assist in channeling subordinates' motivation and behavior toward the values of the department. The understanding and consideration of organizational values enable a leader to be more effective in supporting these appropriate values. This will certainly be noticed by upper management in a positive way and appreciated by all members of the department.

Ethics

Knowing what is right or wrong and behaving accordingly are the essence of ethical behavior. **Ethical behavior** can influence one's success, and it is essential for people to have trust in business and personal relationships. For example, in a survey of chief financial officers, they ranked ethical behavior as the most important personal attribute of graduates.[5] With all that is currently happening in the public

ethical behavior
▪ an individual's character traits, such as honesty, compassion, and courage, and as a standard of moral behavior that is accepted by society as being right.

and in government, some confuse not getting caught at doing something illegal with ethics. The "ethical" question unfortunately seems to have become, "Was it legal?" Ethics and legality are two different concepts. Although following the law is an important first step, ethical behavior requires much more. Many unethical acts fall well within our laws.

ethics
■ from the Greek word *ethos* meaning character. It is a composite of good moral qualities, self-discipline, and judgment.

Ethics comes from the Greek word *ethos,* meaning character. *Character* is defined as a mark or sign of distinctive quality. It is a composite of good moral qualities, typically including moral excellence, self-discipline, high ethics, and judgment. Ethical behavior has been described as an individual's character traits, such as honesty, compassion, and courage, and as a standard of moral behavior that is accepted by society as being right. Ethics refers to the "principles of conduct governing an individual or a group" and specifically to the standards one uses to decide what one's conduct should be. Ethical decisions almost always involve morality; in other words, society's accepted ways of behavior. Moral judgments tend to trigger strong emotions. People bring to their jobs their own ideas of what is morally right and wrong, so the individual must shoulder most of the credit (or blame) for the ethical decisions he or she makes. Every decision an individual makes and every action he or she takes reflects, for better or worse, the application of personal moral standards to the question at hand. Situations are judged as ethical based on group dynamics and societal values. At times defining ethics is highly subjective; it describes the relationship between character traits and the moral aspects of relations with others. As such, the definition of ethics includes elements of both character and morality.

Ethics and ethical theories can be traced back to ancient philosophers such as Aristotle and others. The concepts of ethical behavior have withstood the test of time and have existed for centuries. Three basic ethical theories guide many of the ethical decisions and situations everyone faces virtually every day. Most people think of "ethical" in accordance with one of these three philosophies or some variation depending upon many circumstances.

ETHICAL THEORY	ETHICAL GUIDELINE
Utilitarianism	Do that which leads to the best consequences for everyone involved.
Deontology	Do that which you are motivated to do for the sake of duty.
Virtue ethics	Do that which is consistent with the type of person that you wish to become.

The theory of utilitarianism focuses on the consequences of ethical situations. Utilitarians would state that the only thing that matters when one is making decisions is the consequences of the action. They tend to believe that the proper course is the one that does the most good and generates long-term happiness. This theory seems to be intuitively correct to most people; however, it can be very difficult to practice. It is very complicated to measure and compare the benefits and costs of each consequence and the interaction between multiple consequences. Therefore, this theory is difficult to apply in many situations.

Deontological theory focuses on morality and believes ethical behavior results from doing one's duty. It goes well beyond consequences and in fact in some cases the consequences of doing one's duty ethically may have severe ramifications. To make an action right, it must be done for the sake of some duty or ethical law. Many actions are taken in paramilitary organizations, such as fire departments, because they are duties to be performed. The effect on the individual person is important as are the principles involved. Therefore, the motivations behind the actions are very important in this theory.

Both the utilitarian and the deontological theories focus on actions, whereas the virtue theory focuses on character. According to Aristotle, virtues are character traits that are habitual and good for people to have because they lead to successful human living. These can be such character traits as honesty, courage, generosity, and others. Ethical decisions emanate from the virtues and guide one's actions toward becoming a better person by virtuous acts.[6]

Ethical behavior begins with the individual person, as each has the ability to affect his or her own actions the most. However, ethical decisions about behavior take into account not only one's own interests but also equally the interests of those affected by the decision. Do not expect others to behave in an ethical manner when you have chosen not to do so. Fire department officers have a clear responsibility to act in an ethical manner and demonstrate integrity at all times. It is important to set the example and promote proactive leadership regarding ethical decision making. This is one area in which it is clear that actions speak louder than words.

It is always proper when discussing ethics to remind oneself that ethical behavior begins with the individual person. Minor ethical misjudgments often lead to more serious ethical issues over time. Although one cannot change the past behavior of subordinates, fire officers must demand ethical behavior from themselves to set an example. When fire department members act unethically, they generally have a basic gut-level feeling through intuition and conscience that their actions are wrong. Decisions made against one's sense of right and wrong make an individual feel bad and erode self-esteem. A fire service personnel manager has an obligation to define and publish a code of ethics for the organization to guide employees when they may have ethical questions or concerns.

COMPONENTS OF ETHICAL LEADERSHIP

Ethical leadership begins with the way leaders perceive and conceptualize the environment in which they work. Ethical leadership, organizational ethics, and social responsibility are inseparable concepts. How ethical leaders relate to and come to understand the world around them involves judgment and action. The leader's role is to guide the human potential of the organization to achieve organizational goals and objectives in ways that are both purposeful and ethical.

Ethical leadership should be effective and constantly striving toward excellence. It is not enough just to be ethical in one's individual actions to be an ethical leader. Four major components of ethical leadership must be understood and developed: purpose, knowledge, authority, and trust. The relationship among these four components can be visualized as interrelated components acting and reacting with each other constantly. Attention to any one part alone is incomplete and misleading. These four major components of ethical leadership follow:

- *Purpose*—The ethical leader reasons and acts with organizational purposes firmly in mind. This provides focus and consistency.
- *Knowledge*—The ethical leader has the knowledge to judge and act prudently. This knowledge is found throughout the organization and its environment, but must be shared by those who hold it.
- *Authority*—The ethical leader has the power to make decisions and act, but also recognizes that all those involved and affected must have the authority to contribute what they have toward shared purposes.
- *Trust*—The ethical leader inspires and is the beneficiary of trust throughout the organization and its environment. Without trust and knowledge, people are afraid to exercise their authority.

MODELING ETHICAL BEHAVIOR

Modeling is a powerful leadership strategy whereby one demonstrates to one's employees through one's own behavior how one wants them to behave with others. It has been used successfully to demonstrate flexibility, politeness, decisiveness, compassion, and numerous other desirable traits. It also works extremely well as an ethics-teaching tool. In fact, one of the principal findings of the Ethics Resource Center's 2000 National Business Ethics Survey was that the modeling of ethical behavior by organizational leaders, managers, supervisors, and coworkers establishes benchmarks for other to emulate. When employees perceive formal and informal leaders are ethical, they:

- Feel less pressure to compromise ethical standards
- Observe less misconduct on the job
- Are more satisfied with their organization
- Feel more valued as employees.

In other words, organizational ethics become real for employees when they see good ethics being applied. One issue with modeling is that all too often the audience present when ethical dilemmas arise is small, such as in the fire station environment. A leader wants to multiply the effectiveness of modeling and seek out opportunities to do so. Make it acceptable to talk about ethics by placing the subject on the agenda for staff meetings and training sessions. This can be accomplished by some of the following techniques:

- Take time during a staff meeting or other group event to commend someone on demonstrated ethical conduct.
- Talk about ethical conduct routinely and encourage employees to seek guidance whenever they question whether or not an act is ethical.
- Keep the lines of communication open. When someone wants to talk about an ethical issue, the time to talk about it is then, not at some future date. Pushing the discussion off until later diminishes the importance the organization is trying to place on ethics.
- Capture organizational stories of proper conduct and use them to illustrate desired behavior.
- Use organizational newsletters and other publications to spread the word about ethical successes.

- Make expectations for ethical conduct well known.
- Capture examples of ethical conduct to use when delivering performance evaluation.
- Provide some award or incentive for ethical performance (one organization awarded certificates titled "Caught you . . . doing something good").[7]

When a leader reaches the point of talking about ethics routinely and the employees know where the leader stands on the need for good conduct and behavior, the effect is multiplied throughout the organization and many positive benefits result. As good as discussing and modeling ethical behavior is, the organization will also want to measure its ethical effectiveness. Two basic methods to do so are process measures and outcome measures. Process measures tend to look at events and activities and include such things as number of employees attending ethics training, number of ethics cases requiring investigation, and the like. These are good facts to know, but what conclusions can one draw from this type of activity, other than who attended and what they were exposed to? Although this is interesting information, what one really wants to know is how employees will behave or act when confronted with ethical issues. Outcome measures indicate the true effectiveness of the ethics program as it is directed at the results of behavior. Outcome measures would include the results of investigations, records of substantiated ethical lapses by employees, disciplinary actions as the result of unethical behavior, and others. It is important to measure the ethical compliance of an organization from both perspectives.

ETHICS POLICIES AND CODES

The leader's actions may be the single most important factor in fostering corporate behavior of a high ethical standard; surveys also rank ethics as very important. A policy signals that top management is serious about ethics and wants to foster a culture that takes ethics seriously. Many firms have ethics codes. One study that surveyed corporate accountants found that 56 percent of the respondents had corporate codes of conduct. Following are some conclusions from the survey.

Top Manager's Role.   Top management must make it clear that it is serious about code enforcement. Top management also must ensure that customers, suppliers, and employees are aware of the firm's stance on ethics.

Approval of Code.   It is important that both the top executives of the organization and the employees of the organization endorse the code.

Communication of Code.   To influence employee behavior, the ethics code must be properly communicated.[8]

Increased workforce diversity may make ethics codes even more important in the future. One expert contends that, with the flow of immigrants across national borders, it may become more difficult to rely on a shared organizational culture to control ethical behavior. In other words, because it is harder to infuse common values and beliefs in a diverse workforce, it may become more necessary to emphasize explicit rules, expectations, and ethics codes.

The following code of ethics has been in place in the city of Los Angeles since 1959 and was amended in 1979. The document's message is just as appropriate today as when it was developed. Good ethics are timeless, and this code has provided excellent direction regarding ethics for more than 50 years.

CITY OF LOS ANGELES

Statement of Approved Principles for Public Service in the Government of the City of Los Angeles

I General Rule with Respect to Conflicts of Interest

Persons in the public service shall not engage in nor shall they have an interest, direct or indirect, in any business or transaction, nor incur any obligation which is in substantial conflict with the proper discharge of their official duties in the public interest or which impairs their independence of judgment in the discharge of such duties.

II Actions and Conduct Designed to Build Public Confidence

Persons in the public service shall not only be ever conscious that public service is a public trust but also shall be impartial and devoted to the best interests of the City, and shall so act and conduct themselves, both inside and outside the City's service, as not to give occasion for distrust of their impartiality or of their devotion to the City's best interests.

III Acceptance of Favors and Gratuities

Persons in the public service shall not accept money or other consideration or favors from anyone other than the City for the performance of an act which they would be required or expected to perform in the regular course of their duties; nor shall such persons accept any gifts, gratuities or favors of any kind which might reasonably be interpreted as an attempt to influence their actions with respect to city business.

IV Use of Confidential Information

Persons in the public service shall not disclose confidential information acquired by or available to them in the course of their employment with the City, or use such information for speculation or personal gain.

V Use of City Employment and Facilities for Private Gain

Persons in the public service shall not use, for private gain or advantage, their city time or the City's facilities, equipment or supplies, nor shall they use or attempt to use their position to secure unwarranted privileges or exemptions for themselves or others.

VI Contracts With the City

Persons in the public service shall not exercise any discretionary powers for, or make any recommendations on behalf of the City or any department or officer thereof with respect to any contract or sale to which the City or any department thereof is a party and in which such persons shall knowingly be directly or indirectly financially interested.

VII Outside Employment Impairing Service to the City

Persons in the public service shall not engage in outside employment or business activity which involves such hours of work or physical effort that it would or could be reasonably expected to substantially reduce the quality or quantity of work or interfere with such persons giving a full day's labor for a full day's pay.

VIII Outside Employment Incompatible With Official Duties

Persons in the public service shall not engage in any outside employment which involves the performance by them of any work which will come before them as officers or employees of the City, or under their supervision, for approval or inspection; provided that nothing in this paragraph shall be taken to limit in any manner the outside employment of such persons where the interests of the City are protected under Section 28.1 of the Charter and ordinances adopted there under.

IX Personal Investments

Persons in the public service shall not make personal investments in enterprises which they have reason to believe may be involved in decisions or recommendations to be made by them, or under their supervision, or which will otherwise create a substantial conflict between their private interests and the public interest. If, however, persons in the public service have financial interests in matters coming before them, or before the department in which they are employed, they shall disqualify themselves from any participation therein.

X Discussion of Future Employment

Persons in the public service shall not negotiate for future employment outside of the city service with any person, firm, or organization known by such persons to be dealing with the City concerning matters within such person's areas of responsibility or upon which they must act or make a recommendation.

XI Conduct With Respect to Performance on the Job

Persons in the public service shall perform their duties earnestly, economically and efficiently.

XII Activities Incompatible With Official Duties and the Reporting of Improper Government Activities

Persons in the public service shall not engage in any improper governmental activity or in any actions or practices which would interfere with the proper performance of the duties of others. Persons in the City service are strongly encouraged to fulfill their moral obligations to the City by disclosing to the extent not expressly prohibited by law, improper governmental activities within their knowledge. No officer or employee of the City shall directly or indirectly use or attempt to use the authority or influence of such officer or employee for the purpose of intimidating, threatening, coercing, commanding, or influencing any person with the intent of interfering with that person's duty to disclose such improper activity.

XIII Loyalty

Persons in the public service shall uphold the Federal and California State Constitutions, laws and legal regulations of the United States, the State of California, the City of Los Angeles, and all other applicable government entities therein.

XIV Affirmative Action

Persons in the public service shall not, in the performance of their service responsibilities, discriminate against any person on the basis of race, religion, color, creed, age, martial status, national origin, ancestry, sex, sexual preference, medical condition, or handicap and they shall cooperate in achieving the equal employment opportunity and affirmative action goals and objectives of the City.

Source: Reprinted with permission of the City of Los Angeles.

Although ethics codes vary greatly, they can be classified into two major categories: compliance based and integrity based. **Compliance-based** ethics codes emphasize preventing unlawful behavior by increasing control and penalizing violators. These codes are mainly concerned with avoiding legal punishment. **Integrity-based** ethics codes define the organization's guiding values, create an environment that encourages and supports ethical behavior, and stress a shared accountability among employees. Integrity-based codes move beyond legal compliance to support a climate that promotes core values such as honesty, fairness, commitment, and moral excellence. Although these values are ethically desirable, they are not necessarily legally mandatory.[9]

To be effective, all ethics codes must be enforced and all employees must be held responsible for their behavior. This starts at the top of the organization and extends down through each rank. Expectations for ethical behavior begin at the top, and senior management expects all employees to act accordingly. It is the responsibility of a fire service personnel manager to be a role model and discuss leadership situations that will help other officers and firefighters better understand how to meet their responsibilities regarding ethical behavior. It may be a good idea to establish an ethics office or officer to whom individuals can go for advice if questions or conflicts arise. It is also important to back any ethics program with timely action if any rules are violated. These actions provide the clearest message to all that the code is serious and that it will be upheld.

Two types of conflict generally present themselves within organizations:

- **Conflict of interest** typically involves potential financial gain or personal benefit to an employee or related individual. An example would be a situation in which a firefighter receives money for overlooking fire code violations.
- **Conflict of commitment** typically arises when otherwise positive activities may compromise the employee's basic job responsibilities. An example would be a firefighter operating a private business while on duty at the fire department.

Personal responsibility, integrity, and high ethical standards are the principal factors in avoiding conflicts such as these. Situations in which such conflicts may arise can be sufficiently complex that opinions may differ as to whether they actually exist or are likely to arise. Most government agencies have an ethics panel or such body to review and make decisions regarding these activities. To be most effective, these bodies must become engaged prior to the acts being committed. To ensure that this is accomplished, most organizations require full and prompt disclosure. This disclosure enables potential conflicts to be reviewed and, if appropriate, properly managed to the benefit of all parties concerned. Many progressive organizations also make access to advice and legal counsel available to individuals and supervisors in order to provide them with guidance and clarification when necessary. Ethical conflicts are always easier to resolve when they are disclosed before the act occurs. If disclosed after the fact, many complications result for both the individual and the organization.

In most situations the clear-cut cases rarely become a problem. Everyone knows it is wrong to take a bribe or to steal from a patient on the emergency scene. These situations can be dealt with in a forthright manner and with great clarity within the organization. Often unethical actions result because of the perception that superiors want an employee to reach an objective or an end to a

compliance based
- ethics codes that emphasize preventing unlawful or unethical behavior by increasing control and penalizing violators.

integrity based
- ethics codes that define the organization's guiding values, create an environment that encourages and supports ethical behavior, as well as stress a shared accountability among employees.

conflict of interest
- typically involves potential financial gain or personal benefit to an employee or related individual.

conflict of commitment
- when otherwise positive activities may compromise the employee's basic job responsibilities.

problem. As long as only a limited number of people are adversely impacted and the organization as a whole benefits, then everyone just looks the other way. This scenario presents a morally difficult dilemma. A fire officer who acts unethically, thinking that the greater good of the organization is served, engages in a rationalization that serves no one well. When firefighters observe this conduct in their officer, they may assume that unethical behavior is a way of doing business, and they too will begin to decide which rules to follow and which to overlook. Eventually such behavior ends up in scandal and on the evening news.

After a review of the ethics programs at eleven major firms, one study concluded that fostering ethics at work involved five main steps. They are listed here as a guide for fire service organizations to improve their ability to deal with ethical situations in a more proactive manner.[10]

Emphasize Top Management Commitment. To achieve results, top management needs to be openly and strongly committed to ethical conduct and provide constant leadership in tending and renewing the values of the organization.

Publish a "Code." Organizations with effective ethics programs set forth principles of conduct for the whole organization in the form of written documents.

Establish Compliance Mechanisms. Pay attention to values and ethics in recruiting and hiring, emphasize ethics in training, institute communication programs to inform and motivate employees, and audit to ensure compliance.

Improve Personnel at All Levels. Use roundtable discussions among small groups of employees on the topic of corporate ethics and surveys of employee attitudes regarding the state of ethics in the organization.

Measure Results. Organizations with effective ethics programs used surveys or audits to monitor compliance with ethical standards.

The following "Code of Ethics for Fire Chiefs" was developed by the International Association of Fire Chiefs to guide its membership regarding ethical conduct.

CODE OF ETHICS FOR FIRE CHIEFS

The purpose of the International Association of Fire Chiefs (IAFC) is to actively support the advancement of the fire service, dedicated to the protection and preservation of life and property against fire, provision of emergency medical services and other emergencies. Toward this endeavor, every member of the International Association of Fire Chiefs shall represent those ethical principles consistent with professional conduct as members of the IAFC:

- Recognize that we serve in a position of public trust that imposes responsibility to use publicly owned resources effectively and judiciously.
- Not use a public position to obtain advantages or favors for friends, family, personal business ventures, or ourselves.
- Use information gained from our positions only for the benefit of those we are entrusted to serve.
- Conduct our personal affairs in such a manner that we cannot be improperly influenced in the performance of our duties.

(continued)

- Avoid situations whereby our decisions or influence may have an impact on personal financial interests.
- Seek no favor and accept no form of personal reward for influence or official action.
- Engage in no outside employment or professional activities that may impair or appear to impair our primary responsibilities as fire officials.
- Comply with local laws and campaign rules when supporting political candidates and engaging in political activities.
- Handle all personnel matters on the basis of merit.
- Carry out policies established by elected officials and policy makers to the best of our ability.
- Refrain from financial investments or business that conflicts with or is enhanced by our official position.
- Refrain from endorsing commercial products through quotations, use of photographs or testimonials, for personal gain.
- Develop job descriptions and guidelines at the local level to produce behaviors in accordance with the code of ethics.
- Conduct training at the local level to inform and educate personnel about ethical conduct and policies and procedures.
- Have systems in place at the local level to resolve ethical issues.
- Orient new employees to the organization's ethics program during new employee orientation.
- Review the ethics management program in management training experiences.
- Deliver accurate and timely information to the public and to elected policy makers to use when deciding critical issues.

Source: IAFC, 2008.

One can ask a number of questions when faced with an ethical dilemma. These questions may help guide an individual to the proper decision.

1. Is it legal? Check to see whether the action violates any law or company policy. Everyone needs to be aware of the legal implications of his or her actions. This is the most basic question regarding ethics, but only the first one.
2. Is it fair and balanced? Would I want to be treated this way? Do you win everything at the expense of someone else? Every situation cannot be balanced, but it is important to look for imbalances and make decisions when possible that benefit all parties involved.
3. How will it make me feel about myself? Would I feel proud if my family were to learn of the decision? If a private matter became public, would it make anyone uncomfortable or embarrassed? The Golden Rule of "Treat others as you would want to be treated" makes a lot of ethical sense.

Rarely do ethical problems come with easy solutions. No set formulas guide the decisions that must be made, especially in what could be considered "gray areas" or subjective issues with unclear paths. Occasionally managers must be able to defend and justify their decisions. Often, it just comes down to doing what one thinks is right. However, others may not agree, even when faced with the same set of circumstances. In these situations, it is necessary to be careful and review actions to ensure that an actual ethical issue is at stake and not to overreact or assume certain circumstances to be true when they are not. Additionally, some people confuse not agreeing with a decision and ethics. One may not agree with or like a decision that may affect him or her in a negative way, but those feelings do not necessarily make the decision unethical. Many times it comes down to the motives, which are difficult to determine at times, of the individuals involved. My advice is to always take the high road.

To illustrate how ethical issues occur in the real world, I use the following personal example. During my first week as fire chief for the Prince George's County Fire Department, I was faced with a situation that concerned me, even though I knew what to do. A Hispanic citizens group in the county was conducting a fundraiser and reception for the county executive. The event was advertised and tickets were sold for this important political event. I learned that the event was to be held in a vacant store that had significant fire code violations. Several hundred people were scheduled to attend. When I first contacted the people who had arranged the event, they were extremely upset and wanted the event to go forward. At one point I felt as though I might have the shortest tenure of any fire chief in the history of the department.

Some members of my staff advised me on how to avoid the situation or get around the code for such a one-time event. Others advised me not to allow the event to be held. With little time to bring the building into compliance before the event, I contacted the county executive and explained that I could not allow the event to be held in that building. His response really impressed me; he thanked me for keeping him from being in a potentially embarrassing situation and told me he understood the position. The event was subsequently held in an outside tent close to where the original event had been scheduled. I learned a valuable lesson in ethics and an even more important lesson about the ethics of the person I worked for. It made me have a great deal of respect for the county executive. This example demonstrates the respect that lies within ethical decision making and its importance to an officer's career and to those under one's employ.

The rewards of an ethical approach to management are evident by the self-satisfaction of having exercised leadership in a positive way and in knowing that such actions fall within the highest tradition of the fire service. The importance of high ethical standards figures prominently in the success and respect that any fire department hopefully enjoys.

Relationship with Personnel Office

As organizations grow and become more complex, the personnel management function necessarily becomes more complex and takes on greater importance. The basic purpose and application of personnel management remain the same, but its structure and location within the organization often change, in most cases with relation to the size of the organization in terms of employees.

According to Wayne Casio in his text *Managing Human Resources* (2006), there are five activities that organizations have to deal with regarding personnel management regardless of their size or complexity: staffing, retention, development, adjustment, and managing change.[11] All of these activities can be carried out at the individual, team, or organizational level, and salary is not an issue because volunteer fire departments go through the same activities. Many people tend to think that a number of these activities such as staffing and retention are the sole purview of the larger personnel department. But every officer in a fire department has an individual degree of responsibility regarding these issues, and they all will have to work closely with the personnel department to be successful. For example, fire officers are responsible for the retention activity by treating employees fairly, resolving conflicts, promoting teamwork, and recommending pay increases based on merit. The following chart summarizes the responsibilities of the fire department and the personnel department.

ACTIVITY	FIRE DEPARTMENT	PERSONNEL DEPARTMENT
Staffing	Data for job analysis, interviewing candidates, selection for hire and promotion	Job analysis, recruitment, compliance, testing, background and references
Retention	Fair treatment, communications, conflict resolution, respect, working conditions	Compensation and benefits, employee relations, services, health plans
Development	Training, motivation, performance evaluation, professional development	Performance management system, career planning, counseling, human resource (HR) research
Adjustment	Discipline, transfers, termination, demotion	Retirement counseling, policy development, equity
Managing change	Department vision, policies, research	Expertise for change process, resources, data analysis

Source: *Managing Human Resources*, 2006.

In many smaller-sized fire departments the personnel management functions are often located in the government's personnel department. Fire department officers coordinate their activities with the personnel department, but this department performs the major personnel functions and most likely sets the policy for the fire department in this process. The majority of the personnel work and decisions are made within the personnel office with the fire department in an advisory role. Because of the relatively small number of personnel actions, no one within the fire department holds a full-time role regarding personnel management. Supervisors within the fire department obviously continue to have significant personnel management responsibilities as they evaluate, counsel, and perform other personnel management functions, whereas the policy and main personnel functions are performed in another department of government. Normally this is because the fire department is not of sufficient size to generate an appreciable number of personnel actions, such as hiring, retirements, separations from service, and other actions. This system will work well until the fire department reaches a certain size and complexity requiring a higher level of personnel services.

In larger fire departments, generally those with 200 or more employees, the personnel management functions need to be included more formally within the fire department structure. Individuals in the fire department personnel section can play the key role in the design and implementation of the personnel management system within the fire department. They also coordinate necessary liaison activities with the larger personnel office of the city or county government. In essence, the city or county personnel office delegates much of its responsibility to the fire department. This structure (see following chart) provides the fire department with personnel experts who understand and work within their system on a daily basis. It has been my experience that this type of structure generally works best for the management of the fire department because it is much more responsive to the special needs of the department. In order for this format to work, two important requirements must exist. One is that the personnel office of the city or county government must be willing to delegate much of its responsibility to the fire department personnel office. Second, the fire department must accept this responsibility and have the fire department personnel office staffed with knowledgeable personnel experts.

CENTRALIZED PERSONNEL MANAGEMENT STRUCTURE

DECENTRALIZED PERSONNEL MANAGEMENT STRUCTURE

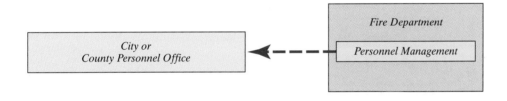

The structure and responsibility for this type of system must be clear and well communicated. The personnel office of the city or county government must be satisfied that the fire department is operating within the personnel law and general requirements of that jurisdiction at all times. This operation requires a good deal of understanding and communication between the two offices. The fire department needs to staff its personnel office properly with fully qualified individuals, which may include a fire department officer with a personnel management background or, more likely, a professional personnel manager specifically hired for that position. Within this situation, the detail and quality of the personnel services will be much improved because they will be concerned solely with the business of the fire department. When a general personnel office of the city or county government handles these functions, the fire department is just one of many agencies competing for services and attention to their issues. Often this arrangement results in personnel services of a lower quality and significant time delays on decisions and personnel actions. As a fire chief, one should attempt to have as much control over the personnel management functions as possible.

Regardless of the structure of the personnel management system, the functions of personnel management must be performed well if the fire department is to be successful. The key to success in this area is that everyone in the organization has some degree of shared responsibility for the personnel management of the fire department and an obligation to operate within the guidelines of its jurisdiction. Managers within the fire department must understand effective personnel management concepts and practices and use them in the discourse of their business. They need to work closely with the personnel management office of the general city or county government at all times. The policies issued within the fire department must be acceptable regarding personnel management and be the result of an inclusive process. All employees of the fire department must accept responsibility for managing their own behavior and contributing toward the betterment of their career in particular and the department in general.

Summary

Managing human resources is a critical task in any organization, large or small, regardless of industry or service. Every personnel management decision made will have consequences and send a message to employees throughout the organization. As organizations grow and become more complex, changes are required in the way personnel management is implemented. However, the basic purpose of personnel management remains the same; only the application is performed in a different manner. The elements of personnel management explored in this book must be well coordinated and properly implemented throughout the organization.

Governmental organizations have general values that are formed from a number of sources and modified over time. The interaction of and, at times, the conflict between these values can be very difficult to manage. The importance of ethics in one's personal and organizational life will provide a sound framework and will command the respect of those with whom one is associated. Proper personnel actions within a fire department that are communicated and coordinated with the overall personnel office of the city or county government will improve this relationship and allow the department to be more effective. Personnel management is a shared responsibility involving everyone within the fire department.

Review Questions

Use these questions to review the material in this chapter and for discussion purposes.

1. Identify and discuss the ramifications of at least three major trends in society that have affected the implementation of personnel management.
2. Discuss how the conflict among different values can affect a fire department, and provide at least two examples from your fire department. How were these situations resolved?
3. Why is ethical behavior important to the fire department, and how would you describe this behavior?
4. Describe the structure of your fire department personnel functions and that of the local government personnel office. Who is responsible for specific personnel functions? How can you improve the relationship with your city or county personnel office?
5. How can a fire chief demonstrate his or her commitment to good ethics?

Case Study

You are the officer in charge of an engine company that is returning from a fire incident. The crew decides to stop in a local restaurant for lunch to take back to the station. The owner of the restaurant says that he will not charge the firefighters for their sodas. He offers the sodas in appreciation for what firefighters do every day to assist the community. You say that it is not necessary, but you appreciate the offer. He becomes somewhat offended in that his generosity has been refused. The crew would just like to have their sodas for free. How do you handle this situation?

Two weeks later your company is operating at the scene of a two-alarm fire in the downtown section of the city. It is July and the temperature is very hot. After the fire is under control an employee of the local McDonald's shows up with sodas and cheeseburgers for everyone on the incident scene. You accept these and thank the employee for his generosity.

How are these situations different? Are the ethical issues the same or not? How would you handle a situation like this in your department?

Fire Service Professional Qualifications

This chapter contains information that will assist fire officers in meeting the job performance requirements (JPRs) of NFPA 1021, Standard for Fire Officer Professional Qualifications:

4.2 Human Resource Management—Fire Officer I

4.2.5 Apply human resource policies and procedures given an administrative situation requiring action, so the policies and procedures are followed.

Endnotes

1. Wendell French, *Human resources management,* 6th ed. (Boston: Houghton Mifflin, 2006).
2. Gary Dessler, *Management: Leading people and organizations in the 21st century,* 2nd ed. (Upper Saddle River, NJ: Prentice Hall, 2001).
3. Donald E. Klingner and John Nalbandian, *Public personnel management: content and strategies,* 5th ed. (Upper Saddle River, NJ: Prentice Hall, 2002).
4. The Teal Trust, *The value of organizational values* (South Croydon, UK: The Teal Trust, 2007).
5. Klingner and Nalbandian, *Public personnel management: content and strategies.*
6. M. Neil Brown et al., *Practical business ethics for the busy manager* (Upper Saddle River, NJ: Pearson Prentice Hall, 2004).
7. Jerry Brown, *Setting the stage for modeling ethical behavior* (Washington, DC: Ethics Resource Center, 2003).
8. Robert Sweeney and Howard Siers, Survey: Ethics in corporate America. *Management Accounting,* June 1990.
9. William G. Nickels et al., *Understanding business,* 5th ed. (New York: McGraw-Hill, 1999).
10. Alan Rowe et al., *Strategic management: A methodological approach* (Reading, MA: Addison-Wesley, 1994).
11. Wayne Casio, *Managing human resources,* 7th ed. (New York: McGraw-Hill, 2006).

Suggested Readings

Brown, Jerry. 2003. *Setting the stage for modeling ethical behavior.* Washington, DC: Ethics Resource Center.

Brown, Neil M., et al. 2004. *Practical business ethics for the busy manager.* Upper Saddle River, NJ: Pearson Prentice Hall.

Cascio, Wayne F. 2006. *Managing human resources,* 7th ed. New York: McGraw-Hill.

Dessler, Gary. 2001. *Management: Leading people and organizations in the 21st century,* 2nd ed. Upper Saddle River, NJ: Prentice Hall.

French, Wendell. 2006. *Human resources management*, 6th ed. Boston: Houghton Mifflin.

Gibson, James L., et al. 2006. *Organizations*, 12th ed. New York: McGraw-Hill Irwin.

Klingner, Donald E., and John Nalbandian. 2002. *Public personnel management: context and strategies*, 5th ed. Upper Saddle River, NJ: Prentice Hall.

Lazere, Cathy. 1997. Ethically challenged. *CFO*, April.

Nickels, William G., et al. 1999. *Understanding business*, 5th ed. New York: McGraw-Hill.

Noe, Raymond A., et al. 1997. *Human resource management*, 2nd ed. Chicago: McGraw-Hill.

Rowe, Alan, et al. 1994. *Strategic management: A methodological approach*. Reading, MA: Addison-Wesley.

Sweeney, Robert, and Howard Siers. 1990. Survey: Ethics in corporate America. *Management Accounting*, June.

The Teal Trust. 2007. *The value of organizational values*. South Croydon, UK: The Teal Trust.

University of Maryland. 1999. *MFRI Fire Officer III program*. College Park, MD: Maryland Fire and Rescue Institute.

Workforce Issues of the Twenty-First Century

Source: MFRI

OBJECTIVES

After reading this chapter, you should be able to:

- Understand general population and fire service demographics.
- Possess the ability to manage diversity effectively.
- Understand fire service workforce issues.
- Describe fire service organizational benefits of diversity.
- Create a fire service positive work environment.
- Utilize the concepts of alternative dispute resolution.

The changing composition of today's workforce creates new challenges for managers, employees, and the community. As significant shifts in demographics, changing job skills, educational requirements, and the dynamics of different cultures converge, a much different workforce will characterize the twenty-first century. **Diversity** affects many issues. To some, diversity is about sex

diversity
- a society consisting of people of different ethnic and cultural backgrounds who possess different attitudes, values, and norms.

and race, but it is much more than that. It includes values, age, disabilities, and education, in addition to the many issues that revolve around a person's sex or race. Organizations that expect to prosper and survive in the future must work to attract, integrate, and retain an increasingly diverse, nontraditional workforce.

Formerly, the fire service consisted of a workforce in which large portions were similar, and those who were different were expected to adapt to an era in which the workforce is composed of many different individuals, each of whom wants to be valued and supported. Today's workforce does not look, think, or act like the workforce of the past; nor does it have the same values, experiences, or expectations. If fire service personnel management practices are based on what worked well in the past, then that organization is in for some serious confrontations. This chapter examines these issues and the importance of diversity to a fire department.

Demographics

It has been said that increasing diversity is moving American society from a "mass society" to a **mosaic society**. Organizations in both the public and private sector reflect this trend in terms of a more diverse workforce. This diversity is not only in terms of race, gender, and ethnicity but also in terms of age, sexual orientation, and other demographic variables. Fire service managers of today have to interact and communicate with many people from rich and diverse backgrounds. The skill to motivate and adapt effectively across many different demographic factors will be very important to one's success as a leader.

mosaic society
■ as organizations move toward a more diverse workforce, the diversity not only in terms of race, gender, and ethnicity but also in terms of age, sexual orientation, and other demographic variables.

Diversity in organizations will continue to increase. As indicated by the U.S. Census Bureau National Population Projections, the Hispanic population will increase by 11.2 percent between 2000 and 2025 to become the largest minority group in the United States. All other minority groups will increase by about 9 percent, while the number of Caucasians will decrease by approximately 19 percent.[1] The world population is growing at a high rate in developing countries, while remaining stable or decreasing in the developed world.

The aging American workforce population means more retirees and potential gaps in availability of experienced workers. According to American Association of Retired Persons (AARP), by 2015 nearly 1 in 5 U.S. workers will be age 55 or older. Retirees often want to remain active in the workforce. AARP's research shows that nearly 8 of 10 baby boomers envision working part time after retirement; 5 percent anticipate working full time at a new job or career; only 16 percent foresee not working at all.[2] It is conceivable that many fire department jobs that do not involve emergency services could be filled by a much older workforce in the future.

People of diverse ethnic and cultural backgrounds possess different attitudes, values, and norms. Increasing cultural diversity in both public and private sector organizations focuses attention on the distinctions between ethnic and cultural groups in their attitudes and performance at work. Organizations with greater diversity will be respected for that by the population that they serve. Not only do these organizations embrace a diverse workforce internally, but they are better suited to serve a diverse customer base in both emergency and nonemergency settings. Organizations with greater diversity also display higher creativity and innovation.

Civilian Labor Force by Sex, Age, and Race

(Number in thousands)

	1986	1996	2006	PERCENTAGE CHANGE, 1986–1996	PERCENTAGE CHANGE, 1996–2006
Total	117,834	133,943	148,847	13.7	11.1
Men	65,422	72,087	78,226	10.2	8.5
Women	52,413	61,857	70,620	18.0	14.2
White	101,801	113,108	123,581	11.1	9.3
Black	12,654	15,134	17,225	19.6	13.8
Asian	3,371	5,703	8,041	62.2	41.0
Hispanic	8,076	12,774	17,401	58.2	36.2

Source: Bureau of Labor Statistics, 2007.

Recent data from the Bureau of Labor Statistics of the Department of Labor indicate that the workforce increased by 15 million between 1996 and 2006, reaching a total of about 149 million.[3] For women, the rate of growth in the labor force slowed, but it still increased at a faster rate than for men. Women were estimated to be 47 percent of the workforce in the year 2006. Also in 2006, the Hispanic labor force was projected to be larger than the African American labor force, primarily because of faster population growth. The youth labor force (aged 16 to 24) grew more rapidly than the overall workforce for the first time in 25 years. At the same time, the number of persons in the workforce aged 25 to 44 decreased, as the baby boom generation continued to age.

Workforce growth over the 1986 to 1996 period (16 million) was significantly slower than the rate of growth over the 1976 to 1986 period (22 million), when larger numbers of the baby boomers caused rapid rates of workforce growth. Workforce growth for men was less than that for women in both the 1976 to 1986 and the 1986 to 1996 periods, whether measured by numbers of persons or rates of change. White non-Hispanics were the largest group in the labor force in 1986, accounting for 80 percent of the total. However, from 1986 to 1996, this group had the lowest growth rate. All minority groups increased their share of the workforce during this period.

There is no question that the year 2006 workforce had a greater proportion of women and Hispanics than did the 1996 workforce. The Bureau of Labor

Statistics reports that between 1996 and 2006, 40 million workers entered the workforce, with 25 million workers expected to leave and 109 million workers expected to remain in the labor force. As a result, the workforce in 2006 was approximately 149 million, up 15 million from the 1996 level.[4] These figures represent both a rate of growth as slow as the growth experienced in the 1950s and a great deal of change and an increasingly diverse workforce as we have entered the twenty-first century.

As evidenced by these noted statistics, managing diversity will be one of the most challenging aspects of fire department management. The complexity of the differences is wide ranging and there is not simply one way of perceiving the world around us. People are not the same and although they want to be treated equally, this does not necessarily translate to being treated the same. This has major implications, not only with respect to the workforce but also in the delivery of emergency services. Managers that have the understanding and ability to manage a more diverse workforce will be the most successful in the future.

Fire service personnel managers will have a key role in transforming the organizational culture so that it more closely reflects the values of the increasingly diverse workforce. Some of the skills that will be needed follow:

- An understanding and acceptance of managing diversity concepts
- Recognition that diversity is threaded through every aspect of management
- Self-awareness, in terms of understanding one's own culture, identity, biases, prejudices, and stereotypes
- Willingness to challenge and change institutional practices that present barriers to different groups

It is probably natural to want a cookbook approach to diversity issues so that one knows exactly what to do. Unfortunately, given the many dimensions of diversity, there is no easy recipe to follow. Advice and strategies given for one situation may not work given the same situation in another context. Managing diversity means acknowledging people's differences and recognizing these differences as valuable; it enhances good management practices by preventing **discrimination** and promoting inclusiveness.

Managing diversity focuses on maximizing the ability of all employees to contribute to organizational goals. **Affirmative action** focuses on specific groups because of historical discrimination, such as people of color and women. Affirmative action emphasizes legal necessity and social responsibility; managing diversity emphasizes business necessity. In short, while managing diversity is also concerned with underrepresentation of women and people of color in the workforce, it is much more inclusive and acknowledges that diversity must work for everyone. Ignoring diversity issues costs time, money, and efficiency. Some of the consequences can include unhealthy tensions between people of differing genders, races, ethnicities, ages, and so on; loss of productivity because of increased conflict; inability to attract and retain talented people of all kinds; complaints and legal actions; and inability to retain women and people of color, resulting in lost investments in recruitment and training.

The following chart summarizes the primary aspects of diversity, which are those a person is born with. It also depicts the secondary aspects of one's

discrimination
- refers to taking specific actions toward or against a person based on the person's group.

affirmative action
- refers to the policies intended to promote access to education or employment aimed at an historically nondominant group, typically minorities or women.

diversity, which represent the culture and other factors that are acquired from one's environment.[5]

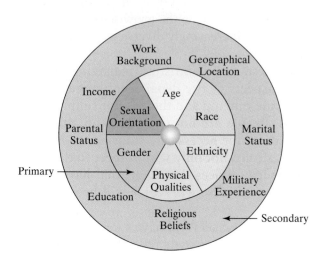

Fire Service Demographics

The effects of increasing diversity on the population and the workforce in general present a variety of issues for the fire service profession. What are the demographics of the fire and rescue service of the United States, and how will they affect employees and fire service personnel managers? The National Fire Protection Association (NFPA) recently conducted a study of fire departments regarding certain demographic factors. Some of the information from this report, titled *U.S. Fire Department Profile Through 2006*, is listed next.[6] Managers need to be aware of these statistics and understand their implications. Let us look at the number and type of fire departments within the United States.

- In 2005, an estimated 30,635 fire departments operated in the United States. This includes 52,900 fire stations, 68,950 pumpers, and 6,600 aerial apparatus.
- Of that total, 70 percent were all volunteer and another 17 percent were mostly volunteer, approximately 26,583 fire departments. A total of 8 percent were career departments and another 6 percent were mostly career, approximately 4,052 fire departments.
- In other words, 14 percent of all departments are all career or mostly career but protect 60 percent of the U.S. population, whereas 86 percent of the departments are mostly volunteer or all volunteer and protect 40 percent of the population.

Obviously a very large number of individual fire departments serve our country. This large number of independent departments makes it difficult to reach consensus on issues that affect the fire service, which often inhibits progress. It is clear that the fire service has an added issue of diversity: whether one is a career or volunteer member of the fire department. This dimension of the diversity situation often creates managerial challenges, especially when both career and volunteer members work within the same department. In reviewing firefighter demographics in the United States, trends are similar to the departmental statistics.[7]

- In 2006, the United States had a total of 1,140,900 firefighters.
- Of that total, 72 percent were volunteers (823,950).
- Of that total, 28 percent were career firefighters (316,950).
- Ninety-five percent of volunteers were in departments that protect populations of fewer than 25,000 people, and more than half are located in small, rural departments that protect fewer than 2,500 people.

In looking at the statistics, one can see that the fire service of the United States is not as diverse as society as a whole. A recent analysis of the career fire service by the National Fire Protection Association shows that 96 percent of the fire service is male, of which 83 percent is white. There were on average 24,000 African American firefighters, 19,000 Hispanic firefighters, and 9,000 female firefighters annually during the 2000 to 2004 period. The minority percentages are the highest in the larger metropolitan areas that have higher percentages of minorities in the labor pool and where residency requirements may be in effect. Women firefighters continue to be severely underrepresented in both the career and the volunteer fire service.[8]

Racial–Gender Demographics of Career Firefighters

YEAR 2000–2004	TOTAL	WOMEN	AFRICAN AMERICAN	HISPANIC
Annual average	251,400	9,000	24,000	19,000

Source: National Fire Protection Association, 2007.

In April 2008 *A National Report Card on Women in Firefighting* was published, and this report supported by the Ford Foundation included many revelations regarding the utilization of women in the fire service.[9] Clearly the fire service significantly lags behind other occupations with respect to the employment of women firefighters. The report states that 2000 census reports indicate that women career firefighters number approximately 3.7 percent of the total firefighter workforce. One startling statistic was that among 291 metropolitan areas in the 2000 census, 51.2 percent had no career women firefighters. The report developed a benchmark for expected female representation using the 2000 census. The benchmark included the nation's labor force of the typical active firefighter age of 20 to 49, educational background of high school graduate, and working in one of 184 occupations that resemble firefighting in terms of strength, stamina, dexterity, and involving dangerous work. The result showed that women occupied 17 percent of the positions in like occupations compared to 3.7 percent in the fire service. Obviously the significant barriers to women in the fire service need attention and resolution.

Distribution of Firefighters by Age Group

AGE OF FIREFIGHTERS (YEARS)	NUMBER OF FIREFIGHTERS	PERCENTAGE
16–19	44,650	3.9
20–29	240,900	21.1
30–39	327,550	28.7
40–49	300,800	26.4
50–59	169,750	14.9
60 and over	57,250	5.0

Source: National Fire Protection Association, 2006.

Title VII
■ part of the Civil Rights Act of 1964 that prohibits employment discrimination based on race, color, religion, sex, or national origin.

EEOC
■ U.S. Equal Employment Opportunity Commission; an independent federal agency created by Congress in 1964 to eliminate discrimination in employment; provides oversight and coordination of all federal equal employment opportunity regulations, practices, and policies.

Title VII requires that organizations categorize their employees according to race and gender and report this information to the U.S. Equal Employment Opportunity Commission **(EEOC)**. This report is the basis for further analysis to determine whether employment patterns show evidence of discrimination. Equal employment opportunity analysis has four basic steps:

1. *Analyze the workforce* to determine representation of minorities and women in each job group in the organization.
2. *Analyze availability* in the relevant labor force to determine the proportion of minorities and women qualified and interested in the job opportunities.
3. *Establish goals* by comparing the present workforce to availability. Goals are the percentage of job opportunities to be shared with women and minorities to increase their representation in the targeted jobs.
4. *Choose programs* specifying how the goals are to be achieved and a proposed timetable for achievement.

Any fire department that has not performed this analysis yet, should. This will assist in identifying whether the current employment pattern within the fire department is inappropriate and needs attention. The personnel department within the overseeing government will generally have the demographic information for the particular area necessary to make these assessments.

Employment in the fire service is very competitive. When positions open, the number of applicants is normally substantially higher than the number of jobs available. Additionally, the turnover of the workforce in fire departments is low, because when people get the job of firefighter they tend to stay until retirement. Most fire departments have restrictive single-point entry systems, whereby one has to begin at the entry level and then attempt to advance up the ranks. Some volunteer departments still operate as social clubs, with members voted in by the current membership. Add to this a tradition-bound history and paramilitary organizational structure, and it is easy to see why the fire service in the United States is not very diverse. In fact, in numerous legal actions, fire service organizations have been found to employ discriminatory hiring and promotional practices that exclude women and minority groups.

This segment of fire service organizational history is not one of which to be proud, but progress over the past decades has been made in being more inclusive. Diversity of the workforce is an issue to be faced in a proactive and forthright manner. The strength and success of fire departments in the twenty-first century will depend on their ability to attract and manage a diverse workforce.

Managing Diversity

One of the foremost issues of the workforce of the twenty-first century will be the management of diversity. Managing diversity requires an acute awareness of characteristics common to a culture, race, gender, age, or sexual preference, while at the same time managing employees with these characteristics as individuals. Diversity management presents new challenges in the workplace. Managers must be knowledgeable about common group characteristics in order to manage diversity effectively. Fire service managers need to recognize that people with common characteristics, but different from those of the mainstream, often think and act differently as individuals. One cannot stereotype people regardless of culture.[10] This task requires training and a degree of sensitivity that may be new to many fire service officers.

One thing that managing diversity does not mean is lower standards of performance. Being sensitive to an individual's needs and being flexible in managing people are quite different from not requiring adherence to established standards. This assumes that the standards are valid and do not contain inherent bias toward a certain group of people. Job-related performance standards that are properly designed and implemented create a level playing field for all. It is very important that the firefighters and officers in the department understand this.

Many diversity programs have difficulties because fire departments are more concerned with increasing diversity and counting numbers and percentages as opposed to managing diversity in a meaningful way. The goal is not just to count numbers and percentages of people but to benefit from the diversity and the best mix of people within an organization. Diversity management needs to become a real part of the fire department culture and not just a part of the mission statement that is on display. This view is very important to the success of managing diversity in any organization, and organizations need to understand the following with respect to managing diversity:

- Managing diversity is not affirmative action.
- Managing diversity is not a program, but a process.
- Managing diversity is not only a moral and social issue but also an economic one.
- Managing diversity increases performance and productivity.

Many organizations experience a constant mix of values at any given time. Some firefighters may value their career, whereas others place a higher value on family or sporting events. Some are loyal to their assigned company but not as much so to the department. In many cases people place a high value on what they perceive they lack, be it money, status, or control. Values evolve over time. Although an individual's values vary greatly, research has helped to identify widely held values. If fire service personnel understand these values, they are in a better position to motivate and acknowledge employees for their contributions.

In a research study, a diverse group of employees were asked to identify the work-related values they believed were most important to a large number of people in their workforce. The majority of the respondents identified the following nine values[11]:

1. Recognition for competence and accomplishments
2. Respect and dignity
3. Personal choice and freedom
4. Involvement at work
5. Pride in one's work
6. Lifestyle quality
7. Financial security
8. Self-development
9. Health and wellness

From a diverse group of races, cultures, and sexes, these nine values resound of Main Street America and would be welcome in any fire department in the United States. This study is a good example of different cultures expressing the same organizational values as those of the majority group in the workforce. One has to be careful not to make unwarranted assumptions about people and not to confuse personal values with organizational or work-related values. Even though managers cannot assume that any one individual has these values, they can be used for many personnel actions such as setting policy, designing work schedules, and managing people. The key is to be aware of these values and work to blend them into policy.

Any attempt at managing diversity must begin with understanding the barriers that may prevent an organization from taking full advantage of the potential in its diverse workforce. These barriers include the following:

stereotyping
- a process in which specific behavioral traits are ascribed to individuals on the basis of their apparent membership in a group.

prejudice
- a bias that results from prejudging someone on the basis of some trait.

Stereotyping and Prejudice.   **Stereotyping** is a process in which specific behavioral traits are ascribed to individuals on the basis of their apparent membership in a group. **Prejudice** is a bias that results from prejudging someone on the basis of some trait. Many people form stereotyped lists of behavioral traits that they identify with certain groups. Unfortunately, many of these stereotypes not only are inaccurate but also tend to be negative. When someone allows stereotypical traits to bias him or her for or against someone, then we say the person is prejudiced.

Ethnocentrism.   Ethnocentrism is prejudice on a grand scale. It can be defined as a tendency for viewing members of one's own group as the center of the universe and for viewing other social groups less favorably than one's own. Ethnocentrism can be a very significant barrier to managing diversity.

Discrimination.   Whereas prejudice means a bias toward prejudging someone based on that person's traits, discrimination refers to taking specific actions toward or against the person based on the person's group.

Tokenism.   Tokenism occurs when a company appoints a small group of women or minority-group members to high-profile positions, rather than more aggressively seeking full representation for that group. Tokenism is a diversity barrier when it slows the progress of hiring and promoting more members of minority groups.

Organizations deal with diversity issues in many ways, both negatively and positively. The perspective that the organizational culture presents is very important because it sends constant messages to employees, both direct and implied. The formal position of the organization relating to diversity issues can be undermined by the way the management of the organization manifests itself on a daily basis. The people within the organization will not be fooled by a false program that portends to deal with diversity issues, but in reality does not. The following chart depicts several of the various organizational approaches to diversity.[12]

The authors Herbert Z. Wong and Aaron T. Olson in *Multicultural and Diversity Strategies for the Fire Service* discuss the concept of cultural competence. Cultural competence is a developmental process that evolves over time in both individuals and organizations. The two major values within cultural competence follow:

- Developing a set of principles, attitudes, policies, and structures that allows all individuals to work effectively and equitably with all cultures
- Developing the capacity to apply cross-cultural knowledge in order to respond to and communicate with all of the cultural contexts within the organization

Cultural competence means acquiring the skills and sensitivity to build rapport within one's group but not at the risk of alienating other groups. It recognizes the difficulty that people may have in working in culturally diverse groups. It can be a balance within race, culture, and ethnicity with the concept of equal treatment and at the same time not diminishing one's individuality and culture. Sensitivity and training in this area are very important to the success of implementing cultural competence.

Organizational Diversity Approaches

	RESISTANCE	FAIRNESS	ACCESS LEGITIMACY	DIVERSITY CULTURE
Viewpoint	Diversity is not important and does not affect management	Diversity creates conflicts and problems	Diversity provides opportunities for employers and organizations	Diversity culture permeates organizations
Action	• Resistant to change • Denial of problems	• Diversity training • Affirmative action compliance • Focus on protected groups	• Build "diversity acceptance" culture • Reduce conflicts in multicultural workforce	• Proactive efforts on diversity • Employees seen as resources
Consequences	• Protect status quo • Increased possible legal liabilities	• Discrimination addressed through internal responses • Minimize legal exposure	• All employees are valuable to recruit and retain • Acceptance leads to internal problem solving	• Effective relations among all employees • Diversity access throughout organization

Source: Human Resource Management, Tenth Edition.

The predominant management style of the twentieth century can perhaps be summed up as the "one-size-fits-all" approach. There are standard, precise policies and individuals are to be subordinate to the policies and fit into the organization. We now know that much more flexibility is necessary in the workplace, as evidenced by increased job-sharing arrangements, home offices, and other strategies that give employees more freedom. In their book, *Managing Workforce 2000*, David Jamieson and Julie O'Mara offer a system entitled "flex management," which they believe will assist in managing the workforce of the future. Flex management consists of the following four basic strategies that can enable managers to develop organizations in which people are valued, supported, and satisfied as individuals and contribute in a high-performance way to the organization.

Matching People and Jobs.   The focus of this strategy is to match the individual skills and work preferences to real job characteristics and demands. It involves paying closer attention to both the subjective and the objective sides of people and work. More information is gained through communication and assessment, which is then related to policies and procedures within the organization. A fire service application may be the careful assessment of people who wish to be assigned to certain jobs to ensure a good match for the firefighter and the department, such as an assignment as an arson investigator.

Managing and Rewarding Performance.   This strategy recognizes that people do not work in the same ways and are not motivated by the same methods and incentives. Different approaches to planning, motivation, appraisal, and managing the process of work are valuable. A fire department should have a number of ways to recognize and reward performance not limited to valor awards. One suggestion may be to send letters of commendation to the home address so that the firefighter's family is more aware of the firefighter's success on the job.

Informing and Involving People.   This strategy recognizes the significant desire among people to be more informed about and involved in their work. Policies may deal with the use of work time or the establishment of and participation in groups. One example is the use of focus groups consisting of officers and firefighters to review policies of the fire department before they are implemented.

Supporting Lifestyle and Life Needs.   This strategy identifies people's needs and interests and where possible creates supportive options, such as child care or substance abuse counseling, flexible work hours, and leave options. The fire department could provide a list of child-care providers who will provide their services on the same shift-work schedule that the department works.[13]

These general strategies are not strict rules and regulations to be applied without regard to individual circumstances. The management of an increasingly diverse workforce requires sensitivity and flexibility, which means neither that an organization applies different rules for different individuals nor that the regulations of the department will change based on personal circumstance. No organization can be managed in that manner, especially public safety organizations, which certainly have special needs regarding order and discipline. It means that everything is not "either/or" or "one way"; one needs to be receptive to methods that get the job done with the required degree of performance, while at the same time being flexible and sensitive where appropriate. Managers do not allow people to abuse these privileges, but practice flexibility that includes a positive outcome for the employee and the organization. A commitment to diversity is a commitment to all employees, not an attempt at preferential treatment for any individual or single group.

DIVERSITY, EEO, AND AFFIRMATIVE ACTION

As discussed, a diverse organization is one that invites, embraces, and manages difference in the changing workforce to gain the best from people and to provide the best service to its customers. It encourages the breaking down of barriers of the past to look at issues in a new way. Many organizations view their diversity as a competitive advantage, especially in the private sector. The success of a diversity program can be measured through several factors, such as an inclusive workplace culture, and growth in the hiring, retention, and promotion of women and minorities. To be successful it is not enough to have leadership that is supportive; it must be committed leadership. Committed leadership understands that acknowledging and addressing diversity issues takes a substantial effort and unwavering commitment to long-term change. It involves the need to change policies, practices, and procedures of the formal organization as well as the informal organization. Denial is one of the major issues that will have to be dealt with in a progressive manner. Fire departments in general have been publically criticized for their reluctance about and denial of diversity issues resulting in lawsuits, political consequences, and poor public relations.

Affirmative action refers to policies intended to promote access to education or employment aimed at an historically nondominant group, typically minorities or women. Motivation for affirmative action policies is to redress the effects of past and current wrongful discrimination and to encourage organizations to be more representative of the population. This is commonly achieved through targeted recruitment programs, by preferential treatment given to applicants from disadvantaged groups, and in some cases through the use of quotas. Opponents of affirmative action policies argue that it is based on collectivism and is merely another form of discrimination because it can result in qualified applicants being denied entry to employment opportunities because they belong to a particular racial or gender group.

Focusing in particular on education and employment, affirmative action policies require that active measures be taken to ensure that minorities enjoy the same opportunities for employment, promotions, salary increases, and career advancement as the dominant group. From the outset, affirmative action was envisioned as a temporary remedy that would end once there was a "level playing field" for all Americans. By the late 1970s, however, flaws in the policy began to show up amid its good intentions. Reverse discrimination became an issue, epitomized by the noted *Bakke* case in 1978. The Supreme Court outlawed inflexible quota systems in affirmative action programs, which in this case had unfairly discriminated against a white applicant. In the same ruling, however, the Court upheld the legality of affirmative action per se.

The Supreme Court in a landmark 2003 case involving the University of Michigan's affirmative action policies upheld the right of affirmative action in higher education. Two cases, first tried in federal courts in 2000 and 2001, were involved: the University of Michigan's undergraduate program *(Gratz v. Bollinger)* and its law school *(Grutter v. Bollinger)*. The Supreme Court upheld the University of Michigan Law School's policy, ruling that race can be one of many factors considered by colleges when selecting their students because it furthers "a compelling interest in obtaining the educational benefits that flow from a diverse student body." The Supreme Court, however, also ruled that the more formulaic approach of the University of Michigan's undergraduate admissions program, which uses a point system that rates students and awards additional points to minorities, had to be modified. The undergraduate program, unlike the law school's, did not provide the "individualized consideration" of applicants deemed necessary in previous Supreme Court decisions on affirmative action. In the Michigan cases, the Supreme Court ruled that although affirmative action was no longer justified as a way of redressing past oppression and injustice, it promoted a "compelling state interest" in diversity at all levels of society.

Many believe that the forced diversity affirmative action plans of the past contained both good and bad features that made their implementation difficult, as just explained. The country and most organizations have evolved toward the concept of diversity in a positive light, not with the punitive aspects of affirmative action plans. Fire department leaders need to embrace diversity and understand the difference between diversity management and affirmative action. The confusion in these concepts has caused misunderstanding and has impeded the legitimate movement to more diverse organizations.

Differences in affirmative action (AA) and diversity follow:

- AA is government initiated, whereas diversity is voluntary and organizationally driven.
- AA is legally driven, whereas diversity is driven by organizational performance.
- AA is quantitative and diversity is qualitative.
- AA is problem focused, whereas diversity focuses on opportunities.
- AA is reactive, but diversity is proactive.

A fire department diversity program needs to include goals and benchmarks to measure success, but preferably not quotas or strict affirmative action plans. Diversity issues need to be recognized and dealt with because the worst strategy is to do nothing and allow the court system to intercede. A diverse workforce in the fire department will be reflective throughout the organization and at all levels within the organization.[14]

EEOC

The U.S. Equal Employment Opportunity Commission (EEOC) is an independent federal agency created by Congress in 1964 to eliminate discrimination in employment. The various statutes enforced by the Commission prohibit employment discrimination on the basis of race, color, sex, national origin, religion, retaliation, age, and disability. EEOC has authority to receive, initiate, and investigate charges of discrimination filed against employers who have a statutory minimum number (generally fifteen or more) of employees. EEOC's role in an investigation is to evaluate the charge allegations fairly and accurately in light of all the evidence obtained.

EEOC has five commissioners and a General Counsel appointed by the president of the United States and confirmed by the Senate. Commissioners are appointed for 5-year staggered terms. The term of the General Counsel is 4 years. The president designates a chair and a vice chair. The chair is the chief executive officer of the Commission. The five-member Commission makes equal employment opportunity policy and approves most litigation. The General Counsel is responsible for conducting EEOC enforcement litigation.

The federal laws prohibiting job discrimination follow:

- *Title VII of the Civil Rights Act of 1964* (Title VII), which prohibits employment discrimination based on race, color, religion, sex, or national origin
- *Equal Pay Act of 1963* (EPA), which protects men and women who perform substantially equal work in the same establishment from sex-based wage discrimination
- *Age Discrimination in Employment Act of 1967* (ADEA), which protects individuals who are 40 years of age or older
- *Title I and Title V of the Americans With Disabilities Act of 1990* (ADA), which prohibit employment discrimination against qualified individuals with disabilities in the private sector and in state and local governments
- *Sections 501 and 505 of the Rehabilitation Act of 1973*, which prohibit discrimination against qualified individuals with disabilities who work in the federal government
- *Civil Rights Act of 1991*, which, among other things, provides monetary damages in cases of intentional employment discrimination

The U.S. Equal Employment Opportunity Commission (EEOC) enforces all of these laws. EEOC also provides oversight and coordination of all federal equal employment opportunity regulations, practices, and policies.

The following chart summarizes the differences between diversity, equal employment opportunity, and affirmative action in terms of focus and practice.

DIVERSITY	EEO	AFFIRMATIVE ACTION
Is inclusive of all groups	Eliminates discrimination	Targets underutilized groups
Maximizes the potential of all employees	Provides equal access opportunity	Measures good faith efforts for minorities and women
Is broader than ethnicity, race, and gender	Excludes no one from participation	Helps to prevent discrimination
Not legally mandated	Legally mandated	Legally mandated
Grounded in individuality	Grounded in assimilation	Grounded in assimilation
Changes the system or operation	Changes the mix of people	Changes the mix of people
Equality	Preference	Preference
Focuses on all elements of diversity	Focuses on race, gender, and ethnicity	Focuses on race, gender, and ethnicity

Fire Service Workforce Issues

The issues within the current workforce and beyond will be substantially different from those of the past, mostly because the composition of the workforce will be significantly diverse. It was projected that in the years from 1985 to 2000, minorities, women, and immigrants would compose 85 percent of the net growth in the workforce. The largest minority growth is that of Hispanics in the workforce and in the population in general. Women will also continue to be an expanding factor. It is important to realize that these projections refer to workforce growth. It does not mean that the percentage of white males in the labor force will change dramatically. In 1985, for example, white males comprised 49 percent of the labor force; in 2006, they constituted approximately 45 percent.[15]

Because the demographics within the fire service include fewer minorities and women than in the general workforce, it will remain predominantly white male for many more years. This trend is also affected by the relatively slow turnover in positions and the general downsizing experienced by most fire departments in the last decade. The fire service, then, faces an increasingly greater challenge because other organizations and departments of government are showing increased diversity at a faster rate than the fire service. Many local governments set goals to reflect the diversity of their political jurisdiction within their workforce. If the fire department is vastly white male in a government and a community that are not, then political and other consequences may follow.

This section reviews the fire service work environment, discusses what makes it different from most other occupations, and covers the challenges and built-in advantages due to the nature of the fire service work environment. The specific issues of the workforce that arise from increased diversity such as sexual harassment and discrimination are covered in subsequent chapters.

Source: MFRI Archives.

Within the fire service environment, several unique factors make managing diversity and people more challenging as compared to many other occupations.

One of the biggest factors is the length of the work shift and the closeness of the fire station environment. Many fire departments are now working a 24-hour shift arrangement, which places individuals in close contact with coworkers for long periods of time during each shift. Differences in work styles or personalities are much more likely to surface during this extended time period. In many occupations, 8 hours is the most time one has to put up with someone he or she does not like or agree with. The extended duration of fire service shift work is a factor to be considered in managing diversity.

Teamwork is critical to successful fire service operations; all firefighters learn it from their first day on the job. Not only do firefighters work long shifts, but they also do it together in the fire station. Firefighters work, eat, and even sleep together, and in this regard the only real comparison it has is the military. Even law enforcement does not generally entail such a close working relationship; in fact, many police officers spend most of their day on patrol by themselves. This close working relationship, sometimes called the *fire service family*, has the potential to greatly magnify differences in people, including the differences often found among races, genders, and cultures.

Additionally, the fire service team members depend on each other for critical tasks; one team member who fails to do an assigned job can impose severe consequences on other members of that team. Therefore, if one member of the fire service team is viewed as weak or not up to the task, then great pressure is applied on that individual. If a firefighter already disrespects someone because of culture or sex, then this effect is exaggerated to an even greater extent. These types of situations cause dysfunction and can damage the company and the firefighter who is subjected to such treatment. In most cases the individual can do the work, but do it in a way different from what other members of the crew are used to. Company officers need to be particularly aware of and sensitive to these situations because they can rapidly escalate and create major personnel problems and legal actions.

The increased emphasis on teamwork also offers a major advantage within the fire service. One of the basic needs that people have is to belong and to identify in a significant way with other people and groups. The fire service work environment provides such an asset to its employees and reinforces it throughout their careers. People in many other occupations would love to have the natural teamwork advantages enjoyed in the fire service. Many of the quality management concepts of the 1990s cite teamwork as a major factor in the success of an organization. When managed properly, fire departments have some natural advantages in this area.

In most fire service organizations, a diversity of assignment options is available to personnel within the department. One can work in fire suppression, emergency medical services, fire prevention, code enforcement, arson investigation, training, and other areas. This variety not only allows for continued professional development but also offers the flexibility to match people with jobs in an effective way. It gives the fire service personnel manager more tools with which to work and the ability to reenergize employees with new and challenging assignments throughout their career. Even rotating firefighters among various companies has real advantages toward enhancing their abilities with new experiences. It can be as simple as transferring someone from an outlying fire station to a downtown station.

Fire service personnel managers need to review the assignments of department members on a regular basis to ensure that they reflect the diversity that is desired. For example, if the department has one rescue squad company and it is deemed to be an elite assignment, then what is the makeup of that company? If it is all white male, then what does this say to the rest of the department? It speaks volumes, and other firefighters will understand its message quite well. The fire department can have all the diversity and nondiscrimination policies in the world, but no one will believe it until these policies are manifested in the way the department conducts business and manages its people. Therefore, diversity should be reflected both vertically throughout the chain of command and horizontally through assignments to the various companies and divisions within the fire department.

Perhaps one of the most important advantages within the fire service is its clear mission, which is easily understood by those within the fire service and the public. Fire service personnel save lives and protect property and the environment. The mission is simple and direct, and the citizens appreciate it. In virtually every citizen survey poll conducted as to which government agency provides the best service or whom citizens trust, the fire service is generally at the top of the list. Fire service history is abundant with acts of courage and unselfishness by firefighters to the degree of sacrificing their lives for others. Most private companies would gladly spend millions of dollars to have the name recognition and reputation with the public that the fire service has. This is something to be very proud of and protect at all times.

In August 2008 the *USA Today* newspaper conducted a survey of the most prestigious occupations. The crux of the survey was that "high prestige does not always equal high pay." The most prominent occupations were the following:

Firefighter	57%
Doctor	53%
Nurse	52%
Teacher	52%
Police officer	46%

Source: Harris Poll Survey of 1,010 adults 18 and older, *USA Today*, August 21st Edition. Gannett Co. Inc. Weighted to represent actual population.

This recognition and clear mission encourage the workforce to strive to uphold this special place in the minds of the public. Firefighters really want to do their job well and enjoy the excitement and satisfaction inherent in this profession. Many of the workforce issues that hinder other occupations are not an issue in the fire service. Therefore, these real, built-in motivators make the fire service profession somewhat easier to manage in certain aspects. It also means that careers in the fire service are highly sought after, and the exclusion of any segment of society will not be easily accepted. Additionally, because firefighters are role models in the community, it is important that others, especially children, see people of their own race and sex performing these jobs.

Fire Service Organizational Benefits of Diversity

As discussed in the preceding paragraphs, many of the specific issues within the fire service personnel environment are not found in numerous other occupations. These include both positive and negative factors, which combine to provide both wonderful opportunities and dangerous pitfalls. A good diversity program recognizes similarities and differences while bringing a balanced approach to common goals. The following organizational attributes are necessary in order to manage diversity in a proactive and successful manner:

- Allowing fire departments to attract and retain the best people for firefighter and fire officer positions
- Reducing the exposure to and risk of legal sanctions, the organizational costs in terms of financial settlements, and the negative publicity that these types of actions involve; removing barriers that restrict people from realizing their potential and ensuring that all requirements are job related
- Improving the quality and quantity of the performance of the workforce via the strength of diversity; allowing all employees to contribute and take full advantage of different viewpoints and methods
- Preparing the fire department to respond to the social and demographic changes in the community and being able to interact better with the community at all levels

A number of methods will create and maintain a positive work environment in the fire service that respects and encourages diversity, but none are easy to build or sustain over time. In addition, other factors in the department or government such as labor negotiations, understaffing, and other negative influences may be creating unrest. Attempts at changing behavior and the culture of a department during periods of general dissatisfaction are extremely difficult. However, whatever one can do as a fire service personnel manager to create a positive work environment will eventually be appreciated and recognized.

A good way to demonstrate the organization's commitment to diversity and to ensure that all within the organization understand the principles is to have a diversity mission statement. John Lee Cook, Jr., in the March 2002 edition of *Fire Engineering*, proposed a "Sample Diversity Mission Statement." This excellent statement should be used as an example for any department's diversity mission statement.

SAMPLE DIVERSITY MISSION STATEMENT

Diversity, as it is understood in the workplace today, implies differences in people based on their identification with various groups, but it is more. Diversity involves the process of acknowledging differences through action. In organizations, this means developing a variety of initiatives at the management and organizational levels, as well as at the interpersonal levels.

The continued excellence of the department is largely dependent on the ability to attract, develop, and retain highly skilled, talented, and motivated members. An essential element in maintaining this quality of service is the recognition of the value of a diverse workforce.

VISION

Characteristics such as age, culture, ethnicity, gender, race, religious preference, sexual orientation, and the expression of unique philosophies and ideas provide the opportunity to better understand each other. This understanding will strengthen the efficiency and productivity of the workforce, whose primary objective is to provide excellent service to the community.

MISSION

The mission of the department is to maintain a high standard of excellence by attaining and fostering a diverse workforce. This will be accomplished by reaching the following goals:

1. Uphold the federal, state, and local laws and the department's rules and regulations regarding employment.
2. Attract and retain qualified individuals from diverse backgrounds who are committed to the continued excellence of the department.
3. Achieve a diverse workforce in terms of culture, ethnicity, gender, race, religious preference, and sexual orientation.
4. Provide all employees with the opportunity for development and growth at every rank in the department.
5. Expect all employees to treat each other with dignity and respect, regardless of perceived differences.

Source: Fire Engineering, March 2002.

Fire service organizations have the opportunity to approach the management of diversity from a number of different perspectives. The diversity of fire department organizations is only going to increase with time. Many members of minority groups perceive that they work in organizational cultures and climates that are hostile to them. They perceive that they have to work harder to be accepted and to move up in the organization. It is logical to have these perceptions in an organization that is not diverse and demonstrates no commitment to be so. For decades many fire departments had a reputation for being the last element of local governments to accept increased diversity within its structure.

A good way to overcome these perceptions is to have an active diversity management program. Organizations can approach diversity from several perspectives. The continuum runs from resistance to the creation of an inclusive organizational culture that is accepted by all within the department. There are many important reasons for an organization to address diversity issues proactively. The organization and each of its members will experience significant benefits because of the diversity efforts, as described next.

Diversity and Organizational Performance. A number of organizations have found that, because they serve a diverse community or set of customers, there are significant business reasons for having a diverse workforce. The greater cultural understanding and knowing how the services can be improved from a number of perspectives are important. If fire departments want to be accepted and respected by the community they serve, then they should reflect that community's diversity.

Recruiting and Retention.   An organization that is composed of diverse members will aid in recruiting and retaining workers from diverse backgrounds. Most people prefer to work in a climate in which they are accepted and have people there who are like them. A recent study concluded that one-third of minority and female job candidates will rule out an employer because of a perceived lack of diversity. This means that the entire organization and not just the recruiters have to be diverse.

Diverse Thinking and Problem Solving.   Any decision or decision-making process will be improved by more diverse thinking and a better range of solutions. Diverse problem-solving groups will more likely see factors and circumstances differently and consider a wider range of options. For fire departments this involves inclusion of women and minorities on committees and work groups, as well as members of different ranks to be a part of these groups.

Reduction in Discrimination Complaints and Costs.   There is an increasing prevalence of protected group members pursuing legal complaints against employers who persist in resisting diversity and engage in discriminatory employment practices. Any organization will benefit from a reduction in the time and legal costs associated with discrimination complaints. In addition, these complaints are public matters and the reputation and perception of the fire department in the community will suffer if the department is found to be guilty of these practices. Organizations that value and promote diversity from a number of perspectives are less likely to face these challenges.

One effective way to improve the organization's ability to deal with increased diversity is to have a well-thought-out diversity training program. There have been many successes and numerous failures in this regard. My experience is that if the training is focused only on the legal liabilities and if the effort is just to say that the training was done, then it is doomed to failure. Diversity training programs that focus on the acceptance and understanding of people, cultural awareness issues, and a degree of sensitivity have the best chance of success. The legal aspects of discrimination and unfairness must be discussed and understood, but should not be the major focus of the program.

Diversity training is in many cases very difficult to perform. The instructor is not only showing firefighters how to use a new piece of equipment but also discussing issues of their personality and deeply held beliefs. Some of the major criticisms of diversity training programs are that they draw attention to differences and build up defenses, rather than break them down. Additionally, much of the content in some programs is viewed as "politically correct" and blames the majority individuals for past wrongs. This turns people off in the first few moments of the program; thus, the ability to make a change is negated. Teaching appropriate behaviors and skills in relationships with others is more likely to produce satisfactory results than focusing only on the attitudes and beliefs among diverse employees.

Another issue to be aware of in diversity training programs is that they will raise the bar for organizational performance and increase the expectations of minority employees. Many minorities and women may see the efforts as inadequate and be disillusioned when they return to their normal work assignment and nothing has changed. Individuals in the majority, such as white males, may see the emphasis on diversity as setting them up as scapegoats for societal problems over which they perceive no control. This results in anger and resentment when they see programs benefiting women and minorities at their presumed expense.

Diversity training is a necessity, but it must be performed carefully and by the proper human resource professionals. The training program needs to be a part of the entire organizational effort and not just exist by itself. Top management of the fire department needs to be committed and demonstrate this by their attendance at the training courses. The diversity training programs should lead to improved effectiveness of multicultural work teams, mentoring programs, and the everyday business of the fire department.

EQUAL EMPLOYMENT OPPORTUNITY COMPLIANCE PLANNING

With little or no training in equal opportunity management, fire departments will continue to see themselves on the evening news involved in discrimination and sexual harassment litigation. Even if only a few departments are accused of such discrimination, all fire service organizations are painted with the same broad brush of exclusion and bias. One way to prevent such difficulties is to have a program of equal employment opportunity compliance planning. Equal opportunity compliance-based planning management in the fire service involves evaluating risks, planning to respond to those risks, and implementing control measures to mitigate risks.

Central to equal opportunity compliance planning is an understanding of what is required by law and a plan that addresses the specific needs of the organization. Compliance planning is the process of preparing, adopting, and implementing a compliance plan to help an organization come into, and remain in, compliance with the laws and regulations applicable to that organization. Compliance planning involves reducing an organization's inherent legal risks. Planning to identify and reduce risks is nothing new to the fire service, but fully exploiting the idea of legal compliance planning is too often overlooked. The Equal Employment Opportunity Commission (EEOC) defines it this way: "EEO compliance refers to a business' responsibility to be compliant with Equal Employment Opportunity Laws. It is also sometimes referred to as EEOC compliance, in reference to the Equal Employment Opportunity Commission, or AAP for Affirmative Action Plan. Failure for an organization to be compliant with the EEOC laws can result in lawsuits, fines and potentially even worse, negative publicity about your business or organization." The failure of organizations to implement **compliance programs** has cost millions of taxpayer dollars as organizations must defend themselves in litigation. Some experts include the cost to a fire service's reputation in their estimates of EEO noncompliance.

compliance programs
■ an understanding of what is required by law and a plan that addresses the specific needs of the organization regarding equal opportunity.

One of the major factors in effective compliance planning is to have the commitment from the top of the organization. A commitment to equal opportunity compliance must start with the fire chief and include the support of the company officers. Further, the organization will have implemented EEO training and education at all levels of the organization. These organizations communicate that they want to bring people together who represent various backgrounds, age groups, talents, and education levels to form one team focused on providing exceptional service to the citizens of their communities. Finally, these organizations do not rely on their programs to work without oversight. Instead, they complete annual or biannual audits to ensure the compliance program is meeting its goals.

DEVELOP A COMPLIANCE PROGRAM

In part, compliance planning is based on the concept of affirmative defense. To invoke the affirmative defense, departments must establish, publicize, and distribute antiharassment policies and complaint procedures. The organization should provide every employee with a copy of the policy and complaint procedure, and redistribute them periodically. The policy and complaint procedure should be written in a way that all department employees will understand. Other measures to ensure effective dissemination of the policy and complaint procedure include posting them in central locations and incorporating them into general orders. If feasible, the department should provide training to all members to ensure that they understand their rights and responsibilities. The antiharassment policy and complaint procedure should contain, at a minimum, the following elements:

- A clear explanation of prohibited conduct
- Assurance that members who make complaints of harassment or provide information related to such complaints will be protected against retaliation
- A clearly described complaint process that provides accessible avenues of complaint
- Assurance that the department will protect the confidentiality of harassment complaints to the greatest extent possible
- A complaint process that provides a prompt, thorough, and impartial investigation
- Assurance that the department will take immediate and appropriate corrective action when it determines that harassment has occurred

Equal opportunity compliance planning is a cultural change that affects the organization at its deepest level. It creates a more professional work environment that values all members of the organization. The standing of the fire department in the community will be enhanced and the entire department will benefit from this. But most importantly, it is the right thing to do.[16]

Fire Service Positive Work Environment

Creating an environment in which everyone can realize his or her potential while being included and valued is the ultimate goal. Many factors contribute to a positive work environment, especially the following:

Fairness. Ensure that access to training and career opportunities, promotions, and employment are based on job-related knowledge, skills, and abilities. The principles of equity should guide all actions.

Respect. Demand a work environment in which employees are treated with dignity and respect, free from mistreatment and harassment, such as unwelcomed remarks, jokes, sexual innuendo, and so forth.

Trust. Maintain a work environment where employees can raise issues and make suggestions without fear of reprisal, one in which they feel fully comfortable contributing at all times.

Flexibility. Consider the changing and different needs of employees as balanced against the mission of the department. Be flexible where appropriate and when it is warranted.

Sensitivity. Foster awareness of the differences in people and situations. Active efforts to ensure inclusion of people of all races, both sexes, and all cultures are welcomed and appreciated.

If organizations set out to create an atmosphere in which everyone can realize his or her potential through inclusion, then a positive work environment is needed. One of the most important aspects is personal awareness and commitment toward diversity and its benefits. In many cases managers, through their actions and attitudes, have the single most important impact on diversity. A keen awareness of personal work style and a personal comfort level with ambiguity, conflict, change, and newness will help managers understand how diverse people may experience a supervisor in different work situations. It will allow for a better understanding of how a manager's style impacts others, which in turn will assist in replacing old thinking with new ideas and information and allow managers to set the example for others in the organization.

When, as a supervisor, one understands oneself, one places oneself in a better position to help others. The book *Managing Workforce 2000* identifies five basic skills that enable successful management of a changing workforce. These skills build on basic good management practices and judgment. The fire service personnel manager of today and of the future will have to understand and utilize these five basic skills for managing a diverse workforce.[17]

Empower Others. A fire officer shares power and information, solicits input, and rewards people based on performance. He or she attempts to manage more as a colleague and encourages participation by maintaining accountability and providing a framework for success of those who work with him or her.

Develop Others. A supervisor's main tasks in development include coaching, setting the example, mentoring, and providing opportunities for growth for all employees. A supervisor conscious of development will delegate freely and counsel employees when necesssary as well as provide training and educational opportunities that are individualized if necessary.

Value Diversity. A manager must understand his or her own strengths, weaknesses, and biases. Diversity is seen as an asset in attempting to understand different cultures, people, and situations. A manager readily helps others identify their needs and options.

Work for Change. Through supporting employees by adapting policies, systems, and practices when practical to help meet their needs, a supervisor is a key player in identifying and influencing organizational change.

Communicate Responsibility. Clear communication of expectations, asking questions for clarification, listening, and demonstrating empathy when appropriate help a supervisor develop clarity across cultures and languages and provide sensitivity toward individual differences.

A new type of fire service personnel manager who is fully competent at basic management skills and who can blend the preceding five skills when appropriate will be needed in the future. Managing with diversity requires new skills and vision. As a fire service personnel manager, one must use personal power and responsibility to respond to the changing workforce by initiating change. This challenge will take managers beyond the limitations of using only one approach; they will appreciate and use many options and make decisions that are the most effective for the employee and the organization.

Alternative Dispute Resolution

Across the United States, both large and small organizations have found it advantageous to establish in-house dispute resolution programs to resolve conflicts between and among employees and managers. These programs are proactive, are future directed, and have been designed to avoid both the legal expenses and the workplace stigma associated with litigation. As a by-product, in-house alternative dispute resolution systems increase productivity and lessen the chance of workplace violence, which often occurs when anger and hostility among employees are not effectively addressed.

Workplace conflict is not uncommon in the fire service, and organizations need to be better prepared to deal with these issues. Conflicts may arise from ethnic differences, sexual harassment, and performance and behavior issues, among others. Due to the fire service work environment and its increasing diversity, new methods to resolve disputes that are effective, timely, and less costly are needed. Managers often spend a great deal of their available work time on employee conflict issues. Conflicts that are not resolved within the fire department can often result in the overuse of transfers, in an employee leaving the service, or in a lawsuit—all costly and protracted outcomes. One of the most effective alternative dispute resolution techniques is peer mediation. This process addresses disputes between employees and managers through teams composed of staff members trained to act as peer mediators. When disputes arise, the involved parties are invited to participate in a mediation conducted by the team. The parties adhere to voluntary settlement agreements reached through this process a majority of the time. Such positive results reflect the advantages of peer mediation.

Control of the Outcome. Settlement agreements are decided by the disputants and are not imposed by outside sources, such as a judge or jury.

Control of the Process. Mediation can occur at any time during a dispute, even during litigation. Disputants have control over which issues will be discussed and can choose which peer mediation they feel most comfortable with.

Avoidance of Legal Fees. Peer mediation is free to all parties because it uses existing organizational resources and people. Settlement agreements are enacted through the power of mutual agreement.

No Unwanted Publicity. Mediation avoids public disclosure of organizational problems, as well as protecting employees from uncomfortable publicity.

Preservation of Work Relationships. Because decisions are based on consensus and as such are long-lasting and meaningful, work relationships often outlast the dispute. Protracted litigation destroys working relationships and trust.[18]

Peer mediation training is relatively easy to learn and has powerful results. It is a future-oriented technique that builds trust between workers and their organization. Mediation provides a safe and controlled opportunity for the expression of perspectives, an effective way to de-escalate hostility and resentment. A fire department that embraces alternative dispute resolution creates a culture of peacemaking, thus establishing itself as a committed, compassionate, and proactive organization. Organizations of the twenty-first century will need these types of programs in-house to reduce conflict and improve working relationships, thereby increasing productivity and in general creating a better place to work.

Summary

Diversity is defined by who we are, how we live and work, and our different ideologies. It goes beyond race, gender, and ethnicity. It goes to the root of how we communicate, listen, share information, and interact with each other. It impacts our openness to new opportunities and challenges our past practices. The representation of diversity at all levels of the organization reflects commitment and forms a strong bond with the community served by the fire service. The issues involved with diversity are not management fads or trends; they are the reality of the future.

Review Questions

Use these questions to review the material in this chapter and for discussion purposes.

1. Describe the diversity of your fire department as well as the diversity of the community you serve.
2. Identify and discuss the four basic strategies of flex management.
3. Discuss at least three issues in the fire service work environment that may complicate the management of diversity.
4. Name two fire service organizational benefits of increased diversity.
5. Identify four basic skills that a fire service personnel manager will need in order to manage an increasingly diverse workforce effectively.

Case Study

You are the company officer and have a member of your crew who is frequently late for duty. After many steps trying to rectify this problem, you have decided to place him on formal disciplinary charges. It is common knowledge within the battalion that the lieutenant in another station also is frequently late for duty and is often seen out of his regulation uniform, yet no formal action has been taken against him.

When you present the charges to the firefighter, who is African American, he accuses you of racial discrimination and points out the example of the lieutenant, who is white but has not been disciplined. He says that he is being mistreated and that he will file an EEOC complaint.

In the past, you have permitted, in your presence, several racially derogatory remarks and jokes to be made at the expense of this firefighter. You know that this will not look good in an EEOC hearing.

What action do you take and how?

Fire Service Professional Qualifications

This chapter contains information that will assist fire officers in meeting the job performance requirements (JPRs) of NFPA 1021, Standard for Fire Officer Professional Qualifications:

4.2 **Human Resource Management—Fire Officer I**

　4.2.5 Apply human resource policies and procedures given an administrative

situation requiring action, so the policies and procedures are followed.

4.4.1 Recommend changes to existing policies and/or implement a new departmental policy at the unit level, given a new departmental policy, so that the policy is communicated to and understood by unit members.

6.2 Human Resource Management—Fire Officer III

6.4.7 Review current policies, then evaluate the need for and supervise the development of a new or revised policy or procedure, so that the recommended policy or procedure addresses the need.

7.2 Human Resource Management—Fire Officer IV

7.2.1 Appraise the department's human resource demographics, given appropriate community demographic data, to determine if the recruitment, selection, and placement of human resources is effective and consistent with law and current best practices.

Endnotes

1. U.S. Census Bureau Website, *U.S. population projections,* retrieved November 15, 2008, from www.census.gov/population/www/projections/index.html
2. American Association of Retired Persons (AARP) Website, retrieved October 20, 2008, from www.aarp.org/research
3. U.S. Department of Labor Bureau of Labor Statistics Website, retrieved October 20, 2008, from http://stats.bls.gov/data/#employment
4. Ibid.
5. Marilyn Loden, *Implementing diversity* (New York: McGraw-Hill, 1996).
6. Michael J. Karter, Jr., and Gary P. Stein, *U.S. fire department profile through 2006* (Quincy, MA: National Fire Protection Association, 2007).
7. Ibid.
8. Ibid.
9. Marc Bendick, A fair shake, *Fire Chief* magazine, April 2008.
10. Stephanie Overman, Managing the diverse workforce, *Human Resources Magazine,* April 1991.
11. Thomas R. Roosevelt, Jr., *Beyond race and gender* (New York: American Management Association, 1991).
12. Robert L. Mathis and John H. Jackson, *Human resource management,* 10th ed. (Mason, OH: Thomson Learning, 2003).
13. Ibid.
14. Mauricio Velasquez, Jump starting your law firm's diversity program, Minority Corporate Council Association, *Diversity & the Bar,* July–August 2005.
15. David Jamieson and Julie O'Mara, *Managing workforce 2000* (San Francisco: Jossey-Bass, 1991).
16. Paul B. Brown, Compliance culture, *Fire/Rescue* magazine, October 2007.
17. Jamieson and O'Mara, *Managing workforce 2000.*
18. JoEllen Kelly, Interview on May 25, 1999, in the offices of Greenridge Associates, Dunkirk, MD.

Suggested Readings

Brown, Paul B. 2007. Compliance culture. *Fire/Rescue* magazine. October.

Hulett, Denise M., et al. 2008. *A national report card on women in firefighting.* Washington, DC: Egan Economic Consultants, Inc.

Jamieson, David, and Julie O'Mara. 1991. *Managing workforce 2000.* San Francisco: Jossey-Bass.

Johnson, Kevin B. 2001. *Employment law compliance.* Wayne, PA: Oakstone Legal & Business Publishing, Inc.

Johnston, William B., and Arnold E. Packer. 1987. *Workforce 2000*. Indianapolis, IN: Hudson Institute, Inc.

Loden, Marilyn. 1996. *Implementing diversity*. New York: McGraw-Hill.

Loden, Marilyn, and Judy B. Rosener. 1991. *Workforce America! Managing employee diversity as a vital resource*. Homewood, IL: Business One–Irwin.

Mathis, Robert L., and John H. Jackson. 2003. *Human resource management*, 10th ed. Mason, OH: Thomson Learning.

National Fire Academy. 2005. *Cultural diversity for the fire and emergency services instructors*. Emmitsburg, MD: Author.

National Fire Protection Association. 1998. *Fire department profile overview*. Presented at National Volunteer Fire Summit, Emmitsburg, MD, June.

Overman, Stephanie. 1991. Managing the diverse workforce. *Human Resources Magazine*, April.

Roosevelt, Thomas R., Jr. 1991. *Beyond race and gender*. New York: American Management Association.

Velasquez, Mauricio. 2005. Jump starting your law firm's diversity program. Minority Corporate Council Association, *Diversity & the Bar*, July–August.

Wong, Herbert Z., and Aaron T. Olson. 2009. *Multicultural and diversity strategies for the fire service*. Upper Saddle River, NJ: Pearson Prentice Hall.

Source: MFRI

3 CHAPTER

Legal Issues

KEY TERMS

common law, *p. 64*

doctrine of *stare decisis*, *p. 65*

essential functions, *p. 61*

hostile environment, *p. 67*

prima facie, *p. 81*

protected class, *p. 77*

reasonable accommodation, *p. 62*

sexual harassment, *p. 67*

undue hardship, *p. 62*

writ of certiorari, *p. 84*

OBJECTIVES

After reading this chapter, you should be able to:

- Understand the law and personnel management.
- Identify appropriate federal/state statutes and executive orders.
- Prevent sexual harassment.
- Understand fire service equal employment and the law.
- Describe the Americans With Disabilities Act (ADA).
- Describe the Age Discrimination in Employment Act (ADEA).
- Understand the relationship with legal counsel and the court system.

Given the litigious society in which we live, the legal issues faced by personnel managers in the fire service will be frequent, complex, and challenging. This was not the case in the past when legal challenges in the fire service usually involved union versus management issues, not employment law, civil liability, and other such issues. The fire department was seldom seen in the community except during an emergency, and few people would have even thought of suing the fire department as a result of its actions. This is not the case today, nor will it be in the future.

In today's fire service, because of the diverse mixture of people, cultures, and circumstances, conflicts might occur that need to be resolved through litigation or by other legal means. Currently, employees are much more aware of their rights and are prepared to pursue them if necessary. An effective fire officer must be conscious of and generally understand the legal implications of personnel management. Understanding does not require one to be an attorney, but it is important to know when to call one. Through research and study, managers can be in a position to recognize when circumstances warrant legal assistance for the protection of individuals and their department, and consequently avoid playing the part of a "firehouse lawyer," dispensing personal opinion as if it were legal advice. Every fire station seems to have at least one of these.

The best practice is proactive treatment of people and decisions in order to avoid legal conflict. In many cases by the time legal assistance is provided, the illegal acts or circumstances have already been committed. The situation is then reduced to damage control. By knowing the right action to take in the appropriate circumstance, an officer will be a tremendous asset to the organization and save him- or herself the turmoil of getting involved in these actions. For example, providing appropriate counseling and follow-up action for a male firefighter under one's command who is harassing a female firefighter may prevent the need for legal action at a later date. Like a fire prevented, the effects of being proactive are difficult to measure, but we know that they have tremendous value. A person's respect as an officer will increase because his or her decisions, based on an understanding of legal actions and laws, will be fair, just, and legal.

The Law and Personnel Management

The law does not interfere with good management, but it does prescribe certain limits in the way business is done and it does increase the consequences and the administrative workload of the fire service personnel function. Since the 1960s the law has imposed more restraint on the employee–employer relationship than at any other time in recent history. The personnel policies of fire service organizations are subject to audits by government regulatory agencies and by the legal system via employee actions, which motivate personnel management to be more objective and aware of legal issues in developing job-related policies.

The fire officer must consider possible legal implications before taking action, and therefore more knowledge of the law is required than in the past. The law's effect on personnel relations has greatly improved this area of management, whereas other efforts have failed. Currently employees are treated better and more fairly than they have been at any prior time in history; everyone has benefited from these changes in one way or the other. Yes, the law is sometimes misapplied to an undeserving individual; and yes, sometimes people gain an advantage that they do not deserve. However, more problems in this area come from the political objectives of organizations and governments rather than from application of the law. We need to look at the larger picture and applaud the progress to date. Tests today have to be job related and accurately predict performance on the job. Questions on job applications also have to be job related and not just about one's personal information. Equal work deserves equal pay and much, much more.

The current legal implications will make the job of a fire service personnel manager more difficult and at times more frustrating. Many of the problems occur because some fire departments are poorly prepared to comply with legal requirements and to make objective decisions. Poor leadership is easy to blame on society or on the lawyers who complicate everything. Departments who care about their employees, their most valuable asset, invest the time and money into ensuring that their personnel practices are fair, objective, and not easily subjected to challenge. When the decisions of these agencies are questioned, the department usually prevails. Learning about the law and developing a good working relationship with the legal counsel available to one's organization can greatly improve managerial effectiveness.

Federal Statutes

This section presents a brief description of the most relevant laws pertaining to fire service personnel management. It is not important for managers to understand the details or the exact legal fact pattern of each case; it is, however, crucial to understand the general history and major issues involved in these laws because they have significantly shaped personnel management.

CIVIL RIGHTS ACTS OF 1866 AND 1871

The Civil Rights Act of 1866 is based on the Thirteenth Amendment of the Constitution, which was initially created to abolish slavery and currently prohibits racial discrimination in the making and enforcement of contracts, including hiring and promotion decisions. The Civil Rights Act of 1871 is based on the Fourteenth Amendment and prevents the state from taking life, liberty, or property without due process of law and prevents the states from denying equal protection of the laws. This act granted all citizens the right to sue in federal court if they felt deprived of their civil rights. These statutes establish constitutional power as the foundation of subsequent laws.

CIVIL RIGHTS ACT OF 1964

The Civil Rights Act of 1964 is the major legislation regulating equal employment opportunity in the United States. The act is divided into several sections or titles. The most important title with respect to personnel management decisions is Title VII, which states that it is illegal for an employer to

(1) fail or refuse to hire or discharge any individual, or otherwise discriminate against any individual with respect to compensation, terms, conditions, or privileges of employment because of such individual's race, color, religion, sex, or national origin, or (2) to limit, segregate, or classify employees or applicants for employment in

any way that would deprive or tend to deprive any individual of employment opportunities or otherwise adversely affect status as an employee because of such individual's race, color, religion, sex, or national origin.

Title VII covers employers affecting interstate commerce with fifteen or more employees as well as state and local governments, schools, colleges, unions, and employment agencies. This act also created the Equal Employment Opportunity Commission (EEOC), the agency empowered to enforce compliance with the statute.

EQUAL EMPLOYMENT OPPORTUNITY ACT OF 1972

This act amended the Civil Rights Act of 1964 and expanded its coverage. The Civil Rights Act now covers almost all public and private employers with fifteen or more employees, except certain private clubs, religious organizations, and employment connected with Indian reservations.

CIVIL RIGHTS ACT OF 1991

Prior to this act Title VII limited awards to equitable relief such as back pay, lost benefits, future pay (in some cases), and attorney's fees and costs. Under the Civil Rights Act of 1991, additional compensatory and punitive damages were available in cases of intentional discrimination under Title VII and the Americans With Disabilities Act. The addition of damages to the Civil Rights Act of 1991 increased the potential payoff for a successful discrimination suit and thus has increased the number of suits filed against organizations. The plaintiff also has the right to demand a jury. Title VII now also applies to a U.S. citizen employed in a foreign facility owned by a U.S. company.

FAIR LABOR STANDARDS ACT (FLSA) OF 1938

The Fair Labor Standards Act (FLSA) established minimum wage and maximum hour standards for the first time on a national basis and controls working hours for children. This act originally applied to the private sector only. The FLSA has been amended several times since its enactment and in 1985 was held to be applicable to state and local governments by the U.S. Supreme Court in *Garcia v. San Antonio Metropolitan Transit Authority.* Under the FLSA, all employees are classified as either exempt or nonexempt. A nonexempt employee is entitled to a minimum wage and overtime pay as well as other protections set forth in the FLSA.

Exempt employees are not protected under these rules. To classify an employee as exempt, he or she must be paid on a salary basis. Anyone paid on an hourly basis is automatically considered nonexempt; however, there can be nonexempt employees who are paid a salary. Excluding certain industry-specific exceptions, federal and state law require that nonexempt employees be paid overtime. Under the FLSA, nonexempt employees must be paid one-and-a-half times their normal rate of pay for hours worked in excess of 40 hours during a workweek. A workweek is defined as seven consecutive 24-hour periods.

Public safety agencies have been granted exceptions to the application of FLSA under Section 7-K, which allows for increased workweeks, rotating shift schedules, as well as maximum work hours over a certain period of time.

EQUAL PAY ACT OF 1963

The Equal Pay Act, an amendment to the Fair Labor Standards Act, requires that men and women in the same organization doing equal work must be paid equally. The act defines *equal* in terms of skill, effort, responsibility, and working conditions. Pay differentials are legal if they are based on seniority, merit, quality and quantity of work, or any factor other than gender. In general to correct an inequity under the Equal Pay Act, employers must raise the rate of the lower paid workers, not lower the rate of the higher paid employees.

AGE DISCRIMINATION IN EMPLOYMENT ACT (ADEA) OF 1967

Passed in 1967 and amended in 1986, the Age Discrimination in Employment Act (ADEA) prohibits discrimination in pay, benefits, and continued employment for employees age 40 and over. The ADEA was designed to protect older workers who in times of economic trouble may be singled out for layoffs or termination because of their higher salaries. As with Title VII the EEOC is responsible for enforcing this act. The effect of this act was to remove many of the maximum hiring age limits in public safety agencies.

OCCUPATIONAL SAFETY AND HEALTH ACT OF 1970

The Occupational Safety and Health Act requires employers to provide a safe working environment for employees. It establishes a "general duty" to provide a place of employment free from recognized hazards. Originally federal, state, and local governments were excluded, but that limitation has changed and they are now covered to various degrees. The Occupational Safety and Health Act created three new government agencies to enforce the Occupational Safety and Health Administration (OSHA) regulations. OSHA enforces the safety and health standards of the act. The Occupational Safety and Health Review Commission rules on appeals by businesses regarding the application of the standards by OSHA. The National Institute for Occupational Safety and Health (NIOSH) provides research and analysis on the causes and effects of occupational injury and illness.

REHABILITATION ACT OF 1973

The Rehabilitation Act requires federal contractors (those receiving more than $2,500 in federal contracts annually) and subcontractors to actively recruit qualified people with disabilities and to use their talents to the fullest extent possible. The Rehabilitation Act is the federal legislation that authorizes the formula grant programs of vocational rehabilitation, supported employment, independent living, and client assistance. It also authorizes a variety of training and service discretionary grants administered by the Rehabilitation Services Administration.

The act authorizes research activities that are administered by the National Institute on Disability and Rehabilitation Research and the work of the National Council on Disability. The act also includes a variety of provisions focused on rights, advocacy, and protections for individuals with disabilities.

VIETNAM ERA VETERANS' READJUSTMENT ASSISTANCE ACT OF 1974

Federal contractors and subcontractors are required under the Vietnam Era Veterans' Readjustment Assistance Act to take affirmative action to ensure equal employment opportunity for Vietnam-era veterans. Vietnam-era veterans, veterans with special disabilities, and veterans who served on active duty during a war or in a campaign or expedition for which a campaign badge has been authorized are protected in employment by the Vietnam Era Veterans' Readjustment Assistance Act of 1974, as amended, 38 U.S.C. 4212.

The law requires that employers with federal contracts or subcontracts of $25,000 or more provide equal opportunity and affirmative action for Vietnam-era veterans, veterans with special disabilities, and veterans who served on active duty during a war or in a campaign or expedition for which a campaign badge has been authorized.

PREGNANCY DISCRIMINATION ACT OF 1978

Passed as an amendment to Title VII of the Civil Rights Act, the Pregnancy Discrimination Act prohibits discrimination in employment based on pregnancy, childbirth, or related medical conditions. The basic principle of the act is that women affected by pregnancy and related conditions must be treated the same as other employees on the basis of their ability or inability to work. Pregnancy is a disability and must receive the same benefits and treatment as any other disability.

IMMIGRATION REFORM AND CONTROL ACT (IRCA) OF 1986

The Immigration Reform and Control Act established criminal and civil sanctions against employers who knowingly hire an unauthorized alien. This act applies to every employee, whether full time, part time, temporary, or seasonal. Employers have the responsibility to verify the identity and work authorization of every new employee. Under IRCA, employers may hire only persons who may legally work in the United States; that is, citizens and nationals of the United States and aliens authorized to work in the United States. The employer must verify the identity and employment eligibility of anyone to be hired, which includes completing the Employment Eligibility Verification Form (I-9). Employers must keep each I-9 verification form on file for at least 3 years, or 1 year after employment ends, whichever is longer.

AMERICANS WITH DISABILITIES ACT (ADA) OF 1990

The Americans With Disabilities Act provides people with disabilities the protection from discrimination in employment, transportation, and public accommodation. The ADA defines an individual with a disability as a person who has, or is regarded as having, a physical or mental impairment that substantially limits one or more major life activities. In general the ADA prohibits an employer from discriminating against a qualified individual with a disability. A qualified individual is deemed to be able to perform the **essential functions** of the job with or without accommodation.

essential functions
■ the fundamental job duties performed in a position; function may be essential because the position exists to perform that function, a limited number of employees perform the function, or the function is highly specialized.

Under the ADA, the employer must make a **reasonable accommodation** for the mental or physical limitations of the individual with a disability who is otherwise qualified for the position, unless it would pose an **undue hardship** on the employer's operations. The act has prompted a great deal of litigation involving what a "reasonable accommodation" is and what constitutes an "undue hardship."

The enforcement agency for ADA is the EEOC, and the act applies to private employers, state and local governments, employment agencies, and labor unions with fifteen or more employees.

FAMILY AND MEDICAL LEAVE ACT OF 1993

This act requires employers having more than 50 employees to provide up to 12 weeks of unpaid leave for employees after the birth or adoption of a child; to care for a seriously ill child, spouse, or parent; or in the case of the employee's own serious illness. Eligible employees are those who have worked for the employer more than 1 year and have worked 1,250 hours during the previous year. Some exceptions include "key" employees defined as those in the highest 10 percent of the workforce and whose absence would cause substantial economic harm to the employer.

The employer is required to maintain the health and insurance benefits and to give the absent workers their previous positions back when they return to work. The enforcement of the Family and Medical Leave Act is the responsibility of the Department of Labor.

RELIGIOUS FREEDOM RESTORATION ACT OF 1993

The Religious Freedom Restoration Act (RFRA) is a 1993 U.S. federal law aimed at preventing laws that substantially burden a person's free exercise of religion. The law reinstated the *Sherbert* Test, mandating that strict scrutiny be used when determining whether the Free Exercise Clause of the First Amendment to the United States Constitution, guaranteeing religious freedom, has been violated. In this, the court must first determine whether the person has a claim involving a sincere religious belief, and whether the government action is a substantial burden on the person's ability to act on that belief. If these two elements are established, then the government must prove that it is acting in furtherance of a compelling state interest, and that it has pursued that interest in the manner least restrictive, or least burdensome, to religion.

UNIFORMED SERVICES EMPLOYMENT AND REEMPLOYMENT RIGHTS ACT OF 1994

The Uniformed Services Employment and Reemployment Rights Act (USERRA) prohibits discrimination against persons because of their service in the Armed Forces Reserve, the National Guard, or other uniformed services. USERRA prohibits an employer from denying any benefit of employment on the basis of an individual's membership, application for membership, performance of service, application for service, or obligation for service in the uniformed services. USERRA also protects the right of veterans, reservists, National Guard members, and certain other members of the uniformed services to reclaim their civilian employment after being absent due to military service or training. To be protected the employee must provide reasonable advanced notice and the employer is required to maintain health benefits, but not the difference between military and civilian pay.

HEALTH INSURANCE PORTABILITY AND ACCOUNTABILITY ACT OF 1996

This act requires the Department of Health and Human Services to establish national standards for electronic health care transactions and national identifiers for providers, health plans, and employers. It also addresses the security and privacy of health data. Adoption of these standards is intended to improve the efficiency and effectiveness of the nation's health care system by encouraging the widespread use of electronic data interchange in health care.

Executive Orders

The president of the United States has several methods whereby he or she can influence the legal environment of the nation. Options include approving or vetoing legislation passed by Congress, appointing the justices of the Supreme Court with Senate confirmation, and directing the executive branch agencies of government to take certain actions. The president has the authority to issue executive orders that specify the conditions and rules for government business. These executive orders are issued unilaterally by the president and do not require the approval of Congress. Even though orders do not apply directly to private businesses and contractors, such businesses cannot conduct transactions with the federal government unless they are in compliance. Obviously, a huge incentive to be in compliance comes from the number and dollar volume of federal government contracts.

The Office of Federal Contract Compliance Programs (OFCCP), a section of the Department of Labor, has the responsibility to monitor and enforce many of the executive orders. With respect to affirmative action, contractors must have plans that address three basic components: a proper utilization analysis of the workforce and labor market, goals and timetable, and action steps. The OFCCP annually audits government contractors to ensure that they have been actively pursuing the goals in their plans. If a contractor is found to be in noncompliance, the Department of Justice may institute criminal proceedings or cancel current contracts as well as restrict the right to bid on future contracts.

Several of the most relevant executive orders regarding fire service personnel management are summarized here.

EXECUTIVE ORDER 11246

Executive Order 11246 prohibits discrimination based on race, color, religion, sex, and national origin. It applies to federal contractors and subcontractors with contracts of $10,000 or more. Those with contracts greater than $50,000 must develop a written affirmative action plan for each of their establishments within 120 days of the beginning of the contract.

EXECUTIVE ORDER 11375

With this order, discrimination in employment based on sex was prohibited. In 1967, Executive Order 11375 expanded Executive Order 11246 to include women. After considerable pressure from women's groups, the sex discrimination provisions of the order finally began to be enforced in 1973. The order requires federal contractors and subcontractors to "take affirmative action to ensure that applicants are employed, and that employees are treated during employment, without regard to their race, color, religion, sex or national origin." Specific actions outlined in the order include upgrades, transfers, recruitment, compensation, and training.

EXECUTIVE ORDER 11478

This order requires the federal government to base its employment practices on merit and fitness. The order proposes, to the maximum extent possible, to provide sufficient resources to administer such a program in a positive and effective manner; to assure that recruitment activities reach all sources of job candidates; and to utilize to the fullest extent the present skills of each employee. In addition, the order provides the maximum feasible opportunity to employees to enhance their skills so they may perform at their highest potential and advance in accordance with their abilities, and it provides training and advice to managers and supervisors.

State and Local Laws

Numerous state and local laws affect the management of personnel within a fire department. Even prior to federal legislation a number of states and jurisdictions had passed laws regarding discrimination, fair employment practices, and other issues. State laws must be consistent with federal law, but they are not necessarily the same. State laws often apply to companies and circumstances that are not covered by federal law. For example, federal employment laws such as Title VII apply only to businesses with fifteen or more employees, but similar state laws may apply to businesses with fewer employees. State and especially local laws may apply to volunteer fire departments, whereas federal employment laws usually do not.

In addition, most city and county governments have personnel laws or merit regulations. Because these laws apply directly to members of the fire department, one should be particularly aware of these regulations. Many of the legal actions and grievances managers face originate from the misapplication of these regulations. It is a manager's responsibility to be aware of the regulations and how they are applied within a particular jurisdiction. Being aware of previous actions and how they were resolved may prove helpful in dealing with both current and future disputes. The local office of a law or legal assistance agency can provide great assistance and should be contacted when questions arise.

common law
■ a rule made by judges as they resolve disputes between parties that are recorded and subsequently referred back to as precedent for making future decisions.

Common Law

Common law, or case law, is a rule made by judges as they resolve disputes between parties. The common law system, on which the U.S. legal system is based, began as a system of laws established in England several hundred years ago. To help unify England, William the Conqueror established the King's Court. The purpose of the King's Court was to develop a common set of rules and apply them in a uniform manner throughout the kingdom. Important decisions were recorded and subsequently referred back to as precedent for making future decisions. In this way, earlier cases became precedents for deciding subsequent cases with similar issues and facts. When new types of disputes arose, the judges created new laws to resolve them. This type of system is still in place today in England.[1]

Judges in the United States do not make laws, but they interpret and apply them. Their interpretations establish the precedents that are used to decide new cases. The use of precedents to decide cases is referred to as the **doctrine of *stare decisis*,** which helps litigators anticipate how the courts may rule in cases that are similar to the precedent-setting case. This can be confusing in that depending on the type of court and the section of the country it is in, the decisions may or may not be precedents for other courts. The U.S. Supreme Court serves as the court of final appeal. Decisions made by the U.S. Supreme Court are binding and can be overturned only through legislation.

doctrine of *stare decisis*
■ the use of precedents to decide cases, which helps litigators anticipate how the courts may rule in cases that are similar to the precedent-setting case.

In order to make sense of how precedents apply to situations in the fire department without going to court, one will need legal assistance. All too often today, someone will see a news story of a court verdict with a large monetary award and recognize similarities to his or her own circumstances. Normally, the news provides only a synopsis of the case that may highlight the more provocative points. The next day the organization is faced with several grievances from employees who may, for lack of a better term, be attempting to "cash in." In most cases the issues are not exactly the same. The publicized case most likely included several weeks of testimony and evidence presentations. These arduous details get lost in the media "sound bite," and people sometimes misinterpret their particular situations. It is always best to get professional legal advice on how precedents apply to particular circumstances and subsequently let it guide one's actions.

Of the huge volume of legal interpretations and case law in existence, major decisions by the U.S. Supreme Court are the most important and affect how to manage personnel in the fire department. The following cases have been selected for review because of their importance to the personnel management functions of fire departments and public safety agencies. Please review these cases carefully and be aware of them and of their importance, but remember to obtain competent legal advice for specific circumstances.

GRIGGS V. DUKE POWER COMPANY

After the enactment of Title VII, the company required job applicants to have a high school diploma or pass a written test to qualify for certain jobs. The plaintiffs demonstrated that more whites than blacks in the relevant labor pool had a high school diploma. People already in the jobs were performing well without a high school diploma, and as such no business necessity could be shown for this requirement. Duke Power was unable to defend its use of these requirements. Testimony on the record stated that the company had not studied the relationship between these employment practices and the employee's ability to perform the job.

The U.S. Supreme Court ruling in 1971 against the Duke Power Company stated that employment practices must be related to job performance. This decision meant that when employment practices eliminate substantial numbers of minority or women applicants, the burden of proof is on the employer to show that the practice is job related. The court did note that the company had not intended to discriminate, which highlights the importance of consequences as opposed to motivations in cases of disparate impact on certain groups.

ALBEMARLE PAPER COMPANY V. MOODY

In 1975, the U.S. Supreme Court reaffirmed the requirement that any test used in the selection or promotion decisions must be validated if found to impact adversely the selection of women or minorities. The employer has the burden of proof to show that the test is valid and that it actually measures what it is supposed to measure.

This case and *United Papermakers and Paperworkers v. Moody* were decided in the same opinion. They dealt with two important issues under Title VII of the Civil Rights Act of 1964: (1) the standards a district court should use in deciding back pay awards to employees who suffered monetary loss because of racial discrimination; and (2) the requirements placed upon an employer to establish that preemployment tests that have a discriminatory effect are sufficiently "job related" to survive a legal challenge.

The Court found that tests used by the company were not sufficiently job related to be valid under Title VII and held that the trial court should have enjoined the use of the tests and awarded back pay. In deciding the issue of back pay awards, the Court ruled that such awards should follow closely upon a finding of discrimination and that ordering back pay was an appropriate incentive to employers to comply with Title VII. It concluded that the certainty of the remedy would best effectuate the statute. In all essential aspects this holding continues to provide a major form of relief under Title VII.

WEBER V. KAISER ALUMINUM

The United Steelworkers of America (USWA) and Kaiser Aluminum entered into an agreement that contained an affirmative action plan designed to eliminate a major racial imbalance in its almost exclusively white workforce. The plan reserved 50 percent of openings in the newly created in-plant training program for black employees. Brian Weber subsequently instituted a class action suit in 1974 alleging that the action by Kaiser and the USWA discriminated against him and other white employees in violation of Title VII.

The U.S. Supreme Court reversed a lower court decision and stated that Title VII does not prohibit race-conscious affirmative action plans. Because the company and the union voluntarily agreed to this specific affirmative action plan, it did not violate Title VII.

REGENTS OF THE UNIVERSITY OF CALIFORNIA V. BAKKE

The University of California had reserved sixteen places in each medical school class for minority applicants. Bakke, who was white, was denied admission even though he scored higher on the admission criteria than did some of the minority applicants. The U.S. Supreme Court overturned a quota policy for admissions at the University of California at Davis, while generally approving affirmative action programs. Allan Bakke, a white male, was denied admittance to the medical school at Davis in 1973 and 1974. In both years the school admitted only one hundred students, reserving sixteen seats for minorities. Bakke sued, claiming "reverse discrimination" because some of the minorities admitted had lower grade point averages and lower scores on the Medical College Admission Test than his. Four justices wanted to end the program for violating Title VII of the Civil Rights Act of 1964 without addressing the constitutionality of the medical school's affirmative action program. Four others wanted the Court to rule that the program was constitutional and that the medical school rightfully denied Bakke admittance. Justice Lewis Powell, Jr., essentially split the difference, arguing not only that fixed racial quota programs such as the medical school's violated the Civil Rights Act but also that a public school's admission policy "may" consider race as long as it was not the determining factor. This was constitutionally permissible under the First Amendment, he argued, because it allowed educational institutions to promote cultural diversity. The Court's decision, which affected only affirmative action programs at schools receiving federal funds, did not settle the question of reverse discrimination, because it was decided on narrow statutory grounds.

MERITOR SAVINGS BANK V. VINSON

Ms. Vinson claimed that she was subjected to **sexual harassment** by her supervisor at Meritor Savings Bank. She initially resisted but eventually gave in over the fear of losing her job. She did not file a grievance under the bank's policy.

In 1986, the U.S. Supreme Court ruled that sexual harassment is covered by Title VII and is not limited to discrimination having economic or tangible effects. The court stated that the appropriate test is not whether the sexual activity is voluntary, but that it is unwelcome. Additionally, the court established in this case the concept of a **hostile environment** in the workplace and that management knew or should have known of the offensive activity. This decision stated that in order to find sexual harassment three elements must be present: (1) It must be unwelcome; (2) the employer must have knowledge, either actual or imputed; and (3) either job opportunities must be involved or a hostile environment must be created.

SCHULTZ V. WHEATON GLASS COMPANY

This case in 1970 provided guidelines to define equal work. The female job class at Wheaton Glass Company was paid 10 percent below that of the male job class. The company stated that this discrepancy was because the male class had to perform additional duties. However, the court found that these extra tasks were infrequently performed and not by all men. The court ruled that the equal work standard required only that the jobs be substantially equal, not identical.

GARCIA V. SAN ANTONIO METROPOLITAN TRANSIT AUTHORITY

In 1985, the U.S. Supreme Court ruled that state and local government employees are covered by the FLSA. This decision holds that Congress has the power under the Commerce Clause of the Constitution to extend the Fair Labor Standards Act, which requires that employers provide minimum wage and overtime pay to their

sexual harassment
- unwelcome sexual advances, requests for sexual favors, and other verbal or physical conduct of a sexual nature when submission to or rejection of this conduct affects employment, unreasonably interferes with work performance, or creates an intimidating, hostile, or offensive work environment.

hostile environment
- concept established by the U.S. Supreme Court in the *Meritor Savings Bank v. Vinson* case; refers to a situation in the workplace regarding a pattern of exposure to unwanted sexual behavior from persons and the employer takes no steps to discourage or discontinue such behavior.

employees, to state and local governments. The decision overruled a previous decision of the Court, *National League of Cities v. Usery* (1976), which had held that such regulation of the activities of state and local governments "in areas of traditional governmental functions" would violate the Tenth Amendment to the United States Constitution. When Congress passed the FLSA in 1938, it did not apply either to employees of private transit companies or to employees of state and local governments. In 1976, however, the Court held in *National League of Cities* that Congress lacked authority to regulate the wages and hours of governmental employees performing "traditional governmental functions." The San Antonio Metropolitan Transit Authority, which had been observing the overtime requirements of federal law up to that point, responded by informing employees that it was no longer obliged to provide them with overtime pay. The *Garcia* decision placed the approximately 18 million state and local government employees under the provisions of FLSA, including fire departments.

MCCARTHY V. PHILADELPHIA CIVIL SERVICE COMMISSION

The U.S. Supreme Court upheld the validity of Philadelphia's residency requirement and said that it was not an infringement of the employee's constitutional right to travel in interstate commerce. Philadelphia municipal regulation requiring city employees to be residents of the city was held to be constitutional as a bona fide continuing residence requirement and not to violate the right of interstate travel of the appellant, whose employment as a city firefighter was terminated under the regulation because he moved his residence from Philadelphia to New Jersey. The court stated that the firefighter did not have a protected right in his job and that the refusal to comply with the residency requirement was a valid basis for discharge.

CLEVELAND BOARD OF EDUCATION V. LOUDERMILL

In 1985, the U.S. Supreme Court ruled that when an employee is faced with termination from employment, that employee is to have a hearing and be afforded certain hearing rights including the opportunity to refute the charges. Employers need to conduct a fair hearing process prior to the final decision regarding the termination of employees. The U.S. Supreme Court found that the essential requirements of due process are notice and an opportunity to respond and that tenured public employees are entitled to oral or written notice of charges against them, an explanation of the employer's evidence, and an opportunity to present their side of the story. Because the respondents alleged that they had no chance to respond, the District Court erred in dismissing for failure to state a claim.

JOHNSON V. SANTA CLARA TRANSPORTATION AGENCY

In this 1987 case, the U.S. Supreme Court ruled that in traditionally sex-segregated jobs, a qualified woman can be promoted over a marginally better qualified man to promote more balanced representation. For the first time, the U.S. Supreme Court decided that a voluntary sex-based affirmative action plan can be used to overcome the effects of past job discrimination based on gender. Further, the *Johnson* decision clearly endorsed affirmative action as a remedy for past discrimination, as long as it is temporary. The Court stressed the need for affirmative action plans to be flexible, gradual, and limited in their effect.

HARRIS V. FORKLIFT SYSTEMS, INC.

In this 1993 case, the U.S. Supreme Court ruled that plaintiffs in sexual harassment suits need not show psychological injury to prevail. Whereas a victim's emotional state may be relevant, she or he need not prove extreme distress. In considering whether illegal harassment has occurred, juries must consider factors such as the frequency and severity of the harassment, whether it is physically threatening or humiliating, and whether it interferes with an employee's work performance.

NLRB V. J. WEINGARTEN

In this 1975 case, the U.S. Supreme Court ruled that employees have a right to union representation at investigatory interviews. These rights have become known as the Weingarten rights.

The employee must make a clear request for union representation before or during the investigatory interview. The employee cannot be punished for making this request. After the employee makes the request, the employer must choose from among three options: (1) Grant the request and delay questioning until the union representative arrives and has a chance to consult privately with the employee; (2) deny the request and end the interview immediately; or (3) give the employee a choice of having the interview without representation or ending the interview. If the employer denies the request for union representation and continues to ask questions, it commits an unfair labor practice and the employee has a right to refuse to answer. The employer may not discipline the employee for such a refusal.

BURLINGTON INDUSTRIES, INC. V. ELLERTH; FARAGHER V. CITY OF BOCA RATON

These two 1998 cases were treated together by the U.S. Supreme Court because they presented similar issues regarding the liability of employers for sexual harassment. The U.S. Supreme Court decided that the company or city was liable for harassment by a supervisor even when it had not been informed of the harassment and when no adverse employment action, such as demotion, had been taken against the accusers. The Court ruled that the employer may be held liable, but that employers may avoid liability through certain actions.

Although these two cases were filed and argued separately, they both involved Title VII of the Civil Rights Act of 1964, were decided on the same day, engendered the same voting pattern among the justices, shared important language, and substantially clarified the law of sexual harassment in the workplace. The effect of the two decisions was to make some lawsuits against employers easier to win while rewarding companies with effective antiharassment policies by limiting their legal exposure.

Both decisions contained an identical page-long set of rules to be applied in sexual harassment cases in the workplace. First, employers are responsible for harassment engaged in by their supervisory employees. Second, when a quid pro quo condition can be demonstrated (i.e., a discharge or demotion), the employer is absolutely liable. And third, where there is no tangible action, an employer can defend itself by showing that it took reasonable care to prevent and correct harassing behavior and that the complaining employee had failed to take advantage of preventive or corrective opportunities provided by the company.

GRATZ ET AL. V. BOLLINGER (UNIVERSITY OF MICHIGAN)

Petitioners filed this class action alleging that the University of Michigan's use of racial preference in undergraduate admissions violated the Equal Protection Clause of the Fourteenth Amendment and Title VII of the Civil Rights Act. The university's guidelines used a selection method under which every applicant from an underrepresented racial and ethnic minority group automatically was awarded 20 points of the 100 points needed to obtain admission. The U.S. Supreme Court ruled in June 2003 that the current policy was not narrowly tailored to achieve the asserted interest in diversity and that this policy violated the Equal Protection Clause and Title VII. The court emphasized the importance of considering each particular applicant as an individual, assessing all of the qualities that the individual possesses. However, the policy that automatically awarded 20 points to every minority is not narrowly tailored and therefore illegal.

GRUTTER V. BOLLINGER ET AL. (UNIVERSITY OF MICHIGAN LAW SCHOOL)

The University of Michigan Law School follows a policy that seeks to achieve student body diversity. The policy neither defines diversity solely in terms of racial and ethnic status nor restricts the types of diversity contributions eligible for "substantial weight" in the selection process. The policy does reaffirm the law school's commitment to diversity with special reference to African American, Hispanic, and Native American students who otherwise may not be represented in the student body in meaningful numbers. The U.S. Supreme Court ruled in June 2003 that the law school's narrowly tailored use of race in admission decisions to further a compelling interest in obtaining the educational benefits that flow from a diverse student body is not prohibited by the Equal Protection Clause or Title VII. To be narrowly tailored race conscious, an admissions program cannot insulate each category of applicants with certain desired qualifications from competition with all other applicants. Instead it may consider race or ethnicity only as a "plus" in a particular applicant's file. It must be flexible enough to consider all pertinent elements of diversity in light of the particular qualifications of each applicant. It follows that universities cannot establish quotas for members of certain racial or ethnic groups or put them on a separate admissions track.

Sexual Harassment

The fire service of this country is and has been since its inception a male-dominated workforce. As of 2008, only an estimated 4 percent of the firefighters in the United States are women. They face challenges uncommon to most other occupations due to many of the traditions and the culture of the fire service. This is further complicated by the close working relationships, long duty shifts, and sleeping arrangements in fire stations. Some of these issues have been addressed in most departments since the early 1980s, especially those that relate to physical changes such as personal facilities. But the far more difficult issues to address are the attitudes and actions that cause sexual harassment and the resulting legal issues arising from these actions. Unfortunately, sexual harassment is a predominant issue in the fire service and requires a major effort by fire service personnel managers to address.

Many factors in fire service history and society influence the prevalence of sexual harassment. This section does not attempt to address these issues, but it will review the legal aspects and policies that the management of a fire department can put in place to best serve its employees and to protect the department from legal actions. As with most personnel issues, proactive prevention measures are the most effective way to deal with harassment issues. No one should ever have to endure these types of challenges within the fire service environment.

Unwelcome sexual advances, requests for sexual favors, and other verbal or physical conduct of a sexual nature constitute sexual harassment when submission to or rejection of this conduct explicitly or implicitly affects an individual's employment, unreasonably interferes with an individual's work performance, or creates an intimidating, hostile, or offensive work environment.

Sexual harassment can occur in a variety of circumstances, including but not limited to the following:

- The victim as well as the harasser may be a woman or a man. The victim does not have to be of the opposite sex.
- The harasser can be the victim's supervisor, an agent of the employer, a supervisor in another area, a coworker, or a nonemployee.
- The victim does not have to be the person harassed but could be anyone affected by the offensive conduct.
- Unlawful sexual harassment may occur without economic injury to or discharge of the victim.
- The harasser's conduct must be unwelcome.

In 1980, the EEOC issued guidelines stating that sexual harassment is a form of sex discrimination and violates Title VII. This policy was amended in 1993 when the EEOC issued regulations that broadened the types of harassment considered illegal and included a requirement that employers have a duty to maintain a working environment free of sexual harassment. These regulations clarified that the standard for evaluating sexual harassment is what a "reasonable person" in the same or similar circumstances would find intimidating, hostile, or abusive. The guidelines clearly state that employers are liable for the acts of those who work for them if they knew or should have known about the conduct and took no immediate, appropriate corrective action. Employers who fail to draw up explicit, detailed sexual harassment policies and procedures may place themselves at greater legal risk.

According to the EEOC guidelines, verbal and physical conduct of a sexual nature is harassment when the following conditions are present:

- Submission to such contact is made either explicitly or implicitly a term or condition of an individual's employment.
- Submission to or rejection of such contact by an individual is used as the basis for employment decisions affecting the individual.

- Such conduct has the purpose or effect of unreasonably interfering with an individual's work performance or creating an intimidating, hostile, or offensive working environment.

Sexual harassment can be further defined as offensive behavior by persons in authority toward those who can benefit or be injured in an official capacity. Therefore, it is primarily an issue of abuse of power, not sex. Sexual harassment is a breach of a trusting relationship that should be sex neutral; it is unprofessional conduct and undermines the integrity of the employment relationship. Sexual harassment is coercive behavior, whether implied or actual. It is unwanted attention and intimacy in a nonreciprocal relationship. Sexual harassment is an illegal form of sex discrimination. Whatever definition of sexual harassment is used, it always includes one key element—the behavior is uninvited, unwanted, and unwelcome.

CONDUCT THAT MAY CONSTITUTE SEXUAL HARASSMENT

Repeated sexual innuendoes or sexually oriented comments
Obscene remarks, jokes, slurs, or language related to sex
Letters, notes, e-mail, or graffiti that is of a sexual nature
Sexual propositions, insults, threats, or intimidation
Sexually oriented demeaning names
Persistent, unwanted sexual or romantic requests or attention
Leering, whistling, or other sexually suggestive sounds or gestures
Display of pornographic photographs, magazines, and the like
Coerced or unwelcome touching, fondling, kissing, and the like
Subtle or overt pressure for sex or sexual favors
Coerced sexual intercourse[2]

The U.S. Supreme Court in the spring of 1998 issued several major decisions that dealt with sexual harassment cases and their impact on employers in particular. The case of *Burlington Industries, Inc. v. Ellerth* involved a claim by a former employee for sexual harassment or gender discrimination. Ms. Ellerth alleged that she had been subject to constant sexual harassment in the form of suggestive and offensive comments and advances, including threats to her job, and was forced to resign. Despite being aware of the Burlington Industries policy against sexual harassment, she did not report the conduct to anyone while it was occurring, nor did she suffer any retaliation for refusing the sexual requests. At the time of her resignation, she cited other reasons for leaving Burlington, but shortly thereafter wrote a letter to Burlington complaining of her treatment by her supervisor. She then filed suit against Burlington, but not against her supervisor.

The U.S. Supreme Court treated the *Ellerth* case and the case of *Faragher v. City of Boca Raton* together because they presented similar issues regarding the liability of employers for sexual harassment. Ms. Faragher also sued under Title VII for sexual harassment. She was a lifeguard at the municipal beach and alleged that her immediate supervisor was fostering a sexually hostile environment at the beach. In 1986 the city adopted a sexual harassment policy, but failed to disseminate it to its employees at the beach. Faragher resigned from her lifeguard job in 1990. She filed suit 2 years later against the supervisor and the city.

The issue for the U.S. Supreme Court was not whether sexual harassment had occurred, because in both cases the lower court's decisions were upheld. Nor did the Court address the liability of the individual harassers. The issue the U.S. Supreme Court had to decide was whether the employer was liable for harassment by a supervisor when it had not been informed of the harassment and when no adverse employment action, such as a demotion or reduction in pay, had been taken against the accusers. The court ruled that the employer may be held liable.[3]

The Court issued an identical rule in both cases that employers are vicariously and therefore automatically liable for harassment committed by supervisors. However, if the harassed employee has not suffered a loss of "tangible job benefits" as a result of the harassment, the employer can avoid liability if it exercised reasonable care to prevent and promptly correct sexually harassing behavior or if the employee unreasonably failed to take advantage of any preventive or corrective opportunities provided by the employer. This represents a shift from prior law, where the employee was required to demonstrate the employer's negligence. The Court defined a tangible job benefit as "a significant change in employment status, such as hiring, firing, failing to promote, reassignment with significantly different responsibilities, or a decision causing significant change in benefits."

These precedents place a fire service personnel manager in a sometimes difficult position, in that even in cases in which the sexual harassment was not reported, the fire department may be held liable. Many feel that the Court was saying that supervisors are expected explicitly and proactively to prevent sexual harassment and hostile working environments from occurring. These cases are a wake-up call for employers to review and revise, if necessary, their sexual harassment policies and procedures, to make sure that they are properly disseminated throughout the workforce, and to respond quickly and decisively when a complaint is received.

Preventing Sexual Harassment

Sexual harassment creates an offensive and hostile work environment that prohibits effective performance, leads to expensive legal actions and negative publicity, and ultimately interferes with the ability of the department to attract and retain the best talent. It is neither a desirable situation in any fire department, nor should any firefighter be treated in this manner. Unfortunately, male-dominated work environments such as fire departments have greater difficulty with these types of issues. Having women in the workforce is not new for most organizations, but it is for many fire departments. In fact, it might be the first time that many male firefighters and officers have had women in their work environment. The study *A National Report Card on Women in Firefighting* in April 2008 concluded the following with respect to harassment experienced by female firefighters.

Discrimination or Harassment Experienced by Firefighters

QUESTION	% OF WOMEN	% OF MEN
I have experienced different treatment because of my gender	84.7	12.4
My gender has created barriers to promotion	36.5	7.4
Males and females are not treated the same in training	33.9	17.8
I have experienced:		
Shunning/isolation	50.8	2.3
Verbal harassment	42.9	2.8
Sexual advances	30.2	0.5
Hazing	18.6	1.4
Hostile notes or other written material	18.2	0.9

No matter how one views these statistics, it is clear that women firefighters are still facing significant issues with regard to disparate treatment and sexual harassment in fire departments. The necessary change in firefighter culture will require a great deal of training and sensitivity as well as a "zero-tolerance" policy toward sexual harassment issues. The barriers to the full inclusion of women in the fire service remain in place in many fire departments either consciously or unconsciously to varying degrees. The goal is to have gender inclusiveness and that means more than just numbers. It will require equal treatment and acceptance in all actions and activities of the department. To achieve this goal fully, the culture of the fire service will require change. Changes in the underlying culture of organizations are difficult and require a multitude of strategies and actions that are consistently applied over time. The *National Report Card on Women in Firefighting* recommends the following five strategies for changing fire service culture:

- Commitment by top leadership of fire department and government
- Monitoring and accountability that result in consequences for noncompliance with departmental policy; in addition, reward of positive contributions
- Human resource policy and procedures that embody transparency, objectivity, and performance relatedness
- Zero tolerance and other policies to change individual behavior and control hostile acts
- Sustained effort over time as real culture change may require 5 years or longer[4]

While culture change is a protracted effort, the changing of policies and procedures can be effective in the short term and prevent the more obvious transgressions. In order to be in a better position to prevent and, if necessary, defend against sexual harassment charges, the following policy steps are helpful:

1. Establish a clear antiharassment policy with examples of prohibited actions that are particular to your department.
2. Educate all employees about the nature of sexual harassment, the kind of conduct that is illegal, and the inappropriateness of engaging in that kind of conduct.

3. Distribute the policy to all employees and consider an attachment that they are required to sign indicating their receipt of the policy.
4. Post the policy prominently in all stations and work locations.
5. Clearly identify the persons and offices to which complaints should be directed, with identification of how these individuals are to be contacted.
6. Provide adequate and frequent training sessions to employees and those responsible for receiving complaints as a part of normal recruitment, in-service, and officer training programs.
7. Educate all personnel about their own potential liability when they engage in harassing conduct, particularly conduct that is retaliatory.[5]

If the fire department is to be successful in preventing sexual harassment, supervisors must assume an active role. They have a special responsibility and duty to prevent these types of actions. Obviously, one of the most damaging types of sexual harassment is when the supervisor is involved as the harasser. At all levels in the department, an awareness of and sensitivity to these issues are crucial.

Fire department officers can be better equipped to handle instances of sexual harassment by observing and committing to the following behaviors:

Understand
- Be fully familiar with departmental policy on discrimination.
- Examine your feelings, attitudes, and opinions about sexual harassment.
- Ensure your behavior is proper.

Observe
- Be aware and conscious of potential sexual harassment behaviors or incidents within your area of responsibility.
- Be sensitive to individuals who may be offended by the verbal and nonverbal behaviors of others.
- Watch for subtle forms of sexual harassment.

Model
- Exhibit the type of behavior you expect of your employees.
- Set the tone of zero tolerance within your department.

Resolve
- Inform your employees that you will take immediate and appropriate action in cases of sexual harassment.
- Thoroughly investigate complaints to determine the facts.
- Emphasize that you have an open-door policy and that anyone who has been harassed or thinks that harassment is occurring can seek your confidential advice.
- Use your department's legal services for advice and guidance.

Educate
- Conduct sexual harassment awareness and prevention training programs.
- Discuss the issues of sexual harassment in staff meetings.
- Post policies in a conspicuous place.
- Ensure a work environment free from harassment.[6]

In order to prevent sexual harassment effectively, proper training at all levels within the fire department will be required. Regular training sessions for all managers and supervisors so that they model appropriate behavior and are able to recognize harassment situations as well as respond to complaints is essential. Firefighters will need to be trained so that they understand their responsibilities in maintaining a zero tolerance regarding sexual harassment in the fire station environment. Sexual harassment training needs to be informative, interesting, and relevant. One method to ensure relevance is to tailor the training to the environment in which it is to be conducted. This will ensure that the special circumstances involving the fire station environment, which may be quite different from other work environments, are properly covered. It is also important to document the training that occurred and who attended. These attendance files need to be retained at a central point in the department.

In the fire service work environment, managers will be faced with a sexual harassment issue at some time, even with the best prevention and education program. How it is handled will be important to the individuals involved and the general welfare and perception of the department. If investigated and adjudicated properly within the department, the complaint will serve as an example to the entire department as to the seriousness and implications of this type of behavior.

Fire Service Equal Employment and the Law

Since the early 1950s, numerous laws have been enacted to prevent illegal employment discrimination. This chapter reviews many of these statutes and court decisions. Even so, some fire departments continue to engage in discriminatory practices that may result in large fines and settlements. Fire service personnel managers must be aware of the implications and consequences of such actions and work to implement programs designed to prevent the occurrence of such acts.

At the federal level the Equal Employment Opportunity Commission (EEOC) was established to enforce provisions of Title VII of the Civil Rights Act of 1964. Title VII forbids discrimination in the workplace based on race, age, disability, religion, sex, or national origin. Title VII covers all phases and aspects of employment including but not necessarily restricted to hiring, termination of employment, layoffs, promotions, wages, on-the-job training, and disciplinary action. Businesses covered by Title VII include employers in the private sector with fifteen or more employees, educational institutions, state and local governments, labor unions with fifteen or more members, employment agencies, and, under certain circumstances, labor–management committees.

The Equal Employment Opportunity Commission's enforcement program manages between 75,000 and 80,000 charges annually. In the EEOC system, charges are prioritized into one of three categories for purposes of investigation and resource allocation: A (top priority charges to which offices devote substantial investigative and settlement efforts); B (charges deemed to have merit but needing additional investigation); and C (charges judged to be unsupported or not under the EEOC's jurisdiction and thus not pursued). Under EEOC rules of operation and investigation, settlements between disputing parties are encouraged at all stages of the process. With this in mind, the EEOC maintains a mediation-based alternative dispute resolution program. The mediation program is guided by principles of informed and voluntary participation at all stages, confidential deliberation by all parties, and neutral mediators.

Anyone who feels that he or she has suffered workplace discrimination because of his or her race, age, physical disability, religion, sex, or national origin is eligible to file a complaint with the EEOC. Complaints or charges are generally filed at an EEOC office by the aggrieved party or by his or her designated agent. All charges must be filed in writing, preferably but not necessarily on the appropriate EEOC form, within 180 days of the occurrence of the act that is the reason the complaint is being filed.[7]

To begin with, it is important to understand the terms and definitions that are used in legal challenges involving equal employment opportunity and affirmative action cases.

Equal employment opportunity. A broad concept espousing that individuals should have equal treatment in all employment-related actions.

Protected class. Individuals within a group identified for protection under equal employment laws and regulations. Various laws have been passed to protect individuals who share certain characteristics such as race, age, or gender.

protected class
■ individuals within a group identified for protection under equal employment laws and regulations.

Affirmative action. Process in which employers identify problem areas, set goals, and take positive steps to enhance opportunities for protected class members. Affirmative action focuses on hiring, training, and promoting of protected class members when they are underrepresented in an organization in relation to their availability in the labor market.

Illegal discrimination. Discrimination can be illegal in employment-related situations in which different standards are used to judge different individuals or the same standard is used but is not related to the individual's job or is applied unequally.

Reverse discrimination. When a person is denied an opportunity because of preferences given to protected-class individuals who may be less qualified.

Whereas equal employment opportunity aims to ensure equal treatment of individuals and groups, affirmative action requires employers to make an effort to hire and promote those in a protected class. Affirmative action includes taking specific actions designed to eliminate the present effects of past discrimination. This concept is probably one of the most volatile issues in managing an organization. The issues of equal employment and affirmative action are easy to understand at the concept level, but not so when they are applied at the individual level.

Increasingly, court decisions and legislative action have focused on restricting the use of affirmative action. Many critics of affirmative action point out that they are paying for circumstances that they did not participate in and that may in fact have been legal at the time the acts were committed. Proponents of affirmative action point to the history of discrimination and the positive effects of increased diversity. Affirmative action programs are supposed to be temporary in nature and end at some point. Where that point is continues to be the legal and political debate. In addition, there are many more minority classes that must be recognized in today's society, such as Hispanics and Asians. The effects of affirmative action programs must take this into account.

CONTINUUM OF DIVERSITY MANAGEMENT

Regardless of one's personal views on equal opportunity and affirmative action, as a supervisor or manager it is important to be aware of and implement the laws and policies of one's jurisdiction in a fair and unbiased manner. It is vital to understand the views on both sides of the debate but contain one's action to what is legal and not expose one's organization to unnecessary legal challenges and the expense and exposure of such actions.

Americans With Disabilities Act (ADA)

The ADA of 1990 requires an employer to provide reasonable accommodation to qualified individuals with disabilities who are employees or applicants for employment, unless doing so would create an undue hardship. For the purposes of this act, an individual with a disability is a person who has a physical or mental impairment that substantially limits one or more major life activities, has a record of such impairment, or is regarded as having such an impairment. Since its enactment, the ADA has had a dramatic impact on the hiring and retention of employees with disabilities. Even though employment in the fire and rescue service has high physical and medical requirements due to the nature of emergency services work, fire departments are being increasingly challenged to include more individuals with disabilities in the fire service work environment.

For fire service personnel managers, determining the essential functions of a job is important because state and federal law protect individuals with disabilities who are qualified—meaning that they possess the requisite skill, experience, education, and other job-related requirements of a position and can perform the essential functions of the job with or without a reasonable accommodation. Also, identifying these key functions helps set expectations and defines a job. Thus, doing so from the outset is critical to avoid inconsistent or unfair hiring decisions and to ensure that each incumbent or employee is treated equally and understands the job and expectations.

Essential functions are the fundamental job duties performed in a position. The term *essential functions* does not include the marginal functions of the position. A function may be essential because:

- The position exists to perform that function.
- There are a limited number of employees available who could perform that function.
- The function is highly specialized.

It is also imperative that the employee actually perform the particular function for it to be considered essential. Once it is determined that the employee actually performs the function, the job analysis should examine whether removing the function would fundamentally alter the position; if the preceding criteria are satisfied, the function is likely essential.

A qualified individual with a disability is an individual who, with or without reasonable accommodation, can perform the essential functions of the job in question. Reasonable accommodation removes workplace barriers for individuals with disabilities and may include but is not limited to the following:

- Making existing facilities accessible
- Job restructuring
- Modifying work schedules
- Modifying equipment
- Changing tests, training materials, or policies
- Providing qualified readers or interpreters
- Reassignments[8]

An employer is required to make an accommodation to the known disability of a qualified applicant or employee if doing so would not impose an undue hardship on the operation of the employer's business. Undue hardship means significant difficulty or expense and focuses on the resources and circumstances of the particular employer in relationship to the cost or difficulty of providing a specific accommodation. Undue hardship refers not only to financial difficulty but also to reasonable accommodations that are unduly extensive, substantial, or disruptive, or to those that would fundamentally alter the nature or operation of the business. An employer does not have to eliminate an essential function or fundamental duty of the position. An employee with a disability who is unable to perform the essential functions with or without reasonable accommodation is not a "qualified" individual with a disability within the meaning of ADA. Employers are not required to lower quality or quantity of production standards that are applied uniformly to other employees.

Major difficulties arise for the fire service in seeking reasonable accommodations, because firefighters respond to almost any type of emergency. Reasonable accommodations in the standard workplace, such as an office or a factory, are easier to achieve because the work environment is known and can be altered to meet the requirements of the law and the individual with a disability. It is more difficult to make reasonable accommodations when the work occurs in a variety of unknown environments. However, fire stations, apparatus, and equipment can be modified to make reasonable accommodations. In addition, certain jobs within the fire service do not require the same physical ability as does active firefighting, such as fire inspector, dispatcher, public educator, and others. These modifications may be used if feasible to accommodate individuals if they meet all other requirements for the position.

One major aspect of ADA concerns the hiring process and the modifications or adjustments that may enable a qualified applicant with a disability to be considered for a position within the fire department. Medical examinations or inquiries of job applicants must be job related and consistent with the employer's business needs. Many fire departments use the medical requirements of the NFPA 1582, Standard on Comprehensive Occupational Medical Program for

Fire Departments, the purpose of which is to identify the minimum medical requirements to become a firefighter. The fire department in its hiring process may not ask job applicants about the existence, nature, or severity of a disability. Applicants may be asked about their ability to perform specific job functions. It is a good practice to attach to employment applications a job description with specific information about job duties and requirements. An applicant can then be asked how he or she plans to perform each function of the job. The decision to hire the applicant must be made first on the specified job-related criteria and then a job offer may be conditioned on the results of a medical examination, as long as the medical examination is required of all job applicants. Therefore, in most fire department hiring processes the medical examination is the last step in the employment qualification process.

When an individual decides to request an accommodation, the individual must inform the employer that he or she needs an adjustment or change at work for a reason related to a medical condition. Although an individual may request an accommodation due to a medical condition, this request does not necessarily mean that the employer is required to provide the change. A request for reasonable accommodation is the first step in an interactive process between the individual and the employer designed to address these issues. The employer may request medical documentation of the disability and then work toward a reasonable accommodation, if possible. Employees and applicants currently engaging in the illegal use of drugs are not covered by ADA.

In 1999, in the case of *Murphy v. United Parcel Service,* the U.S. Supreme Court substantially limited the scope of the ADA. The court ruled that employees who can function normally when their impairments are treated do not qualify for protection under the 1990 Americans With Disabilities Act. The court's decision hinged on what Congress had intended when it enacted the original law. The ADA requires employers to hire or make reasonable accommodations for workers with disabilities, but defines *disability* as "a physical or mental impairment that substantially limits one or more major life activities." The ADA does not specifically state whether workers with correctable disabilities are covered by it. The Supreme Court decided in this case that they are not, stating "that a person be presently, not potentially or hypothetically substantially limited" in his or her ability to work. This ruling will have a major effect on the handling of future ADA cases.

Fire departments will continue to provide reasonable accommodations when they can be accomplished without adversely affecting the response to emergencies. Because entry requirements are high for firefighter positions and the work is difficult and dangerous, the likelihood of successful ADA accommodations is not as high as in other occupations. However, good management and fairness dictate that where a reasonable accommodation can be made, the fire department must strive to comply.

Age Discrimination in Employment Act

The Age Discrimination in Employment Act (ADEA) prohibits any employer from refusing to hire, discharging, or otherwise discriminating against any individual because of age. The act covers compensation, terms, conditions, and other privileges of employment including health care benefits. This act specifically prohibits age-based discrimination against employees who are at least 40 years of age. The purpose of the act is to promote the employment of older persons and to prohibit any arbitrary age discrimination in employment.

The ADEA includes a broad ban against age discrimination and also specifically prohibits:

- Discrimination in hiring, promotions, wages, or firing/layoffs
- Statements or specifications in job notices or advertisements of age preference and limitations
- Denial of benefits to older employees (An employer may reduce benefits based on age only if the cost of providing the reduced benefits to older workers is the same as the cost of providing full benefits to younger workers.)
- Mandatory retirement in most sectors since 1978

The roots of the ADEA can be traced back to 1964, when the U.S. government enacted Title VII of the 1964 Civil Rights Act. The core of Title VII was to prohibit discrimination in employment based on race, color, sex, national origin, or religion. This statute provided a way for women and minorities, in particular, to challenge barriers that limited equal opportunities in organizations. One variable noticeably missing from Title VII was age discrimination. In 1967, the Congress enacted the Age Discrimination in Employment Act to rectify this situation.

The ADEA protects against age discrimination in many employment contexts, including hiring, firing, pay, job assignment, and fringe benefits. Since its enactment, the ADEA has aided older employees by diminishing arbitrary age-based discrimination in virtually all aspects of employment, including hiring, firing, promotions, terms of compensation, and job referrals, as well as by banning age-based harassment and retaliation for filing an ADEA charge. Consequently, employers treat older employees more cautiously, carefully document non-age-related justifications for employment actions, and commonly use incentive programs to reward workers for retiring and waiving any potential ADEA claims.

Under the act, employers are forbidden to refuse to hire, to discharge, or to discriminate against anyone with respect to the terms, conditions, or privileges of employment because of a person's age. The act also forbids employees from limiting, segregating, or classifying an individual in a way that adversely affects that person's employment because of age. The act states that all job requirements must be truly job related and forbids employers from reducing the wage rate of an employee to comply with the act. In general, it forbids seniority systems or benefit plans that call for involuntary requirements due to age and also makes it illegal for employees to indicate any issue related to age in advertisements for job opportunities.

The plaintiff in an ADEA case does not need to prove that age was the *only* factor motivating the employer's decision but must establish that age was one of the determining factors guiding the employer's actions. Once the plaintiff establishes a **prima facie** case, the burden of evidence shifts to the employer. The employer must provide a legitimate, nondiscriminatory reason for the employee's demotion, forced retirement, or discharge.

There are several exceptions to the statute's nondiscrimination provisions. First, employers may use age as an employment criterion if they can justify its use. In other words, they must prove that "age is a bona fide occupational qualification (BFOQ) reasonably necessary to the normal operation of the particular business." Under this exception, employers must show that age is a reasonable measure of a

prima facie
■ means "on its first appearance," or "by first instance." In general, that on first examination, a matter appears to be self-evident from the facts or one of the determining factors.

job qualification that is important to the employer's business. The courts have interpreted this as a very narrow exception to the general prohibition of age discrimination contained in the ADEA.

In 1986, Congress amended ADEA to provide state and local governments an exception on age restrictions for firefighters and law enforcement. Basically fire and police departments that had age restrictions in place as of 1986 could keep these in place until 1993. After much discussion and legal challenges, Congress again amended the ADEA in 1996 when it reinstated the firefighter/law enforcement exception, with no future expiration date.[9]

Age discrimination cases will be of increasing concern to businesses in the future as the workforce in the United States and in many developed countries continues to mature. In addition, changes in Social Security laws mean older workers in protected classes will be working longer than before. Organizations need to be very careful in dealing with ADEA issues, because aggrieved employees have a 78 percent success rate in both state and local jury trials. In federal district courts, the median age discrimination verdict is almost $300,000, the largest award average for all types of discrimination.[10]

Relationship with Legal Counsel

One of the most common errors of fire service personnel managers is not seeking legal advice early enough in the decision-making process. The building of a solid and trusting relationship with the fire department's legal representative is important to one's success as a manager. The particular organizational structure will determine how and when to contact legal assistance. Many large fire departments have in-house legal counsel that actually works in the fire department. This arrangement enables the attorney to understand more fully fire department operations and procedures and to be readily available to answer questions and review policies or legal documents. Other fire departments have access to an office of law within the local government for advice and representation. In this arrangement one may or may not work with the same attorney, so the educational process relating to the organization will be more difficult. Small career or volunteer departments may contract for legal counsel only when they are faced with pending legal action.

When one is seeking legal advice, several basic questions are important to ask the attorney: (1) What are the legal consequences of my decision? (2) What are the legal alternatives? and (3) What can I do to prevent my decision from being challenged in the courts?[11] Legal counsel should be expected to give an accurate legal opinion in answer to these questions. Then it becomes a policy decision as to whether to take the exposure to a lawsuit or to comply and eliminate the likelihood of a suit. Policy decisions are under the purview of the fire chief or designee, not of the attorney. Many times fire chiefs will attempt to have the attorney make what is in essence a policy decision in an attempt to share the blame for what might be an unpopular policy decision on their behalf. However, fully prepared chiefs have confidence in their ability; they make the decision and move on.

Sometimes the attorney may attempt to make the policy decision under the guise of legal advice. The fire chief is in charge of the fire department and has the obligation and responsibility to make such decisions. It is important to make sure that one understands the difference between legal advice and policy advice. In my experience, often the attorney is focusing on the legalities but does not understand the policy ramifications because the fire department has not taken the time to brief the attorney properly of the situation. This type of misunderstanding serves neither the fire department nor the legal office well.

The opinion of legal counsel to a fire service personnel manager is invaluable. Remember to get legal advice as soon as possible. The more lead time before the legal opinion must be rendered, the more time is available for research and a more complete opinion. When requesting a legal opinion or advice, the fire service personnel manager needs to:

1. Give the attorney all known facts. Do not be reluctant to disclose all of the facts, even those that are damaging.
2. Be prepared to provide documents such as memos, letters, and other evidence when giving the facts.
3. Always identify assumptions and state these separately from the facts.
4. Be prompt in returning requested information.
5. Maintain confidentiality of information presented or received.[12]

For the fire service personnel manager, a familiarity with some basic legal terms that may arise during legal actions is helpful in understanding the process more fully.

Interrogatories are sets or series of written questions served on one party in a proceeding for the purpose of a factual examination of a prospective witness.

Deposition is a pretrial discovery procedure whereby the testimony under oath of a witness is taken outside of open court.

Subpoena is an order directed to a person to testify as a witness at a particular time and place.

Summons is an order served on the defendant to appear in court and to give an answer within a specified time to the suit that has begun against the person. The nature of the lawsuit is stated in the complaint. Complaint in a civil proceeding is the first or initial pleading by the plaintiff.

Court System

A general knowledge of how the court system works provides the fire officer with a better understanding of how he or she may be involved in this system. Obviously, the objective is to avoid reaching this stage, but it happens and usually for good reasons. If as a manager you feel that your decisions and actions were proper and the complainant will not withdraw, you may eventually end up in court. The fire department's taking a case to trial may arise from an important legal basis. Beware of the attorney who wants to settle too quickly at the fire department's expense. These actions may set unwarranted precedents within the department. Because it is often a judgment call, the better the fire officer understands the legal process, the better he or she will be able to speak about specific issues.

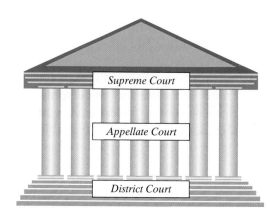

The federal courts are structured at three levels: district court, appellate court, and the U.S. Supreme Court. A case begins in district court, where evidence is presented, witnesses testify, and the judge or jury renders a verdict. Although district court decisions are important, they are not considered to be precedent setting and if not appealed apply only to the area where the court has jurisdiction.

Within a certain time frame after a decision is rendered, either party may appeal the decision to the next higher court. The appellate court hears the appeal and considers whether the decision of the lower court was proper as to the law and the facts presented. The twelve federal circuit courts of appeal in the United States are divided by geographic areas. The appellate court usually will not hear new evidence regarding the case. The decisions of the appellate court set precedent within its circuit and are important decisions in many cases.

The U.S. Supreme Court does not have to accept cases from the appellate court, but chooses which cases it will hear and rule on. The petition for review to the U.S. Supreme Court is called a **writ of certiorari.** If the U.S. Supreme Court refuses to review, the decision of the appellate court becomes law in the circuit where it was decided (see the preceding diagram). The U.S. Supreme Court will

writ of certiorari
■ petition for review to the U.S. Supreme Court by an appellate court. If the U.S. Supreme Court refuses to review, then the decision of the appellate court becomes law in the circuit where it was decided.

not hear new evidence, but will make its decision based on oral arguments, legal reasoning of the appellate court, and evidence presented in district court. Decisions of the U.S. Supreme Court are binding and set precedent for the lower courts and the entire United States.

For the interpretation of state laws, most state courts have the same basic structure as the federal courts. The court where the case starts is the district or circuit court. The next court is an appellate court, followed by a state supreme court. As with the federal system, the appellate court must hear the appeal, but the state supreme court chooses which cases it wishes to hear.

Summary

The significance of the law regarding the personnel management function is not a phenomenon particular to the fire service; it is indicative of the growth of law in all business and social activities. If the fire service personnel manager develops policies and procedures that consider legal implications and obtains competent legal advice, the consequences of legal actions and challenges can be minimized. Knowledge of the law and its proper application will allow managers to be proactive, treat employees fairly, and make the manager a valuable asset within the organization.

Review Questions

Use these questions to review the material in this chapter and for discussion purposes.

1. Why does the fire officer need to have a good understanding of the law and its application?
2. Discuss a legal case that affected your fire department. In your opinion, was it handled properly? What could have been done differently or improved?
3. What is the significance of *Griggs v. Duke Power Company,* and how has it changed the testing of applicants for fire service positions?
4. Describe the difference between a deposition and a subpoena.
5. Which federal agency is empowered to enforce the Civil Rights Act of 1964 and other discrimination legislation?

Case Study

You are the battalion chief of a battalion that consists of six fire stations. You are off duty on a Saturday night and you stop by the fire station to pick up some papers that you left there. As you enter the fire station via the rear entrance, you notice the male lieutenant and a female firefighter in a romantic embrace. They are both married, but not to each other. Both of them plead with you not to report the incident as it will greatly affect their personal and professional lives.

Previously the female firefighter had filed a sexual harassment complaint when she was at a different station. They are both good employees who are professional and generally abide by departmental policy.

It is now Monday morning and you have a previously scheduled meeting with the deputy chief. What do you tell him regarding this incident, and what actions do you take?

Fire Service Professional Qualifications

This chapter contains information that will assist fire officers in meeting the job performance requirements (JPRs) of NFPA 1021, Standard for Fire Officer Professional Qualifications:

4.2 Human Resource Management—Fire Officer I

    4.2.5 Apply human resource policies and procedures given an administrative situation requiring action, so the policies and procedures are followed.

    4.4.1 Recommend changes to existing policies and/or implement a new departmental policy at the unit level, given a new departmental policy, so that the policy is communicated to and understood by unit members.

5.2 Human Resource Management—Fire Officer II

5.4.1 Develop a policy or procedure given an assignment, so that the recommended policy or procedure identifies the problem and proposes a solution.

6.2 Human Resource Management—Fire Officer III

6.2.2 Develop procedures for hiring members, given policies of the AHJ and legal requirements, so that the process is valid and reliable.

6.4.7 Review current policies, then evaluate the need for and supervise the development of a new or revised policy or procedure, so that the recommended policy or procedure addresses the need.

7.2 Human Resource Management—Fire Officer IV

7.3.2 Monitor local, state, and federal legislative activities, given fire service issues, in order to enhance the effectiveness of the fire department.

Endnotes

1. Kenneth L. Sovereign, *Personnel law,* 4th ed. (Upper Saddle River, NJ: Prentice Hall, 1998).
2. J. Curtis Varone, *Fire officer's legal handbook* (Clifton Park, NY: Thomson Delmar Learning, 2008).
3. University of Maryland, System Memorandum of July 14, 1998, regarding "Sexual Harassment."
4. Marc Bendick, A fair shake, *Fire Chief* magazine, April 2008.
5. University of Maryland, System Memorandum of July 14, 1998.
6. University of Maryland, *Sexual harassment education resource manual* (College Park, MD: Author, 1990).
7. U.S. Equal Employment Opportunity Commission Web site, retrieved August 18, 2007, from http://www.eeoc.gov
8. U.S. Equal Employment Opportunity Commission Enforcement Guidance, *Reasonable accommodation and undue hardship under ADA* (Washington, DC: U.S. Government Printing Office, 1999).
9. Lawrence T. Bennett, *Fire service law* (Upper Saddle River, NJ: Pearson Prentice Hall, 2008).
10. P. J. Grossman, Are you ignoring older workers? *HR Magazine,* August 2003.
11. Sovereign, *Personnel law.*
12. Lawrence J. Hogan, *Legal aspects of the fire service* (Frederick, MD: Amlex Inc., 1995).

Suggested Readings

Bennett, Lawrence T. 2008. *Fire service law.* Upper Saddle River, NJ: Pearson Prentice Hall.

Grant, Nancy, and David Hoover. 1994. *Fire service administration.* Quincy, MA: National Fire Protection Association.

Hall, Douglas T., and James G. Goodale. 1986. *Human resources management.* Glenview, IL: Scott Foresman.

Hogan, Lawrence J. 1995. *Legal aspects of the fire service.* Frederick, MD: Amlex, Inc.

Hulett, Denise M., et al. 2008. *A national report card on women in firefighting.* Washington, DC: Egan Economic Consultants, Inc.

Schuler, Randall S., and Susan E. Jackson. 1996. *Human resources management: Positioning for the 21st century.* New York: West Publishing.

Sovereign, Kenneth L. 1998. *Personnel law,* 4th ed. Upper Saddle River, NJ: Prentice Hall.

University of Maryland. 1990. *Sexual harassment education resource manual.* College Park, MD: Author.

Varone, J. Curtis. 2008. *Fire officer's legal handbook.* Clifton Park, NY: Thomson Delmar Learning.

4

CHAPTER

Job Analysis and Design

Source: MFRI

KEY TERMS

job analysis, *p. 89*   job specification, *p. 90*   validity, *p. 91*

job classification, *p. 100*   KSAs, *p. 90*

job description, *p. 89*   subject matter experts, *p. 93*

OBJECTIVES

After reading this chapter, you should be able to:

- Describe job analysis techniques.
- Understand job analysis methods and information sources.
- Describe both job descriptions and job classifications.
- Employ the concepts of job enrichment techniques.
- Understand the relationship with the personnel office.

Who does what in your organization and what are their responsibilities and qualifications? Many fire departments have difficulty responding to this question in a detailed and precise way. Organizations fill jobs based on tradition; often what appears to make sense regarding qualifications is taken for granted. Then one day you are in court facing an Americans With Disabilities Act (ADA) challenge attempting to describe to the judge and jury just what it is that firefighters do. What exactly are the knowledge, skills, and abilities necessary to be safe and productive in this profession? Your best professional opinion may not consist of the necessary facts and processes involved to make these decisions. How can the fire department prevent a potential court order mandating acceptance of an employee who may not be fully qualified for the position of firefighter?

Such a scenario has occurred many times in fire departments throughout the United States and suggests a common problem in human resources management. The job description did not adequately define the duties and skills needed to perform the job. In today's rapidly changing work environment, a sound job analysis system is critical to the success of the fire department. Even if the skills required to suppress fires are essentially the same as they were 20 years ago, many other job functions have been added in most fire departments. Responsibility for emergency medical services, response to hazardous materials incidents, and the use of higher level technology are but a few of these functions. Numerous human resource decisions are driven by the job analysis system; therefore, it must change and be kept current.

Job Analysis

A job consists of a group of tasks that must be performed in order for an organization to achieve its goals. A specific job may require the services of one person, such as the fire chief, or of many people, such as firefighters. A **job analysis** is the systematic process of determining the knowledge, skills, and abilities required for performing jobs in an organization. It provides a summary of a job's duties and responsibilities, its relationship to other jobs, the knowledge and skills required, and the working conditions under which the job is performed. Relevant facts are collected, analyzed, and recorded as the job exists, not as the job should exist.[1]

Much of what happens to people at work depends on the job they currently hold or the ones to which they aspire. For example, recruitment procedures, selection methods, performance measurement, training programs, and compensation are all designed around the demands of the job. For these personnel management functions to be effective, the job must be clearly defined and understood by both the organization and the employees. The job analysis system provides a way to achieve this understanding. In this process the difficulty and dangers involved with firefighting and rescue work become evident.

Two basic types of information are most useful in job analysis: job description and job specification. A **job description** is a document that provides information regarding the tasks, duties, and responsibilities of the job. These tasks, duties, and responsibilities are observable actions. For example, the job of firefighter requires the use of a self-contained breathing apparatus (SCBA). Observation of someone using an SCBA allows an evaluation of its importance to the duties of a firefighter. The job description is a written summary, usually one to two pages long, of the basic tasks associated with a particular job.

job analysis
■ the systematic process of determining the knowledge, skills, and abilities required for performing jobs in an organization. It provides a summary of a job's responsibilities, its relationship to other jobs, and the working conditions under which the job is performed.

job description
■ a document that provides information regarding the tasks, duties, and responsibilities of the job. These tasks, duties, and responsibilities are observable actions.

job specification
■ a list of the knowledge, skills, abilities, and other characteristics that an individual must have in order to perform the job successfully. The KSAs are characteristics that are not directly observable; they are observable only when the person is doing the job.

KSAs
■ stands for knowledge, skills, and abilities. The knowledge refers to factual information necessary to perform a task, a skill is an individual's level of proficiency, and ability refers to a more general enduring capability that an individual possesses.

A **job specification** is a list of the knowledge, skills, abilities, and other characteristics that an individual must have in order to perform the job successfully. Knowledge refers to factual or procedural information necessary to perform a task. A skill is an individual's level of proficiency at performing a particular task. Ability refers to a more general enduring capability that an individual possesses. These knowledge, skills, and abilities (**KSAs**) are characteristics about people that are not directly observable; they are observable only when the person is doing the job. Using the preceding example regarding an SCBA, one cannot just look at a person and determine whether he or she can use an SCBA properly. However, if a supervisor were to observe a person using an SCBA, then he or she could make an assessment of the level of skill the person has in that area.

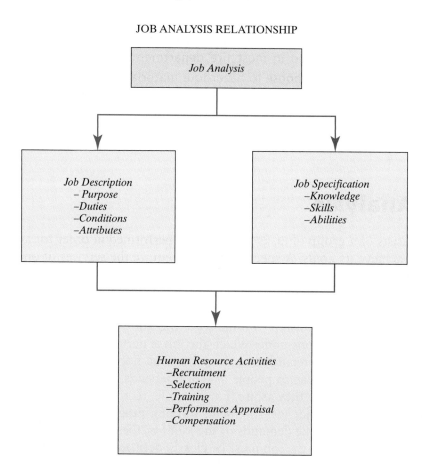

JOB ANALYSIS RELATIONSHIP

The job analysis and the resulting job description and job specification are the basic building blocks for many subsequent personnel management actions and decisions. For example, firefighter recruitment and selection would be haphazard if the recruiter did not know the qualifications needed to perform the job. In many cases job analysis data are necessary to support the legality of employment practices, such as decisions involving promotion and job assignments. Information derived from the job analysis is valuable in identifying safety and health considerations. Compensation decisions are often made based on the results of the job analysis. These few examples indicate the necessity for a valid and reliable job analysis system.

The job analysis serves as the basis for many fundamental personnel decisions and is likely to be scrutinized by legal and professional experts if the organization is sued over its employment practices. Consequently, a large body of legal opinion and regulatory guidelines has evolved around it. By using these guidelines to direct the job analysis processes, an organization can accomplish two things: legal compliance and defense to lawsuits, and a base of information on which to build an effective system of personnel management.

The legal issues in conducting a job analysis (see chart) have been articulated in the federal government's 1978 Uniform Guidelines on Employee Selection Procedures and numerous court decisions. The primary purpose of the Uniform Guidelines is to assist employers in selecting employees using job-related and fair practices. Section 14.C2 of the Uniform Guidelines states that "there shall be a job analysis which includes an analysis of the important work behaviors required for successful performance." The courts also have established that the only way to ensure that selection procedures as well as many other personnel actions are valid is by a proper job analysis. If a fire department is sued by an employee and the department does not have a valid job analysis on record, the defense of a lawsuit becomes more difficult.

Characteristics of Job Analysis

The process of job analysis must be conducted in a logical manner that follows appropriate personnel practices and standards. Regardless of the actual job analysis method, the information and results need to be proven and consist of the appropriate information, because the decisions that are based on the job analysis are very important and have to be defended. An appropriate job analysis consists of five key elements that are discussed next: reliability, validity, quantification, adaptability, and defensibility.

RELIABILITY

Reliability refers to the extent to which a process or a result can be replicated. A reliable job analysis procedure is one that provides essentially the same results when it is applied to the same job by various personnel analysts or is applied at different times. Reliability of a job analysis procedure does not guarantee its validity.

Standardized job analysis methods produce acceptable to high levels of reliability. Particularly high levels of reliability can be demonstrated for methods that rely on structured questionnaires and processes that are carefully planned and implemented.

VALIDITY

Validity is the extent and accuracy to which the analysis measures what it was designed to measure. The validity of a job analysis is of critical importance to its use and justification of the results of the analysis. The validity of a procedure depends on its purpose. For example, a job analysis can be valid for examination development and selection only if it identifies those factors that differentiate among levels of performance in the particular job. A job analysis method that simply describes

validity
- the extent and accuracy to which the analysis measures what it was designed to measure. The validity of a job analysis is of critical importance to its use and justification of the results of the analysis.

what is done and does not evaluate levels of performance is not valid for development of an examination. The job analysis method selected has to have validity regarding its purpose in the end-use work environment.

QUANTIFICATION

Quantification is an important aspect of most major job analysis methods. It involves assigning a numeric value to the data in order to differentiate relative importance of items, analyze large amounts of data, and provide a defensible basis for selection decisions. This is utilized to determine the level of importance to assign to certain aspects of the job analysis.

ADAPTABILITY

Adaptability refers to the extent to which the results of an analysis are adaptable or useful in other contexts. Generally, procedures that are narrative and subjective are low in adaptability. Structured instruments that use a similar format and process for a variety of applications tend to be high in adaptability.

DEFENSIBILITY

Documentation is the key to defensibility. The analysis should provide as much documentation as is practical given the demands for validity and the limitations of time and resources. Reliability also contributes to the defensibility of a job analysis.

The amount and extent of the documentation produced for a job analysis depend on the method used. Listed here are the key elements that need to be included in the documentation at a minimum, regardless of the method:

- A description of the job analysis method used
- The work behaviors, associated tasks, and if the behavior results in a work product, a complete description of the work product
- Criticality and/or importance of the work behaviors
- For each key work behavior, a link to the corresponding KSAs and the level required[2]

To summarize, let's look at what a job analysis is and what it is not.

What Job Analysis Is

- It is a systematic method for gathering information.
- It focuses on work behaviors, tasks, and outcomes.
- It identifies the personal qualifications necessary to perform the job and the conditions under which work is performed.
- It reports the job as it exists at the time of analysis, neither as it was in the past nor as it exists in another organization.

What Job Analysis Is Not

- It is not an analysis of thought processes, attitudes, traits, or aptitudes.
- It is not a time and motion study.
- It is not an analysis of an individual position.

Job Analysis Methods

Considerable information is needed for the successful accomplishment of a job analysis. The job analyst identifies the job's actual duties and responsibilities and properly interprets these points in the overall context of the position studied. In many professions this analysis is not difficult to perform due to the standard nature of the job, which can be easily observed. For example, a job on the assembly line of a manufacturing plant is relatively easy to observe and describe by its functions. However, in the fire and rescue service, this is not the case. Much of the work of fire departments is dangerous and cannot be easily observed by laypersons. When firefighters are advancing a hose line into a second-floor apartment, their work is not apparent to those in the street. It is obviously unsafe for personnel analysts to observe this kind of job function up close.

Most fire departments will turn to their state or local personnel department to provide the professional expertise necessary to perform the job analysis. At that level, experts can provide the fire department with the technical services necessary to perform a proper analysis. A blending of talents is required for this work. The personnel experts develop the methods for the analysis, and the fire department provides the **subject matter experts** to fully describe and demonstrate the critical aspects of being a firefighter.

subject matter experts
■ job incumbents that have direct knowledge of the requirements of the jobs that they perform.

Job analysis can be conducted in a number of ways based on an organization's specific needs and resources for conducting job analysis. The three basic categories of job information collected for analysis are task data, behavioral data, and abilities data.[3] Job analysis methodologies vary depending on which type of data they emphasize.

Task Data. Task data are subparts of a job that describe the purpose of each task. Task data emphasize the content of the job. They reveal the actual work performed and why it is performed, and include information such as the time it takes to perform the task and the degree of importance and prior experience required. An example would be for a firefighter to "read and properly interpret fire code regulations."

Behavioral Data. Behavioral data use verbs to describe the behaviors that occur on the job. If an individual needs to "communicate" well on the job, it will be described as speaking, advising, listening, persuading, and so forth. The emphasis of behavioral data is on the behaviors of the persons performing the job.

Abilities Data. Abilities data assess the underlying knowledge or skill a worker must possess for satisfactory job performance. If the mechanical ability of a firefighter is important to the analysis, then he or she may have to demonstrate an understanding of the operation of fire pumps. The emphasis of abilities data is on the qualifications of the person performing the job.

In this chapter, we examine several of the many different types of job analysis methods that use the job information data as their basis. Although it is not important for the fire service personnel manager to be an expert in this subject, he or she does need to be aware of some of the more standard methods and know what the role and responsibilities are as the coordinator of the subject matter experts (SMEs) for the analysis. If a manager can do a good job explaining and preparing the firefighters to work with the personnel analyst, then the final product is more likely to be accurate and complete.

Position Analysis Questionnaire. The Position Analysis Questionnaire (PAQ) is a structured job analysis tool that uses a checklist approach to identify job elements. Each job element is evaluated on a specified scale, such as extent of use, amount of time, importance of job, possibility of occurrence, and applicability. The PAQ is completed by an employee or employees who are familiar with the job being studied, typically the job incumbents and their supervisors. Using the PAQ, job descriptions can be based on the relative importance and emphasis these employees place on various job elements.

Interviews. The interview method involves a personnel analyst that goes to the location where the work is performed and an interview is conducted with those who perform the work. This method can provide information regarding the standard tasks as well as the mental effort that are required. Usually both the worker and the supervisor are interviewed and in some circumstances group interviews are conducted. The worker may be able to provide the interviewer with information that is not available from other sources. The interview process is time consuming and somewhat subjective.

Job Element Method. The job element method involves asking SMEs to generate a list of the knowledge, skills, abilities, and personal traits required to perform a job. These lists are then evaluated based on a scale. This method provides specific information about the characteristics an employee needs to be successful on the job. This type of information is most relevant to developing selection, training, and performance appraisal systems.

Critical Incident Technique. With the critical incident technique, employees who are knowledgeable about their job describe the critical job incidents that represent effective or ineffective performance. They describe the events leading to the incident, the consequences of certain job behaviors, and whether the behavior was under the employee's control. After the critical incident reports are collected, the job analyst develops task statements. For certain jobs many incident reports are generated in this process. The critical incident technique is well suited to public safety job analysis.

Observation Method. The observation method of job analysis consists of on-site visits in which the job analyst will observe the job being performed by job incumbents, as well as conduct a review of the equipment, materials, and facilities used in the job. Following the observation stage, a group interview process is used to develop a list of work behaviors and of KSAs necessary for effective performance. The work behaviors are described using a standard format. This method can be used for some, but not all, job elements in the fire service.

Job Analysis Information Sources

Information about a job can be obtained from many sources, but a job analyst needs to be careful as to how these sources are used and their validity to the process at hand. Current incumbents of the position to be analyzed, their supervisors, trained personnel analysts, consultants, and others can be utilized in this process. Because each person may see a given job from a different perspective, the results of a job analysis may differ, depending on the information source. Various methods can be used to ensure that these different perspectives are filtered out as the process moves forward and that the personnel experts end up with the essential information.

JOB INCUMBENTS

Job incumbents have direct knowledge of the jobs they do every day. They are often referred to as subject matter experts (SMEs) and can be the most valuable source of information concerning a particular job. They usually provide data through participation in an interview, being observed at work, or responding to a questionnaire.

One major concern in job analysis is selecting specific job incumbents to include in the analysis. For example, if a fire department has 250 firefighters, is it necessary to obtain information from them all? In most cases it would be inefficient to survey everyone, so a sample is used for the analysis. The key point is to ensure the use of a representative sample of the workforce. The sample of firefighters used in the job analysis should include males and females, members of different ethnic groups, senior and junior incumbents, firefighters who work in different battalions and regions, and so forth. They need to be fully briefed and understand the process so that they can contribute to the maximum of their ability.

Managers and incumbents usually agree about whether an incumbent performs specific tasks and duties. However, incumbents tend to see their jobs as requiring greater skill and knowledge than do managers or outside job analysts. One reason for this difference is that job-specific information is more important to incumbents who perform the work than it is to others. The difference may also result from a self-enhancement bias. Because job analysis is related to many human resource outcomes—for example, performance appraisal and compensation—incumbents and, to a lesser extent, their supervisors may exaggerate job duties in order to maximize organizational rewards and self-esteem.[4]

SUPERVISORS

Like incumbents, supervisors have direct information about the duties associated with a job and may in fact have previously performed the job. Therefore, they are also considered as subject matter experts. However, because they are not currently performing the job, supervisors may find it more difficult to explain all of the tasks involved with as much detail as the job incumbent can. This difficulty is more pronounced if incumbents perform their work where the supervisor cannot directly observe them on a routine basis. In the fire department, the company officer would have good direct knowledge of what tasks a firefighter is performing, but the battalion chief would have much less direct knowledge.

Supervisors do have the advantage of observing the work of more than one job incumbent and therefore may be in a good position to consider additional perspectives to the job analysis process. Whereas the incumbent provides information about what he or she does in the job, supervisors can supply information about the tasks typically associated with the job through a number of actual observations. It is important to include this information in the job analysis process.

PERSONNEL ANALYSTS

Some methods of job analysis require input from trained job or personnel analysts. The advantage of enlisting the help of personnel analysts is that they can observe many different incumbents working under different supervisors in diverse locations. This broad exposure leads to more consistency in the data collection for the job analysis process. Trained personnel analysts can also provide research of indirect sources of information such as organizational records and technical documentation. Additionally, they are better trained to appreciate the legal issues that are a part of the job analysis process and can ensure that each component of the process is properly carried out.

If a personnel analyst is used in the job analysis process for a fire department, that person must be fully briefed on the operations and understand the nature of the job he or she is reviewing. Some very critical aspects of emergency service jobs are infrequently performed, such as the rescue of a victim from a fire. The analyst may not observe this action, but will need to understand its importance and the complexity of all tasks that must occur for successful job performance.

SELECTING INFORMATION SOURCES

Job incumbents, supervisors, and personnel analysts all contribute associated advantages and disadvantages to the job analysis process. Therefore, good strategy includes some of each of these sources in the process. Public safety jobs are often more difficult to analyze because of the nature of the work. Firefighters may be called on to respond to anything at any time, from a cut finger to a hurricane. It is necessary to weigh the more critical but seldom performed tasks such as rescues in comparison to the everyday responses and duties, and the analyst will need to clearly understand this.

It is also beneficial to have the analyst observe the job as it is performed in the field. Many fire departments take the analyst to the training academy and perform certain tasks that are reflective of the job. This is a good practice for orientation, but should not stop there. Make sure that the analyst is able to respond to emergencies and observe from a safe vantage point as someone explains what is occurring on the scene. Observations can take place at different times of the day and night and under varying weather conditions. Analysts need to see the job as it is performed under emergency conditions.

Job Description

As noted earlier, a job description is a document that provides information regarding the tasks, duties, and responsibilities of the job. These tasks, duties, and responsibilities are observable actions. It is a written summary, usually one to two pages long, of the basic tasks associated with a job as identified by the job analysis. Job descriptions usually have a label, called a job title, and frequently include a section describing the qualifications needed to perform the job, called job specifications.

Job descriptions are useful for a number of purposes within an organization, particularly in the recruiting and selection processes. Fire departments use them as the basis for writing job announcements that will advertise and explain the position to prospective job candidates, which is particularly important in the fire service due to the nature of the physical demands and the danger associated with this profession.

Typical job descriptions contain three sections that identify, define, and describe the job.

1. Identification—This section may contain the job title, number of incumbents, which department the job is assigned to, a unique job number, and the date of the job analysis. The purpose of this section is to identify the job clearly and to differentiate it from other jobs within the organization.
2. Definition—This section explains the purpose of the job, why it exists, and how it relates to other jobs within the organization. It provides a statement that describes the job in sufficient detail and may include the level of performance that is satisfactory. This section may also include the number of people supervised, the budget responsibility, and the reporting relationships within the organization.
3. Description—This section includes the major job duties, specific work performed, how independent from supervision the work is, and whether certain factors will limit the actions of the job incumbent. An in-depth description of the tasks performed, training requirements, and experience of the job are listed in this section.[5]

Job descriptions are important in orienting and training new employees. They provide an overview of the activities that need to be carried out, which can be explained to the employee in more detail by the training staff. A job description does not list all the duties of the job, but includes the essential and most important ones. It is a waste of everyone's time and effort to have a recruit firefighter quit after 1 week on the job because he or she did not have a clear understanding of what the job entailed. This can be avoided by a clear, up-front understanding of what the job involves. In fact, many progressive fire departments use a videotape that graphically shows the work of a firefighter and what can be expected in firefighter training school.

Job descriptions are also used in the development of performance standards. Performance standards expand the job description and establish, in measurable terms, how well the job is to be performed. Job descriptions are also relevant to the development of performance appraisal systems, which are designed to include some of the job description's major categories.

Job classification and job evaluations are also derived from job descriptions. Job evaluation is the process of determining the relative worth of jobs within an organization to establish wages and salaries, clearly a subject of importance to those who are performing the jobs.

Here is an example of a complete job description for the position of "Firefighter/Rescuer I" within the Montgomery County, Maryland, Department of Fire and Rescue Services.

MONTGOMERY COUNTY GOVERNMENT, ROCKVILLE, MARYLAND

CLASS SPECIFICATION

Firefighter/Rescuer I (Recruit)
Code No. 3172 Grade: F1

DEFINITION OF CLASS

This is entry-level work in fire suppression, fire prevention, rescue, communications, and emergency medical care. Personal contacts are with fire/rescue and emergency medical supervisors and instructors to receive assignments and instructions and with the public to provide information and emergency assistance as required.

Under the immediate supervision of designated fire/rescue officers, an employee in this class receives training in all aspects of suppressing and extinguishing fires, fire prevention, conducting emergency rescues and providing emergency medical assistance. The work performed contributes to the rescue and safety of individuals and the preservation of structures and physical property. Work is performed in accordance with specific instructions and established policies and procedures which must be adhered to, and is closely supervised. Work, which is performed on a rotating shift, requires periods of strenuous physical effort and involves exposure to heat, dirt and other very unpleasant conditions. An employee in this class is exposed to the risk of serious injury, requiring the use of protective clothing and the strict observance of safety precautions and procedures. Work is closely and frequently reviewed for accuracy, adequacy and adherence to instructions and established procedures.

EXAMPLES OF DUTIES: (ILLUSTRATIVE ONLY)

- May periodically drive and/or operate emergency fire/rescue vehicles.
- Responds to and participates in firefighting, rescue, auto extrication, hazardous material incidents, medical emergencies, and other emergency operations as required.
- Lays and connects fire hoses, nozzles, and other firefighting appliances; directs water streams; uses specialized firefighting tools and equipment; raises and climbs ladders; makes forcible entry into buildings.
- Performs salvage operations at scenes of fires; performs overhauling operations to ensure extinguishment and to prevent structural collapse; replaces sprinkler heads at the direction of a superior.
- Performs rescues from burning buildings and other hazardous situations; performs heavy-duty rescues utilizing cutting torches, hydraulic jacks, air compressors, and other specialized rescue equipment.
- Performs immediate and efficient emergency medical care as required at the scene of an incident or during transport; reports vital patient information to medical personnel; operates emergency medical equipment.
- Maintains accurate records of medical treatment administered.
- Participates in cleaning and preventive maintenance, activities concerning fire and rescue apparatus, equipment, facilities, and grounds.
- Attends required fire/rescue training sessions to attain required level of proficiency.
- Attains and maintains required level of emergency medical care certification.
- Completes and maintains accurate records, forms, and incident reports as directed.

- Participates in fire prevention and fire safety programs as directed.
- Performs related duties as assigned consistent with this class specification.

MINIMUM QUALIFICATIONS:

Experience: None

Education: High School graduate or equivalent acceptable to the State of Maryland.

Age: Employee must be 18 years of age as of the first day of employment. Applicant may be permitted to take entrance examination six months before eighteenth birthday.

Knowledge, Skills, and Abilities: Ability to read and understand oral and written policies, rules, instructions, and procedures of the Montgomery County Department of Fire/Rescue Services. Ability to learn and to perform firefighting, rescue, and emergency care practices, techniques, and methods and to apply such information to specific situations.

Ability to analyze emergency situations and to select effective courses of action, giving regard to surrounding hazards and circumstances.

Ability to understand and follow oral and written directions.

Ability to type 25 w.p.m. (Communications appointment only)

Ability to pass an approved diction test. (Communications appointment only)

Ability to think clearly, respond immediately, and act quickly, calmly, and effectively in emergency situations.

Ability to make minor equipment repairs and adjustments under supervision.

Ability to communicate effectively with the public and employees.

Ability to speak clearly and distinctly to the public and other employees.

Ability to successfully complete periodic physical testing and examination for the appropriate assignment.

Ability to drive some of the apparatus and equipment in the fire/rescue corporation to which assigned.

Ability to wear breathing apparatus, climb ladders, and work in confined places, toxic atmospheres, and extreme heat.

Ability to work from heights.

Ability to engage in strenuous physical effort for prolonged periods as required.

Ability to write clearly and legibly.

Ability to cope with stressful situations.

LICENSE AND CERTIFICATION REQUIREMENTS:

Successful completion, within one year from date of appointment, an Emergency Vehicle Operator's course and State of Maryland certification as Emergency Medical Technician-Ambulance (EMT-A).

A recruit must possess and maintain at all times a valid motor vehicle operators license, as required by the Department of Fire and Rescue Services.

A recruit shall successfully complete and be certified to the level of NFPA Firefighter II as approved by the Montgomery County Department of Fire and Rescue Services before attaining permanent status.

(continued)

PROBATIONARY PERIOD:

For new career employees a probationary period of one (1) year from date of hire is required before attainment of permanent status, during which time performance will be carefully evaluated.

MEDICAL GROUP: 1

Successful completion of physical examination, physical agility test and drug screening as required.

Class Established: July 1987
Revised: April 1989,
                      July 1991,
                      June 2001,
                      June 2006

Source: Job description courtesy of the Montgomery County, Maryland, Department of Fire and Rescue Services.

The need for job descriptions that accurately reflect job content is clear in light of the weight given to job descriptions by personnel experts and job applicants. Inaccurate job descriptions may hinder recruiting efforts, reduce the effectiveness of training, and cause the development of unrealistic job standards or inappropriate compensation for classes of work. The job description must be reviewed and updated on a regular basis so that information is always as current as possible. The fire department also needs to be aware that any job descriptions based on standards from national organizations may change. For example, NFPA standards are subject to change on a 5-year cycle.

Job Classification

job classification
■ the process of categorizing positions according to the type of work performed, the type of skill required, or any other job-related factor.

Job classification is the process of categorizing positions according to the type of work performed, the type of skill required, or any other job-related factor. Classification simplifies job analysis, because a standardized job description and qualifications standard can be written for an entire group of positions. A classification, therefore, is actually a group of positions sufficiently alike with respect to their duties and qualifications to justify their being covered by a single job description. For example, a fire department may employ several hundred firefighters. Although some details of their positions may differ (engine company or squad company), the job duties are sufficiently similar to be classified as one occupation (firefighter) or several types of firefighter positions based on the difficulty of work or level of responsibility (Firefighter I or Firefighter II).

As indicated in the preceding paragraph, two types of classification are commonly used. First, jobs are divided into generic occupational classifications (firefighter, fire apparatus operator, or fire officer). Second, within each generic occupational classification, jobs are differentiated according to the level of responsibility (Firefighter I or Firefighter II).

Classification systems for most jobs, in both the public and the private sectors, are based on a variety of factors[6]:

Information input. Where and how does the worker get information used in performing the job?

Mental processes. What reasoning, decision-making, planning, and information-processing activities are involved in performing the job?

Work output. What physical activities does the worker perform, and what tools or devices does the worker use?

Relationships with others. What relationships with other people are required in performing the job?

Job context. In what physical and social context is the job performed?

Work methods. What methods or techniques are used to perform the job?

Worker traits. What personality characteristics or aptitudes are needed for job performance?

Think about the preceding job classification factors and relate them to work within the fire and rescue service. It actually becomes quite obvious how these factors apply to the positions of firefighter, fire officer, and so forth. Again, note that the job analysis process is the basis for these decisions and plays an important and basic role in this process.

Job Enrichment

A planned program for enhancing job characteristics is typically called job enrichment. Job enrichment usually involves increasing skill variety, task identity, task significance, autonomy, and feedback from the job.[7] This type of program includes the addition of some activities previously performed by the supervisory level, including some of the planning. Therefore, job enrichment involves adding tasks from the level above, whereas job enlargement adds tasks from the same level.

Job enrichment is widely advocated as a method to provide greater challenge to the employee. Although job enrichment programs do not always achieve positive results, they often bring about improvements in job performance and in the level of satisfaction of workers in many organizations.

According to Frederick Herzberg, a noted proponent of job enrichment, five principles guide its implementation:

1. Increasing job demands—The job should be changed in such a way as to increase the level of difficulty and responsibility.
2. Increasing the worker's accountability—More individual control and authority over the work should be allowed while the manager retains ultimate accountability.
3. Providing work scheduling freedom—Within limits, individual workers should be allowed to schedule their own work.
4. Providing feedback—Timely periodic reports on performance should be made directly to workers rather than to their supervisors.

5. Providing new learning experiences—Work situations should encourage opportunities for new experiences and personal growth.[8]

A good question to address is whether job enrichment can work in fire service organizations. Because fire departments are hierarchical in nature with a definite rank structure, they tend to be bureaucratic, traditional, and difficult to change. Additionally, it has been my experience that many officers invest a lot of energy in protecting the "turf" of their rank and their place within the organization. They are reluctant to give up their duties, because in their mind only someone of their rank or higher could possibly perform this job. Unions also tend to want a very definite and structured assignment of jobs and tasks in an organization. They expend substantial effort to protect positions, acting pay, and the like. All these tendencies work against the concept of job enrichment.

Managers may spend a lot of time making assignments by title. How often have supervisors said, "I need a captain to do that job"? Do they really need a certain rank, or someone with certain qualifications? I would much rather hear someone say, "I need a person with a fire data background and statistical experience for this assignment." The person who has these qualifications may be a captain, or may not be, but by confining the request to a rank, one may never know.

There are legitimate job enrichment opportunities in fire service organizations. Special projects in particular allow individuals throughout the organization to enrich their jobs. They step up to new levels of difficulty and responsibility and thrive in the freedom to control the destiny of their project or project team. The result will be a more creative and professional approach to the job at hand, and better motivated and trained employees when it is concluded.

Another way of enriching a job is by job rotation, which is the process of switching a person from job to job. This method of enrichment develops the employee's capability for doing several different jobs, which will increase his or her value to the organization. In addition, this technique can break the monotony of an otherwise simple, routine job. This rotation can be a motivating factor in that the employee will face new challenges and a new work environment. However, some studies have shown that these effects are limited and that they tend to wane with time in the new position.

Fire departments are particularity well suited to rotate jobs in this manner. It is to the benefit of the individual firefighter and the department to have personnel capable of performing numerous jobs. Regular rotation among engine companies, truck companies, and rescue squads should be accomplished for all firefighters and officers. In addition, rotation to staff and administrative jobs is beneficial. Nothing creates a better understanding within divisions or bureaus of a fire department than moving between these assignments. For example, a fire officer who once worked in the Communications Division will understand its functions and how to achieve maximum cooperation with it in his or her current position.

Job rotation should be required for fire officers, especially as they advance in rank. This increases their understanding and provides them with a better and more broad-based feel for the entire fire department. As employees progress within the fire department, they need to have a variety of work experiences; otherwise, they become limited and narrow-minded. When I was a fire chief, I required our three deputy chiefs and twelve bureau chiefs to rotate assignments on approximately a 3-year schedule. This caused some difficulty due to the transitions, but in the long run it was very beneficial for the overall management of the department. It provided a broader perspective and understanding from which the officers could make better decisions because they understood the entire department. Some organizations go even further and require that fire officers serve some time in other departments such as the police or emergency preparedness department.

Relationship with Personnel Office

The success of a fire service personnel manager often depends on a strong and viable relationship with the personnel office of the town, city, or county. The professionals in these departments can assist the fire department in developing and implementing the personnel management techniques necessary to ensure it has all the elements in place for a good personnel system. In small fire departments, one may have to contact the personnel department for most actions. Larger fire departments are most likely to have their own personnel section to coordinate their activities with the personnel department of the government entity. Regardless of the structure, it is important that the fire department and the personnel department have regular communication and a clear understanding of what must be accomplished.

The personnel department or human resources department typically plays a major role in the planning and coordination of the systems used; that is, the department determines what job analysis procedures to utilize, making sure that the information obtained through job analysis is used to write appropriate job descriptions and job specifications, and then using this information appropriately within each agency of government. The personnel department has to be concerned with many different government departments. Therefore, it needs to ensure a degree of consistency between departments so that comparisons can be made, especially with regard to compensation and classification decisions.

One of the main jobs of the fire service personnel manager working in cooperation with the personnel department is to coordinate activities between both agencies. The fire chief needs to be informed when job analysis should be performed or reviewed and updated, when job descriptions are revised, how the testing and evaluation techniques need to reflect the job analysis, and of many other issues. The fire department also ensures that the personnel office knows when assistance is needed and under what conditions. The fire service personnel manager must communicate a clear understanding within the fire department as to the necessity of these actions. The fire department must understand that it needs to participate and support the job analysis system because it will benefit from this process being professionally done. This fact is often overlooked given the myriad of challenges that fire departments currently face.

One of the most important things that the fire department can do to assist in the job analysis process is ensure that its personnel are ready to participate. An announcement to a group of firefighters that they have to go off-line for 2 days to assist in developing a job analysis is not normally met with great applause. Often they are told just to do it and not ask any questions. This mandatory compliance sets up a negative situation from the beginning and often complicates the entire process. Fire service personnel managers should have the opportunity to brief the participants thoroughly as to what they are doing and why, and to make sure they understand the importance of the process both to them as employees and to the department as a whole. When they know that compensation, examination, recruiting, and classification decisions will be made from this information, they tend to be much more willing participants. The fire service personnel manager must also ensure that the process is completed as honestly as possible and that individuals or groups do not attempt to circumvent the procedure because they think they can influence pay increases or some other benefit.

The fire department can also prepare the personnel experts to understand the very specific work environment of the fire and rescue service. Many of them may not understand the nature of an emergency service, shift work, dependency on teamwork, or the other factors that make the firefighting occupation unique. Personnel experts will need to be thoroughly briefed and informed about these issues so that the study can proceed as efficiently as possible and, most importantly, truly reflect the work of a firefighter, or whatever position is to be reviewed. My experience is that most personnel analysts seem to enjoy the opportunity to work with the fire department because it is different and exciting for them and offers a break from their more mundane personnel work. Use this to your advantage and make the most of the situation. The primary goal centers on defining the system and recording the results based on the real work of the subject matter experts, those doing the job being studied.

Summary

Job analysis is clearly one of the basic building blocks of personnel management. Because it is the basis of many personnel functions and decisions, it deserves a high priority in any organization. Job analysis is especially critical in the fire service due to the nature of the work and the importance of highly qualified individuals necessary to perform the often dangerous duties. Job descriptions and job specifications further the job analysis process and allow for the development of compensation and classification plans.

Numerous methods and techniques are utilized to perform a job analysis. The relationship that the fire department has with the general government personnel office is important in the selection and administration of the proper techniques to ensure accurate results. If other personnel-related decisions are to be valid, then a proper job analysis system must be in place first. The job of the fire service personnel manager is to ensure that such a system is established and maintained.

Review Questions

Use these questions to review the material in this chapter and for discussion purposes.

1. What are the objectives of job analysis?
2. Discuss what is meant by the following statement: "Job analysis is the most basic personnel management tool."
3. What items are typically included in a job specification, and how does it differ from a job description?
4. What is job enrichment and how could it be implemented in a fire service organization?
5. Can proper job analysis help you defend against legal actions taken by employees?
6. Within the organization, who will provide the most accurate information in the development of a job analysis?

Case Study

Your fire company has been selected to participate in the revisions to the job analysis for the rank of firefighter. As the company officer, it is your responsibility to ensure that your company is prepared to complete the analysis process. Several of the firefighters have expressed disinterest in participating in the process that the city's personnel department will conduct. They intend to inflate all aspects of the job so that they will be in

a better position to receive a pay increase. They think that this will be a good strategy to benefit them and that the personnel department analyst will not know the difference.

It is your responsibility to ensure that a fair and reliable job analysis process is conducted. How do you do this, and what will you say to the firefighters to convince them to follow the proper course of action?

Fire Service Professional Qualifications

This chapter contains information that will assist fire officers in meeting the job performance requirements (JPRs) of NFPA 1021, Standard for Fire Officer Professional Qualifications:

4.2 Human Resource Management—Fire Officer I

4.2.5 Apply human resource policies and procedures given an administrative situation requiring action, so the policies and procedures are followed.

6.2 Human Resource Management—Fire Officer III

6.2.2 Develop procedures for hiring members, given policies of the AHJ and legal requirements, so that the process is valid and reliable.

6.2.3 Develop procedures and programs for promoting members, given applicable policies and legal requirements, so that the process is valid and reliable, job-related, and nondiscriminatory.

Endnotes

1. Wayne R. Mondy, Robert M. Noe, and Robert E. Edwards, What the staffing function entails, *Personnel*, April 1986.
2. Colorado State Government Division of Human Resources, *Technical assistance on job analysis*, March 2007.
3. Ronald A. Ash and Edward L. Levine, A framework for evaluating job analysis methods, *Personnel*, December 1980.
4. F. J. Landy and J. Vasey, Job analysis: The composition of SME samples, *Personnel Psychology*, April 1991.
5. George T. Milkovich and John W. Boudreau, *Human resource management*, 8th ed. (Boston: McGraw-Hill, 1997).
6. Sydney A. Fine, Functional job analysis: An approach to a technology for manpower planning, *Personnel Journal*, 53, 1974.
7. Richard I. Hackman and Greg R. Oldham, *Work redesign* (Boston: Addison-Wesley, 1980).
8. Frederick Herzberg, One more time: How do you motivate employees? *Harvard Business Review*, September–October 1987.

Suggested Readings

Ash, Ronald A., and Edward L. Levine. 1980. A framework for evaluating job analysis methods. *Personnel*, December.

Colorado State Government Division of Human Resources. 2007. *Technical assistance on job analysis*. March.

Dessler, Gary. 2001. *Management: Leading people and organizations in the 21st century*, 2nd ed. Upper Saddle River, NJ: Prentice Hall.

Fine, Sydney A. 1974. Functional job analysis: An approach to a technology for manpower planning. *Personnel Journal*, 53.

Hackman, Richard I., and Greg R. Oldham. 1980. *Work redesign*. Boston: Addison-Wesley.

Herzberg, Frederick. 1987. One more time: How do you motivate employees? *Harvard Business Review*, September–October.

Landy, F. J., and J. Vasey. 1991. Job analysis: The composition of SME samples. *Personnel Psychology*, April.

Mondy, Wayne R., Robert M. Noe, and Robert E. Edwards. 1986. What the staffing function entails. *Personnel*, April.

5

Fire Service Recruitment

GFD Archives.

KEY TERMS

attrition, *p. 111*

recruiting image, *p. 119*

recruitment, *p. 107*

selective recruitment, *p. 108*

workforce analysis, *p. 120*

OBJECTIVES

After reading this chapter, you should be able to:

- Understand the recruitment process.
- Know how to recruit to improve diversity.
- Describe volunteer fire department recruitment and retention issues.
- Describe executive recruitment strategies.

Recruitment is fairly easy to define: It is the set of activities designed to obtain a pool of qualified job applicants for a certain position. It involves attracting individuals on a timely basis, in sufficient numbers, and with the appropriate qualifications. In the fire and rescue service, the attractiveness of the position of firefighter is strong enough that generally many more people apply for positions than could ever be hired. Many other organizations have to do a lot of work in order to get applicants to fill positions within their organizations. In most fire departments, all the department has to do is announce the opening of vacant

recruitment
- the set of activities designed to obtain a pool of qualified job applicants for a certain position. It involves attracting individuals on a timely basis, in sufficient numbers, and with the appropriate qualifications.

positions and a line will form for people to apply in large numbers. The challenge for most fire departments is to have a diverse pool of applicants that includes women and minorities who may be underrepresented in the current workforce of the fire department.

selective recruitment
■ in order to attract more women and minorities, recruiting techniques designed to reach a specific population of applicants.

The focus of this chapter is on the process of recruitment with an emphasis on **selective recruitment** techniques. The more difficult and challenging situations for the fire service personnel manager involve the decisions that are necessary in the hiring process. This area of personnel management must be handled carefully, because it has been the subject of much litigation over the years. As just stated, the position of firefighter is very much sought after; those who feel qualified and are not selected often challenge the process. Therefore, the optimum hiring process is fair and objective, and selects the best qualified applicants from a diverse pool in a consistent manner.

Recruitment

Recruitment is one of the most critical tasks that any organization undertakes. In essence, the department is selecting the future of the organization, which needs to be done carefully. In most fire departments, employees beyond their probationary period tend to stay for a full career of 20 to 25 years. Recruitment decisions can often have long-range impacts regarding the strength and viability of the department. Recruitment programs need serious attention if they are to be successful and productive for the organization. This chapter focuses on the recruitment and selection process for firefighter as the entry-level position in the profession.

Recruitment is often treated as a one-way process in which organizations search out prospective employees. However, in reality, potential employees seek out organizations just as organizations seek them, especially in the fire and rescue service. The goal is to match the organization's needs and the job candidate's desires. In order for the organization and job candidate to match and proceed with the process, three conditions must exist: (1) A common communication medium is established (i.e., organization must advertises where the job candidate is); (2) the job candidate's desires in the employment relationship are met; and (3) the candidate is attracted and motivated to apply for the job. The job of fire service personnel managers in the recruitment process is to meet these conditions in numbers both large and diverse enough to satisfy the employment needs of the fire department.[1]

The recruitment process consists of two major phases. The first phase is an awareness of environmental and organizational changes that necessitate the need for new employees and define both the jobs to be filled and the type of people necessary to fill them. The second phase is having a plan to make a broad range of potential applicants aware of the firefighter positions that are open and how they can enter the process for selection.

Numerous environmental influences affect recruitment programs. One of the foremost is the economic environment of the recruitment area. The variations in the supply and demand for applicants due to economic conditions can vary the level of the recruitment efforts and the techniques used. During good economic times, the labor supply of job applicants is less than that during economic recessions, which may have less effect on fire department recruiting because of its greater number of applicants. One reason for a large number of applicants is that many people already employed elsewhere will apply for firefighter positions when they become available.

Changes in technology can substantially affect recruitment programs. As technology is developed, many existing jobs change or are eliminated and new jobs that require different skills are created. The United States in the last several decades has evolved from a manufacturing base to an information-processing base; therefore, the employee skills needed today in most occupations are more technical. We in the fire service are still hoping for a major technological change for our profession. After all, fire departments still put out fires essentially the same way that Ben Franklin did, with water applied to the base of the fire.

The fire service has experienced advancing technology and the need for a different set of skills for its employees, primarily in the area of emergency medical services and other services now provided by most departments such as hazardous materials response, confined space rescue, and the like. The firefighter both now and in the future needs to be more versatile and adaptable to a number of challenging assignments. The equipment used in certain types of responses is more technical and requires more advanced skills to be successful. The fire service will have to consider these factors in its recruitment program.

Another major influence on recruitment is the political environment, especially federal and state laws and regulations that affect the employment process. Both equal employment opportunity and affirmative action requirements have dramatically affected the recruitment process. Employment needs must be defined in terms of ability to perform the job, not in terms of race, sex, religion, or other arbitrary factors. Because of affirmative action requirements, fire departments recruit far more widely and more selectively in certain areas that primarily serve women and minorities. One political objective of many jurisdictions is to have the makeup of their workforce be comparable to the population in general. Political leaders expect the fire department to resemble the diversity of the community, and those departments that do not may have difficulty explaining why.

The important aspects of the economic environment, technology, and political influence all affect the recruitment program. As formulation of the recruitment program begins, a careful assessment of these factors will improve the results.

Recruitment Process

A successful recruiting process consists of several steps. The six steps outlined in this section describe the necessary components of a recruitment process for a fire department.

IDENTIFY RECRUITMENT NEEDS

The first step in the recruitment process is an accurate forecast of the recruitment needs of the fire department. Many factors must be considered in this step, but the most prominent will be the budget and the number of positions to be filled.

The operating budget for the department is a plan that in terms of financial resources identifies the number and types of positions that must be recruited for. The two basic categories are new positions and filling vacant positions. If the department is staffing a new fire station, then firefighters must be hired; this is identified in the budget. If the department has experienced several firefighter retirements, then the budget allocates funds to fill these positions. Many departments know, based on historical trends, that they will experience a certain number of retirements, separations from service, and so forth in a 1-year period. Regardless of the type of budget process used, at its conclusion, the department will know how many and what level of positions it will be recruiting for in the next fiscal year.

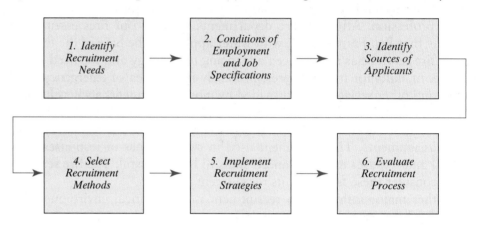

The budget also determines the timing of the hiring process in that the positions are funded for either a full year or part of a year. Fire departments generally have a need to fill firefighter positions in groups because of the extensive training process (16 to 20 weeks) prior to being assigned to duty positions. It would be inefficient to have firefighters on the payroll but not in training or performing firefighting duty. Smaller departments may never hire enough firefighters at one time to form a recruit training school, so they use different strategies. In whatever system the jurisdiction uses, the budget determines the number and timing regarding the identification of recruitment needs.

CONDITIONS OF EMPLOYMENT AND JOB SPECIFICATIONS

Once the department knows the number and type of employees it is recruiting, then it must clearly define the job through a job description created for new positions or updated for positions that may have changed since the last announcement. Prospective employees need to know the conditions of employment such as salary, benefits, work locations and residency requirements, and shift-work schedules. Many fire departments have fitness requirements, no-smoking agreements, and other specific conditions of employment not found in other professions, which also must be identified to applicants so that they understand the conditions of the job. Many departments also provide details of the dangers and types of emergency situations encountered in the fire service work environment. At this point in the process, these descriptions should be general in nature.

Job specifications, which are the qualifications required for successful job performance, are also spelled out at this point, including the level of education, fire service certifications and training, minimum age requirements, class of driver's license, and the like. Medical and fitness requirements, substance abuse testing, criminal background checks, and other requirements must be clear. Detailed and realistic job specifications are important because they help prospective job applicants evaluate their own qualifications for the job. Unclear job specifications may encourage unqualified applicants to apply, whereas job specifications that are too high may turn away qualified applicants.

The job specifications must be valid, job related, and the product of a thorough job analysis for the position being advertised. Such analysis is essential and major legal problems may result if it is not properly conducted. For example, many fire department job announcements used to include height requirements, which were found to be discriminatory toward women and certain races, and were eliminated by the courts. The purpose of the job announcement is to attract people to apply for positions in the fire department; it is not in and of itself a selection device. An effective selection process, and not the job announcement, is used to eliminate unqualified applicants.

Two basic hiring philosophies are prevalent in the fire service. One is that the fire department recruits individuals who have the potential to perform the job of firefighter and then trains the individual after he or she is hired. The other is that the fire department recruits individuals who are already trained and certified for the position and ready to go to work after a brief orientation. Both of these philosophies have merit; the department must decide which is best for its circumstances. If the fire department hires applicants who are already trained and certified, then it will be able to reduce training and recruitment costs significantly, because it will be a smaller market. In addition, the time necessary to fill a position will be shorter, with less concern regarding **attrition** during the training process. Many advantages make it a popular method of hiring firefighters.

attrition
- a reduction in membership or personnel, normally through retirement, resignation, or other factors.

If the department decides to recruit people with the potential for the job and then train them after employment, it will be dealing with a much larger labor pool. One of the major disadvantages of recruiting only trained and certified applicants is that the selection pool is much smaller. In many cases this pool consists of firefighters switching departments for higher pay or benefits, or of volunteer firefighters attempting to get paid positions. Because the demographics of the fire service in the United States are predominantly white males, this pool will consist mostly of white males. If the fire department is underrepresented in women or minorities, then it must use a recruitment strategy that exposes the department to a large number of qualified women and minorities. This objective rules out setting stringent training and certification standards at the application level. In some cases fire departments prefer to train their own employees regardless of the expense, so that they can impart their organizational culture and the specifics of their departments to the recruit firefighters.

It is my experience that most fire departments attempt to recruit and attract as many people as possible to their job announcements and then give credit for specific fire service training and certification somewhere in the selection process. This method will become increasingly difficult in the future as more fire departments provide paramedic services. The training requirements for an employee to be certified to at least the NFPA standards for Firefighter II and also be trained to the National Department of Transportation (DOT) standards for Emergency Medical Technician (EMT)–Paramedic are difficult and time intensive. The question as to who is responsible for providing this training, before or after hire, is interesting. The ever-increasing emphasis on reduced budgets without reducing service will complicate the process even further.

IDENTIFY SOURCES OF APPLICANTS

A key element of recruiting is identifying applicants in sufficient number, quality, and diversity in order to select the most qualified for employment within the fire and emergency services. As stated earlier, it is not difficult for most fire departments to attract applicants. In general, all that needs to be done is to circulate an announcement, and the people will start lining up for the applications. This incidentally says a lot about the profession and the esteem in which it is held within the community. The difficulty for many fire departments, however, is achieving the desired diversity within the pool of applicants.

In order to attract more women and minorities to the fire department, recruiting techniques designed to reach a specific population of applicants are needed. In recruiting minorities the most useful sources, as identified in one study, were referrals by current employees, colleges and universities, advertising in targeted media, community agencies, and private and government employment agencies.[2] Many different types of selective recruiting techniques that have been proven to be effective if conducted properly are available.

Knowing where the best opportunities are for selective recruitment requires a good knowledge of the community. One worthwhile technique is to form a recruitment committee, which consists of firefighters and interested citizens of different backgrounds, races, sexes, and cultures. The committee thinks about and develops a plan for the recruitment effort that identifies the most likely places to recruit minorities and women. In some communities these places may include an historically black college or certain church services, athletic clubs, and community events. The U.S. military was at one time a very good source of firefighter applicants and still is to a degree. The U.S. military is a diverse organization, has many of the same values and attributes that fire departments look for, and members exist in large numbers. The difficulty today is that careers in the military are different. When the draft was in effect, positions within the military experienced constant turnover. Currently the military is considered more of a career, and those within the age range most desirable for entry-level fire service positions who exit the military are far fewer in number. However, it can still be a good source of applicants.

The fire department can do a number of things internally to attract people for employment purposes. Job fairs have a good degree of success as do career-day presentations at schools. Both are directed toward employment opportunities and attract people who are predisposed toward a career in the fire department. One effective approach is to make everyone in the department a "recruiter." When the department members make fire prevention presentations or perform inspections, media demonstrations, or other activities for the public, they can make recruitment brochures and information available. Brochures should be handed out at every opportunity, and every fire station should make them available and know how to assist people who inquire about a job. This recruitment approach is more general than it is specific.

Many fire departments now have "High School Cadet" programs that encourage young men and women to think of a future in the fire service, as either a career or a volunteer member. These programs can be after-school clubs or a more structured program with a full curriculum. Some high school graduates will have completed courses such a Firefighter I/II and EMT-B by the time they finish school. They will have specific job skills for future employment as well as an improved appreciation of emergency services work. This training will serve them well individually, and it allows the fire department to attract perhaps a more diverse pool of future firefighters at a young age. This is a positive situation for both the fire department and the high school cadet.

One effective way to recruit women and minorities to the fire service is by example. If women and minorities hold visible and important positions within the fire department, then this will be known throughout the community. This example helps to form a perspective for those viewing the department from the outside. They see people of their color and gender being successful within the fire department. Conversely, if the department is viewed by the community as sexist or racist, then a successful selective recruitment program will be extremely difficult to accomplish.

An effective recruiting program has a marketing plan to reinforce it. Colorful, well-presented materials should be handed out at recruiting booths. The information contained should have been well thought out in advance and then presented clearly to the public. Fire service personnel are not always the best judges of presentation clarity because of their inherent knowledge about the job. Using a focus group of civilians to review these materials and ask questions about points that are not clear can be helpful toward improving the final product. Also remember that if the department has a desire to improve the recruitment of women and minorities, then the photographs and illustrations used in the handout materials should reflect the diversity that it hopes to achieve.

Many progressive departments have slide presentations, videotapes, and home pages on the Internet dealing with recruitment. The Internet can be especially helpful because of its availability 24 hours a day, every day. Applications should be able to be downloaded, completed, and mailed in or applicants should be able to submit applications online. The use of the Internet as a tool in the recruitment process can be fast, effective, and efficient. These Web sites can be created to target specific recruitment needs and, if properly maintained, include large amounts of relevant and up-to-date information.

Employers have found a number of advantages in using the Internet for recruiting. A primary benefit is that many employers have realized cost savings using Internet recruiting compared to other sources such as newspaper advertising, job fairs, and other traditional methods. Internet recruiting can save valuable time. Applicants can respond rapidly by sending e-mails, and recruiters can respond to qualified candidates more quickly with follow-up information. Another advantage is that an expanded pool of applicants can be generated using Internet recruiting techniques. One effective technique for specialized recruiting is to obtain e-mail databases from predominantly minority organizations and mass e-mail the recruiting material to that list.

Some disadvantages are associated with Internet recruiting. By getting broader exposure, fire departments may also get inquiries from a larger number of unqualified applicants, which will have to be sorted out and responded to. Another issue is that some applicants may not have Internet access, especially individuals from disadvantaged socioeconomic groups and certain racial or ethnic groups. Even in light of these difficulties, it is likely that Internet recruiting will continue to expand.

The following example, which is on the home page of the Phoenix Fire Department, shows the type of information that should be available in a job advertisement or job announcement. Fire department recruiters should strongly consider the use of such technology.

PHOENIX FIRE DEPARTMENT FIREFIGHTER RECRUITMENT GENERAL INFORMATION

Applications for Firefighter Recruit only are accepted once a year, normally for a one-week period in January. However, the length of time for which the eligible list is certified may be extended beyond the one-year limit at the discretion of the Personnel Department. To qualify to begin the testing process, an applicant must satisfy the following requirements:

1. Must live in Maricopa County at the time of application. Applications from individuals who live outside Maricopa County will not be accepted.
2. Must be 18 years of age.
3. Must possess the mental and physical/medical health to adequately perform the duties of a firefighter.
4. Must have a good driving record (fewer than eight points during the previous three years) and be able to obtain and maintain a valid Arizona operator's license.
5. Fingerprints of all applicants will be taken prior to each part of the examination process for identification and validation purposes.
6. May be subject to pre-employment drug screening test.

Medical conditions or disabilities that may impact an applicant's ability to perform the essential functions of the position will be considered on a case-by-case basis. If deemed necessary by the Fire Department physician, or at the request of the applicant, medical specialists will be consulted to assist in assessing the applicant's suitability for the position. If the applicant disagrees with the opinion of the Fire Department physician or the specialists, the applicant can exercise the procedures set forth in the Phoenix Fire Department Administrative Regulations for Medical Arbitration. Certification as an Emergency Medical Technician with the state of Arizona is preferred.

First, pick up an application and a study guide from the city of Phoenix Personnel Department. Information about the time and place for the written exam will be included in the materials. Applicants who pass the written test then must take, and successfully pass, a physical performance test before being placed on the eligible-to-be-hired list. The Firefighter Recruit exam consists of two tests. The first test is a written 100-question, multiple-choice, and the second test is a physical performance test.

1. Written test (Generally administered each year in February). The written test consists of two sections. One section is based on the study guide that is given to each person at the time the application is picked up at the Personnel Department. Questions are asked about material in that study guide. Applicants are not allowed to use the study guide when taking the written examination. The other section is designed to test your ability

to read and comprehend the texts and other written materials used to train Firefighter Recruits at the Emergency Services Institute. Your knowledge of basic math also is tested because you are required to make computations during the training period and in the performance of everyday tasks.

2. Physical Performance Test (Generally given each year sometime in March). The physical performance test is designed to test a person's endurance and ability to perform basic strenuous tasks that professional firefighters must perform. The test is a series of tasks, which are simulations of actual tasks that are performed on the job. These tasks must be performed within seven minutes and 20 seconds. The physical performance test is scored on a pass/fail basis. Applicants perform the tasks wearing a firefighter's helmet, turnout coat, gloves and self-contained breathing apparatus. The total weight of this protective gear is approximately 40–50 pounds. A second physical performance test may be required of applicants as part of the final selection process.

The names of those applicants who receive a passing score on the written test and pass the physical performance test will be placed on the "eligible list" for Firefighter Recruit by the city of Phoenix Personnel Department. When the eligible list is certified, it is forwarded to the Fire Department. The Fire Department uses the eligible list to call applicants for job interviews. After the interviews, the highest qualified individuals will be offered employment pending the results of a background investigation and medical examination.

The Fire Department reserves the right to require a substance abuse screening test. Applicants who meet both medical exam and background investigation standards are scheduled for training at the Emergency Services Institute. A second physical performance test may be required of applicants as part of the final selection process. Periodic drug screening tests may be required at the discretion of the Fire Department.

The Immigration Reform and Control Act of 1986 requires that an applicant provide the city of Phoenix with proof of identification and authorization to work prior to employment. The most common documents that will furnish this proof are an Arizona driver's license and a Social Security card.

All applicants who are selected for hiring then must successfully complete a recruit training program at the Fire Department Training Academy. The Recruit Academy lasts 15 to 17 weeks. Those who successfully complete the recruit training program are placed in the field as probationary firefighters for the remainder of their 12-month probationary period. During the probationary period an on-going structured training program will be provided, along with monthly performance evaluations.

If you have any questions, call (602) 262-7356.

Source: Recruitment announcement courtesy of the Phoenix Fire Department Homepage.

Advertising job opportunities in printed media and/or radio and television is a widespread and versatile method of recruitment. Television is out of the realm for most fire departments due to cost and the lack of availability for public service ads. Radio is useful and can reach a large area of applicants. In addition, if the department uses specific radio stations it can improve the diversity of the recruitment effort. For example, if the department advertises on a Hispanic radio station, the likelihood of reaching that particular population is good. This type of strategy applies to print media as well. Many of the print media are good for advertising positions above the entry level, especially with professional journals and publications. If the department is looking to hire a deputy fire chief, then an advertisement in the local newspaper is not a good choice. This is where the professional publications are important, because their readership is composed of many people with the qualifications required for these positions.

The following chart summarizes the advantages and disadvantages of the major types of media used in a firefighter recruitment program.[3]

Advantages and Disadvantages of the Major Types of Media

TYPE OF MEDIA	ADVANTAGES	DISADVANTAGES
Newspapers	Flexibility Concentrated circulation Well organized and easily accessed by job seekers	Easy to ignore Other competing articles and features Circulation not specialized
Magazines	Specialized magazines by profession or trade Ad size flexibility High-quality printing	Long lead time for placement of ad Not limited to a specific area
Direct mail	Unlimited number of formats and space Can be mailed by zip code Can be personalized	Difficult to obtain mailing lists Cost per prospect is high
Radio/television	Widespread coverage Can be creative Visual medium for advertisement	Only brief, uncomplicated messages Time consuming to make Expensive
Internet	Inexpensive Large volume No geographic boundaries Can be directed to specific groups	May attract unqualified applicants Some may not have Internet access

SELECT RECRUITMENT METHODS

The recruitment methods selected are crucial and will most likely determine the success of the recruitment effort. Making people aware of the positions and encouraging them to apply will not produce the desired results unless the fire department follows through with an effective recruitment method that achieves the goal of appropriate numbers of the desired candidates actually applying to the fire department.

The people chosen as recruiters for the department are critical to the process. They need to be energetic, present a good image, and truly have the desire to be a recruiter. The need for excellent people skills is high because recruiters will be interacting with a large number of prospects in the course of their work. Whether they like it or not, they need to make a quick, good impression on people that they deal with on behalf of the department. The equipment and brochures they use must be professional and readily available. The use of minority and women firefighters as recruiters is appropriate. It establishes a degree of credibility with prospective recruits who are women or minorities and who will want to ask questions of the recruiters regarding their experience and acceptance within the fire department.

In order for the recruitment program to truly be successful, the fire chief needs to be supportive in a visible and meaningful way. Again, the department is not just recruiting firefighters; it is selecting the future of the fire department. The influence of both the fire chief and the top echelon of the department provides direction as well as the budget and personnel assigned to this work. Some fire departments leave the recruitment process up to the personnel department of their jurisdiction. They, however, will not be as committed as is the fire department. Although recruiting is not something that should be delegated to the personnel department, having personnel experts on the fire department recruitment team can be very helpful.

The process that I recommend is the development of a recruitment team that will develop a plan and execute it. In the selection of the recruitment program, a combination of recruiting techniques tends to be most effective. Consider the following factors as the department develops the appropriate plan and approach:

1. Recruitment budget
2. Personnel assigned to the recruitment effort (number and expertise)
3. Number of expected vacancies
4. Projected hire date
5. Affirmative action requirements or goals
6. Analysis of competition—other departments and economic factors

All of the previous factors need to be considered as the fire department develops the methods it will use. Much of this will be driven by the number of positions that the department is attempting to fill. If the department is filling a small number of positions, then the budget and other factors will be proportional. Many large fire departments hire 50 to 100 firefighters or more each year. In this case the recruitment program is a continuous effort needing full-time support. One point to consider strongly is that the department needs to begin early. Recruitment is only the first step in the selection process, and the necessary mental and physical testing, medical evaluations, and background checks can take several months. Therefore, the fire department must start its recruitment efforts early enough to allow for all the steps in the process to occur and still meet the projected hiring date. In addition, the fire department may find that it needs to have six to ten good recruitment prospects for every position that it expects to fill. The attrition experienced through the selection process can be dramatic, which needs to be calculated into the recruitment plan.

IMPLEMENT RECRUITING STRATEGIES

If the fire department has spent the proper amount of time and effort planning the recruitment program and going through the steps as described in the preceding paragraphs, the implementation should go smoothly. It is a matter of actually getting out and doing what was planned. Fire service personnel managers should have goals established as milestones to measure the progress as the department implements the recruiting strategies. For example, one should have a date as to when the recruiting brochures should be mailed, when and where the radio advertisements will be played, the number of women applicants recruited, and other measurements. Keeping track will help guide the department and allow it to make corrections in the plan if necessary and still meet established time constraints.

If the recruitment strategy involves making presentations before groups, and most plans do, then practice the presentation until the delivery is perfect. When a recruiter is called on to speak, it is not the time to be fumbling with a projector or some other aspect of the presentation. The best possible image of the fire department should be presented in a clear and convincing manner. It is very true that the recruiter gets only one opportunity to make a first impression.

A fire department serious about its program works to recruit high-quality applicants. When I was fire chief of the Prince George's County Fire Department in Maryland, we had a standard recruiting program. It was good, but not as good as it could be. The number of applicants was no problem, but the goal was to be selective and to improve the diversity of recruitment efforts, because those efforts were aimed not only at selecting firefighters but also at selecting future officers.

A young officer who was committed to this effort was assigned the project. He selected a team of committed and professional recruiters, and together they developed a new program. One aspect of their program was to recruit at predominantly minority colleges from outside of the state of Maryland. This strategy had not been used before and was somewhat controversial within the department as a waste of resources. They used a multifaceted approach that included visits to nearby military installations while they were on the road visiting the colleges. They were supported with the budget and resources necessary to accomplish their mission.

The result was an exceptional class of recruits that progressed well through firefighter training school. The diversity of the class in terms of women and minorities was excellent and included African Americans, Asians, and Hispanics. In fact the level of college education and other qualifications of the minority recruits exceeded that of the majority recruits. Today, many of those recruited by this effort are now officers having exemplary careers.

The preceding example illustrates an important point. The success of a recruiting program should not be gauged by numbers, but by the quality and the potential of the people who are recruited. In the final analysis, this is what really counts.

EVALUATE RECRUITMENT PROCESS

The evaluation of the recruitment effort should be a continuous process throughout the recruitment program. Management should constantly be measuring the success of the efforts by setting milestones or intermediate evaluation points at certain intervals of the program. If the department needs to make adjustments, then it must be flexible and make them as the program progresses. In this manner, one can correct problems as soon as they arise and keep the momentum of the program moving forward.

One method of evaluation is to develop a record of applicant data. If the department is operating under a consent decree or a mandatory affirmative action program, these statistics might be a requirement. The applicant flow record includes personal and job-related data concerning each applicant. It indicates whether a job offer was extended and, if no such offer was made, an explanation of the decision. Such records enable the department to analyze its recruitment and selection practices and take any necessary corrective action. Sometimes information about those who were not selected is more important than about those who were selected.

After the conclusion of the recruitment effort, it is a good idea to sit down and formally evaluate the performance of the recruitment effort, including a review of the quality, diversity, and management of the program. A good source of information may be those firefighters who were selected for the positions. Ask them what attracted them and why. The recruiters should also have ample input into the critique of the program, because they were the ones who actually did the work.

Make a special effort to bring out all of the positive and negative points during the evaluation. This session can supply a wealth of information to be used when beginning the next recruiting cycle.

Recruiting to Improve Diversity

There is no question that fire departments in general do not reflect the diversity found within the general society of the United States. A recent analysis of the career fire service by the National Fire Protection Association (NFPA) shows that 96 percent of the fire service is male, of which 83 percent is white.[4] Many volunteer fire departments have operated as fraternal organizations and have only recently begun to attract and recruit members from a more diverse pool. Traditional recruitment methods such as employee or member referrals may perpetuate the effects of past discrimination, even after other discriminatory practices have ceased to exist. The result may not be a conscious act, but is an indication of possible systemic discrimination regardless of intent. How many state and local fire service organizations still use the word *firemen* in their titles instead of *firefighters*? Just think of the message this sends to those who may want to apply for firefighter positions.

It is also a political objective of many local governments to reflect the diversity of their community within the public safety forces that protect that community. If the fire department stands out as an organization that rejects or impedes attempts to improve the diversity of the workforce, the results will be negative and political consequences will follow. In addition, the fire department will be deprived of the benefits of a more diverse racial and cultural workforce.

The question for fire department recruiters is how best to achieve success in recruiting strategies that will attract a diverse pool of applicants. Special measures will be required in order to recruit with a more diversity-oriented focus. Two issues of importance are **recruiting image** and recruiting presence. The recruiting image of the fire department is very important to the success of the recruitment program. This reflects the general image of the fire department and the respect it has within the community. In general, the image of the fire service as measured by numerous public opinion surveys is very good and is often rated as the top public service of a given area. However, if the image has been tainted by repeated discrimination/harassment claims or lawsuits or if the department appears to consist of primarily one racial group, this will hamper recruitment efforts. The recruitment presence means that the department engages in a continuous effort to recruit from the community and this presence increases the credibility of the department. This can prove difficult for fire departments that do not hire on a regular basis. If the groups you are attempting to recruit from see you only once every 2 years, do not expect great results.[5]

Unfortunately, organizations sometimes lose qualified applicants in the recruitment process. One such circumstance is when the applicant's potential for the job is underestimated and the applicant is discouraged from applying to the fire department. This can be due to cultural misunderstanding, prejudice, or lack of skill by the recruiter. The second circumstance is when the applicant decides the organization is not a good place to work, and the applicant screens out the

recruiting image
- the general image of the department and the respect it has within the community in the recruiting process.

employer. This can be caused by a misunderstanding of the work environment and how the department presents itself in the job market. Better information to prospective employees will help alleviate this. Obviously, when recruiting for a fire department in which hazardous work conditions do exist, some employees will select out because of this, and that is appropriate.

A National Report Card on Women in Firefighting published in April 2008 includes a great deal of detailed information regarding the underrepresentation of women in the fire service.[6] The authors present several recommendations aimed at increasing the pool of women applicants for fire service positions. The recommendations of this report include the following:

Recruiting Initiatives Targeting Women Are Perceived as More Widespread than They Actually Are. A survey question was "My department recruits and hires women above and beyond its general recruiting efforts." In the response 69.3 percent of men respondents agreed, but only 35.6 percent of the women respondents agreed with this statement. Only 8.8 percent of the women in the survey reported that they were actively recruited by the fire department. This indicates that fire departments need to do a better job of selective recruitment of women applicants.

Personal Relationships Are Very Important in Recruiting. According to the survey 47.5 percent of the female firefighters grew up with a family member or friend who was a firefighter and 60.5 percent were introduced to the occupation by such a person. Many men also reported being aware of fire department opportunities in the same manner. Fire department recruiters need to be aware of this and use it as appropriate.

Men and Women Typically Come to Firefighting with Different Prior Experience. Target-rich recruiting areas for women included medical employment, active sports, and high schools and community colleges. High recruitment areas for men were construction and the military.

Women Are Somewhat Less Likely than Men to Have Pictured Themselves as Future Firefighters when They Were Very Young. As more women firefighters are available to serve as role models, this perception will change. Survey respondents reported first wanting to be firefighters at about age 11.

To improve recruitment diversity, the fire department will need to develop an action plan to enlarge the pool of diverse candidates and attract them to applying for firefighter positions. Some aspects of such plans should include the following:

workforce analysis
- an analysis conducted to determine where the department may be under represented in regard to gender and ethnic group representation as compared to the larger community.

- Conduct a **workforce analysis** to determine where the fire department may be underrepresented in regard to gender and ethnic group representation.
- Use qualified minorities and women as recruiters.
- Recruitment information needs to display diversity in photos and graphics.
- Utilize recruiting trips to minority- and women-oriented colleges/universities.
- Use the resources of local minority and women's organizations to display and promote fire department recruitment.
- Provide training classes (open to all) to prepare people to be successful in the hiring process.
- Measure success by setting goals and monitor the process to ensure it is having the desired effect.

- Have a fire department that reflects diversity in an open and honest manner by valuing all members of society.
- Document the diversity recruitment strategy and efforts.

Fire departments must continually adapt their recruitment strategies to the changes in the economic, social, and political environment of their jurisdiction. There needs to be clear objectives and a defined strategy to recruit a more diverse workforce; it will not just happen by itself. There is the assumption in many fire departments that merit must be sacrificed to gain diversity. Some may believe that seeking diversity means giving preferential treatment—not recruiting on merit or for job suitability, but because certain candidates represent diversity. It actually means paying closer attention to the full range of competencies required for the job and the importance of diversity to the organization's effectiveness.

Second, there is the assumption that any selection processes should be gender and racially blind. Organizational research repeatedly shows that there is no such thing as a gender-neutral selection process. It also demonstrates that we are highly biased toward hiring those most like ourselves. The better alternative is to be more aware and work consciously to avoid assumptions, stereotypes, or generalizations. In many fire departments there is an assumption that final selection is made from a pool of equally qualified candidates. But, there is no such thing as "equally qualified," as all applicants have unique and special qualifications. What one wants to do is establish a list of all candidates who meet the minimum qualifications. Once a pool of qualified applicants emerges, the special attributes that each candidate has to offer can be evaluated against the original list of competencies and standards. One thing is certain: Without a diverse pool of fire-fighter applicants to choose from, one will never have a diverse fire department.

Volunteer Fire Department Recruitment/Retention

As previously discussed, recruitment is difficult in career fire departments that are paying salaries and benefits. Imagine the plight of volunteer fire departments that need to recruit and retain volunteers in today's society. The volunteer emergency services are a long-standing tradition in the United States that often involves generations of families. I know that I joined the fire department at an early age because my older brother and some of our neighbors were volunteers at the local station. If not for that, I do not think that I would ever have considered a career in the fire service.

Many volunteer fire departments are experiencing increased difficulty with recruiting and retaining members than in past years. Although there has been a decline in the number of active volunteer firefighters nationally from a high of 897,750 in 1984, the trend has changed in the last few years. The number of volunteers dipped to a low of 770,100 in 1989. While the number has increased since then, the problem of recruitment and retention is still serious in many areas. The most recent figures indicate that currently there are approximately 800,000 volunteer firefighters.[7]

Volunteer fire departments can no longer count on the children of current members following in their parents' footsteps. Nor can they count on a continuous stream of community people eager to donate their time and energy to their local volunteer fire department. Volunteer fire departments compete for members from a shrinking population that is willing and able to contribute time to community volunteer groups. It is not only the volunteer fire department that may have difficulty in recruiting but also the Boys & Girls Club, Girl Scouts, church groups, and so on. A 2003 report by the Bureau of Labor Statistics indicated the following top three reasons why people do not volunteer:

Lack of time	44.7 percent
Health/medical problems	14.7 percent
Family responsibilities	9.5 percent

Adding to the problem, departments cannot rely on members staying active in the volunteer fire service for long periods of time due to the time constraints and other factors within society.[8]

In May 2007, the United States Fire Administration (USFA) released a report titled *Retention and Recruitment for the Volunteer Emergency Services* indicating that there is no single reason for the decline in volunteers in most fire departments. However, there is a universal consensus that skilled department leadership is a key to resolving the problems. Retention and recruitment problems usually can be traced to several underlying factors: more demands on people's time in a hectic modern society; more stringent training requirements; population shifts from smaller towns to urban centers; changes in the nature of small town industry and farming; internal leadership problems; and a decline in the sense of civic responsibility, among other factors. The problems or challenges involved in the recruiting of volunteers are common in all fifty states.

In order to reverse the trend of declining volunteers and to ensure that volunteer fire departments are open to new strategies and opportunities, the following key points will assist in that effort.

LEADERSHIP

Many studies and discussions have pointed to the leadership factor in volunteer departments. These departments are "big business" controlling large groups of employee volunteers and huge investments in facilities and apparatus. They have to be managed with the same level of expertise and ability as any other large corporation. Strategies to ensure that effective and well-prepared leadership has control of the volunteer fire department will greatly benefit all current and new members. This may involve moving away from the election of officers to a system that promotes merit to a higher degree.

CULTURE

Many of the traditional volunteer departments are still operated as inflexible, fraternally oriented hierarchies rather than as meritocracies. This inhibits efforts to recruit because departments too closely resemble cliques and appear to be unwilling to take advantage of different cultures and ethnic backgrounds within the community. Efforts have to be directed toward recruiting from a broad base as well as exhibiting and modeling the racial and ethnic demographics of the community served.

DIVERSIFICATION

Departments must welcome those who offer their services, even when those services are not for actual firefighting or EMS response. Utilizing people for whatever nonfirefighting services they offer is better than showing them the door. Accepting nonfirefighting volunteers can also open departments to networking opportunities and those with many levels of expertise that may not have the time or talent to volunteer on a regular basis.

MARKETING

Volunteer fire departments have to reach out and bring members into their organizations that have the skill and time to participate. This involves marketing the benefits of being a volunteer, which are numerous, as well as explaining what is required, which can be substantial. Programs that need to attract and retain volunteers should reach out to various opportunities within the community. Many of the positives of being a volunteer, such as self-worth, accomplishment, community service, and others, have to be explained. In addition, tangible benefits such as length of service award programs, uniforms, and other benefits need to be detailed so that prospective volunteers understand what they are getting into.

Managing a volunteer fire department in today's society is a difficult challenge. It really takes an inordinate amount of time and talent that many people choose to use elsewhere. For volunteer fire departments to continue their service, they will have to adapt and change to meet these challenges. Two national fire service organizations are especially helpful in this regard. The National Volunteer Fire Council (NVFC) devotes considerable expertise to extend to and assist volunteer fire departments at the local, state, and national levels. In addition, the Volunteer & Combination Officers Section of the International Association of Fire Chiefs (IAFC) has resources available. Both of these organizations deal with recruitment and retention issues, as well as operational and legislative concerns. They serve as excellent resources for any volunteer fire service organization that needs to recruit and retain volunteer members.

Executive Recruitment

Much of this chapter is devoted to the recruitment of a fire department's entry-level positions. The majority of recruitment occurs at this level because most fire departments tend to promote from within after that point. Generally, promotions within the fire department are open only to those within the system. This type of single-point entry has both advantages and disadvantages for the department. Promoting from within tends to maintain the morale and motivation of those in the department. It also allows one to know more about the applicants prior to selection, and once promoted they can learn the new position faster than can someone from outside the fire department. Promoting from within is less costly and quicker than recruiting external applicants.

However, promoting only from within can stagnate a department and perpetuate the current system, effective or not. New ideas and programs can often be implemented faster by someone from outside the department who is not constrained by the current traditions and relationships. Many times it depends on the perception and condition of the fire department; if things are going well the promotions tend to come from within the department. It is interesting to note that there is a higher degree of mobility that takes place at the fire chief level, but seldom in the lower ranks.

Executive search firms are used to find and attract individuals at the management level within fire departments. The client of the search firm is the employer, rather than the applicant. Search firms locate highly qualified applicants who may or may not be currently employed. Often, people will want to be considered for a position in a confidential way so as not to jeopardize their current position, and executive search firms are adept at this. Search firms now assist organizations in determining their human resource needs, compensation packages, and hiring criteria. This rapidly growing industry is available to assist fire departments with executive searches. However, these firms can be quite expensive, sometimes with fees equal to from 40 to 60 percent of the annual salary of the employee they place.

A more cost-effective solution in many cases uses trade publications and professional associations to advertise a position. This method allows for broad coverage in a specific market in the likelihood that the type and level of person the department is looking for reads these magazines and journals. Once an interest is expressed, one can begin the recruitment and selection process. These types of advertisements have proven to be effective and inexpensive, but require substantial lead time in some circumstances.

Summary

Recruitment is the process of attracting individuals on a timely basis, in sufficient numbers and diversity, and with appropriate qualifications to seek positions in the fire department. Effective planning of the recruitment process is important and involves six basic steps. The implementation of the recruitment program requires top-level support, adequate resources, and committed recruiters to perform this important work. At the conclusion of the recruitment program, managers evaluate it to assess how to modify it for future recruitment efforts.

Review Questions

Use these questions to review the material in this chapter and for discussion purposes.

1. Discuss the factors in the economy, the political environment, and technology that can affect a recruitment program in the fire department.
2. How would you organize a recruitment committee in your department? Whom would you include and why?
3. In your community, name three prominent sources of firefighter recruits.
4. How would you improve the diversity of your recruitment program? Make specific recommendations for your department.
5. Which step do you feel is most important in the recruitment process?

Case Study

You have been selected to direct the recruitment efforts for a firefighter class that will consist of 20 new firefighters. The jurisdiction in which you work is a large county fire department adjacent to a major city. The county's population is approximately 800,000 residents and covers a land area of about 400 square miles. The class is scheduled to begin in 6 months. Your total budget for this project, not including salaries, is $20,000. You can use other firefighters to assist you on an occasional basis, as long as it does not generate any overtime pay.

Explain how you would develop the recruitment program. Identify the recruitment efforts and the media that you would use to advertise for the positions. Contact local recruitment and advertising firms to obtain accurate cost information for your planned recruitment program. Present a complete program, noting the strengths and weaknesses of the plan.

Fire Service Professional Qualifications

This chapter contains information that will assist fire officers in meeting the job performance requirements (JPRs) of NFPA 1021, Standard for Fire Officer Professional Qualifications:

4.2 Human Resource Management—Fire Officer I

4.2.5 Apply human resource policies and procedures given an administrative

situation requiring action, so the policies and procedures are followed.

5.2 Human Resource Management— Fire Officer II

5.4.1 Develop a policy or procedure, given an assignment, so that the recommended policy or procedure identifies the problem and proposes a solution.

6.2 Human Resource Management— Fire Officer III

6.4.7 Review current policies, then evaluate the need for and supervise the de-velopment of a new or revised policy or procedure, so that the recommended policy or procedure addresses the need.

7.2 Human Resource Management— Fire Officer IV

7.2.1 Appraise the department's human resource demographics, given appropriate community demographic data, to determine if the recruitment, selection, and placement of human resources is effective and consistent with law and current best practices.

Endnotes

1. Wayne F. Cascio, *Managing human resources,* 7th ed. (New York: McGraw-Hill, 2006).
2. Bureau of National Affairs, EEO policies and programs, *Personnel Policies Forum 141,* May 1986.
3. Douglas T. Hall and James G. Goodale, *Human resource management* (Glenview, IL: Scott Foresman, 1986).
4. Michael Karter, Jr., and Gary P. Stein, *U.S. fire department profile through 2006* (Quincy, MA: National Fire Protection Association, 2007).
5. Robert L. Mathis and John H. Jackson, *Human resource management,* 10th ed. (Mason, OH: Thomson Learning, 2003).
6. Marc Bendick, A fair shake, *Fire Chief* magazine, April 2008.
7. United States Fire Administration, *Retention and recruitment for the volunteer emergency services* (Washington, DC: Federal Emergency Management Agency, 2007).
8. Ibid.

Suggested Readings

Bureau of National Affairs. 1986. EEO policies and programs. *Personnel Policies Forum 141,* May.

French, Wendell. 1990. *Human resources management,* 2nd ed. Boston: Houghton Mifflin.

Hall, Douglas T., and James G. Goodale. 1986. *Human resource management.* Glenview, IL: Scott Foresman.

Hulett, Denise M., et al. 2008. *A national report card on women in firefighting.* Washington, DC: Egan Economic Consultants, Inc.

Mathis, Robert L., and John H. Jackson. 2003. *Human resource management,* 10th ed. Mason, OH: Thomson Learning.

Mondy, R. Wayne, and Robert M. Noe. 1996. *Human resource management.* Upper Saddle River, NJ: Prentice Hall.

Osby, Robert. 1991. Guidelines for effective fire service affirmative action. *Fire Chief* magazine, September.

United States Fire Administration. 2007. *Retention and recruitment for the volunteer emergency services.* Washington, DC: Federal Emergency Management Agency.

MFRI Archives.

adverse impact, *p. 133*   job relatedness, *p. 132*   **Uniform Guidelines,** *p. 133*

disparate impact, *p. 133*   negligent hiring, *p. 145*

disparate treatment, *p. 133*   reliability, *p. 131*

OBJECTIVES

After reading this chapter, you should be able to:

- Understand appropriate selection standards.
- Utilize selection methods in the fire department.
- Describe concepts of assessment centers.
- Describe the components of a promotional process.

Whether the fire department is selecting firefighters to fill a recruit class or selecting lieutenants to be company officers, the decisions involved are complex and difficult. The decisions made have a substantial impact on those selected and on those not selected for positions within a fire department. Both entry-level positions and higher-level positions in the fire department are very desirable and competitive. A great deal of competition for these opportunities means that those who are not selected can be expected to question the system.

Numerous legal cases have challenged the testing and promotional systems of fire departments throughout the United States.

Most managers admit that employee selection is one of their most complicated and important decisions. Selection is the process of choosing from a group of applicants the individual best suited for a particular position. The decisions are often difficult to make, and the consequences of a poor decision can last a long time.[1] Few fire department officers have an adequate education and background in testing and selection processes. Personnel or human resources departments vary in their ability in this area, especially as it applies to fire and emergency service organizations. This chapter provides background information involving selection, testing, and promotion to increase the cognizance of the important terms and procedures of this process. It will allow fire service personnel managers to work with personnel departments or consultants from a more informed position, thereby increasing their effectiveness in this area.

Selection Standards

The criteria against which an applicant will be evaluated or the particular standards used in the selection process must be chosen with great care. To allow for an accurate and predictable measure of a candidate's success on the job, numerous requirements must be satisfied, from both legal and organizational viewpoints. The importance of each selection decision, combined with the time and cost of administering the selection process, makes the choice of selection tools critical. Several basic standards must be met in the selection procedures; these standards are explained in the following paragraphs.

VALIDITY

An important and basic requirement for a selection device or test is that it be valid. Validity is the extent and accuracy to which the test measures what it was designed to measure. A test is valid if it predicts job performance. If a test cannot indicate ability to perform a job, it has no value as a predictor. For example, many fire departments use examinations that contain mathematical computations. In order to be valid, the type of mathematical computations used in the examination process must be proven to be a valid predictor of job performance. The fire department will also need to substantiate that the level and type of math used is required by the job. So if a firefighter needs a certain level of mathematical ability to perform the job of pump operator in order to calculate friction loss and hydraulic calculations, then it would be a valid measure and could be used as a selection device.

In order to determine job requirements and performance levels accurately, the department needs a job analysis accompanied by an updated job description and job specification. A job analysis provides a summary of a job's duties and responsibilities, its relationship to other jobs, and the knowledge, skills, and abilities (KSAs) required. A job description provides information regarding the observable tasks, duties, and responsibilities of the job. A job specification is a list of the KSAs and other characteristics an individual must have in order to perform the job successfully. (Please refer to Chapter 4 on job analysis and design for more information.) The job analysis is the basic building block of any selection standard and helps make the selection process valid. If the fire department is going to test for a certain knowledge or ability, then it must first have a measurable definition of that knowledge or ability.

The way to determine validity is to compute the statistical relationship between applicant scores on a test and then measure their job performance. This relationship is usually expressed as a correlation. Three general methods are used to determine validity: criterion-related validity, content validity, and construct validity.[2]

As the first method used to assess validity, *criterion-related validity* refers to the correlation between scores on the selection device and ratings on a particular criterion of job performance. If a substantial correlation is found between test scores and job performance scores, then criterion-related validity has been established. In measuring criterion-related validity, a test is the predictor and the desired KSAs and measures for job performance are the criterion variables. Job analysis determines as exactly as possible what KSAs and behaviors are required for each task in a job. Tests or predictors are then devised and used to measure different dimensions of the criterion-related variables. If the predictors satisfactorily predict job performance, then they are legally acceptable and useful. The two basic types of criterion-related validity are concurrent and predictive validity. In *concurrent validity,* the selection tool is administered to a group of current employees, and their scores are correlated with job performance measured at the same time. The logic for this strategy is that if the best employees currently on the job perform better on the test, then the test has validity.

CONCURRENT VALIDITY

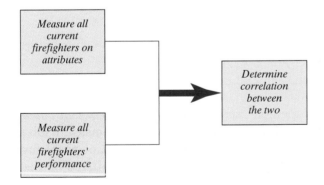

The reason it is called concurrent is because the job performance measures and the test scores are available at the same time rather than subject to a time lag as in predictive validity. Obviously, for this method to be accurate, a good method to measure current success on the job is necessary. For example, if a group of current firefighters all score well on a question regarding the use of math to determine friction loss, then this factor could be used as a selection device with a degree of validity.

The second type of criterion-related validity, *predictive validity*, seeks to establish a correlation between test scores taken prior to being hired and eventual performance on the job. In this method, the performance criterion information is obtained after the person is hired. This method requires the fire department to hire firefighters and then, after some time, measure how they have performed on the job, thereby determining the valid criteria of performance and relating them to the selection test. If determined to be valid, then the selection tool is used in subsequent tests to predict performance. Predictive validity has been preferred by the EEOC because it is presumed to be most closely tied to job performance. However, predictive validity requires a fairly large group of employees as well as a time gap between the test and performance. As a result, predictive validity is not useful in many situations. This type of validity requires the department to hire some people who will not be successful, which would be difficult to perform in a fire department because one could not subject an individual to the dangers of firefighting in these circumstances. However, this type of validity is much stronger than concurrent validity.

PREDICTIVE VALIDITY

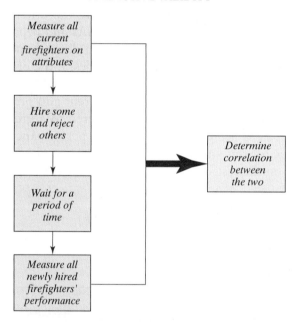

As the second method used to assess validity, *content validity* correlates certain aspects of job performance with test scores to measure job performance. It is demonstrated by showing that the items, questions, or problems posed by the test are representative of the kinds of situations encountered on the job. Content validity is a logical, nonstatistical method used to identify the KSAs and other characteristics necessary to perform a job. A test has content validity if it reflects an actual sample of work done on the job in question. Thorough job analysis and carefully prepared job descriptions are necessary when this form of validation is utilized. The classic example of the use of content validity is a typing test for applicants whose primary job would be to type documents. The content of the test closely parallels the content of the job. Many personnel specialists see content validity as a commonsense way to validate personnel selection decisions and as being more realistic compared to statistically oriented methods.

Whereas criterion-related validity is established by empirical means, content validity is achieved primarily through a process of expert judgment. Subject matter experts (SMEs; people successfully performing the job) are used to judge whether a predictor of performance is job related and its degree of importance to the job. Although this process can be subjective, confidence increases when several SMEs agree that a particular skill is needed.

One assumption behind content validity is that the person to be hired has the appropriate KSAs at the time of hire, as in the example of a typing test. Therefore, it is not appropriate to use content validation in settings where the applicant is expected to learn the job in a formal training school conducted after selection. This scenario would be typical in most fire departments, because firefighter applicants are sent through a recruit school of an extended duration to learn the job. However, at the conclusion of the training program and for promotional examinations, the use of content validation would be appropriate. Additionally, content validity is utilized for many agility and physical performance tests within fire departments. For example, if a person needs to lift a certain-size ladder to be a firefighter, then this type of test is used at the entry level to determine physical performance ability. Care must be used in selecting tests to cover a job's important elements that are used on a frequent basis.

Construct validity as a third test validation method determines whether a test measures certain traits or qualities important in performing a job. It pertains to tests that measure abstract traits in people. In order to be valid, the test must demonstrate that it accurately measures the desired trait, and that this trait corresponds to success on the job. It is a psychological method of validation based on research whereby a psychological trait is identified as essential to the successful performance of the job and a selection device is developed to measure the presence and degree of this trait. For example, if the job of firefighter requires a high degree of teamwork, a test would be developed to measure the applicant's ability to work effectively in teams. These types of tests can be difficult to develop and to correlate to actual job performance in some circumstances because so many factors come into consideration. Because a hypothetical construct is used as a predictor in establishing this type of validity, it is more likely to be questioned for legality and usefulness as a measure of validity. Consequently, construct validity is used less frequently in employment situations than are the other types of validity.

RELIABILITY

The **reliability** of a test or selection device refers to the consistency of the results or measurements that it produces. It is the degree to which a measure is free from random error. Reliability data may reveal the degree of confidence that can be placed in a test. Several different statistical methodologies can measure reliability. Reliable testing and selection methods yield equivalent results, referred to as *test–retest reliability,* time after time, regardless of incidental circumstances. Test–retest estimates of reliability are obtained by finding the correlation between assessments taken at different times. The most reliable correlation is +1.0. The closer the reliability coefficient is to +1.0, the more consistent the results and the more reliable the test. Many times the test is given twice to the same group of individuals for a correlation of the two sets of scores. Another method of assessing reliability is the *split-halves method,* which divides the results of a test within the

reliability
■ regarding a test or selection device, the consistency of the results or measurements that it produces. It is the degree to which a measure is free from random error.

same group into two parts and then correlates the results of the two parts. This method allows for the test to be administered one time, thereby reducing the opportunity for learning or recall, which tends to distort the second score of the test–retest method.

Reliable tools also yield equivalent results, referred to as *interrater reliability,* regardless of who uses them. Unreliability translates into greater measurement error and consequently into greater decision error concerning whom to hire and promote, and other personnel actions.

Summary of Firefighter Selection Devices

METHOD	RELIABILITY	VALIDITY
Application form	Moderate, if verified	Moderate
Written exam	High	Moderate
Physical ability	High	High
Interview	Low	Low
Reference check	Low	Low
Background	Moderate	Moderate
Medical exam	High	High

JOB RELATEDNESS

A great deal of information can be gathered and used to evaluate applicants for a certain job. Whether that information should be used depends on its degree of **job relatedness**. For selection standards to have job relatedness, they must be relevant to actual performance on the job. The objective is to use information that predicts successful performance as it is found in the work environment and not to allow criteria to be used that do not meet this standard. Therefore, all selection and testing items must be directly related to the actual job as demonstrated by the job analysis. It is especially important to avoid using information that may unfairly discriminate against members of certain groups, such as women and minorities.

The law does not discourage the use of testing and measurement procedures as long as they are valid, reliable, and relevant to the job. What has caused so many problems in fire departments is that the testing and selection procedures have not been validated. In many cases, insufficient job analysis coupled with tests that have no proven validity to the job remain a part of the selection and evaluation processes. Fire departments must be able to demonstrate these standards if they expect their testing procedures to be upheld in a legal challenge. The job of firefighter is too important and too dangerous to do otherwise; high standards as well as appropriate testing are essential. It is the duty of a fire service personnel manager to ensure that the department is using selection standards that meet or exceed these requirements.

job relatedness
■ selection standards relevant to the actual performance on the job. All selection and testing methods must be directly related to the actual job as identified in the job analysis.

ADVERSE IMPACT

To assist employers in developing legally acceptable selection standards, the Equal Employment Opportunity Commission (EEOC) cooperated with three other federal agencies to develop the **Uniform Guidelines** on Employee Selection Procedures. These guidelines are designed to assist employers in determining the impact of their selection procedures based on the entire selection process. The guidelines contain much information on reliability and the validity of selection procedures.

To further our discussion and to ensure a clear understanding of the terms utilized in determining adverse or **disparate impact** and discrimination, the following definitions are provided.

Business Necessity.   A practice necessary for safe and efficient organizational operations. If a business necessity is claimed, it must be proven to be essential to the performance of the job.

Bona Fide Occupational Qualification (BFOQ).   A characteristic providing a legitimate reason why an employer must exclude persons on what otherwise may be a discriminatory reason. The BFOQ must be reasonably necessary to the normal operation of the particular business or enterprise.

Protected Class.   Individuals within a group identified for protection under equal employment laws and regulations.

Disparate Treatment.   Occurs when protected class members are treated differently from others. If **disparate treatment** has occurred, the courts generally have said that intentional discrimination exists.

Disparate Impact.   Occurs when substantial underrepresentation of protected class members results from employment decisions that work to their disadvantage.

Burden of Proof.   What individuals who are filing suit against employers must prove in order to establish that illegal discrimination has occurred. The plaintiff charging discrimination must be a protected class member and prove that disparate treatment or disparate impact existed.

A fundamental principle contained within the Uniform Guidelines is the concept of **adverse impact**, a term central to the case of *Griggs v. Duke Power Company* as decided by the Supreme Court in 1971. The *Griggs* case was very important with respect to testing and the job relatedness of selection processes. The following principles came from this landmark decision:

1. A test must be job related.
2. An employer's intent not to discriminate is irrelevant. This principle applies to Title VII cases; other statutes require intent.
3. If a practice is fair in form but discriminatory in operation, it will not be upheld.
4. The defense for any existing program that has adverse impact is business necessity.
5. Title VII does not forbid testing, only tests that do not measure job performance.
6. The test must measure the person for the job and not the person in the abstract.
7. Less qualified applicants should not be hired over those better qualified because of minority origins; no law or precedent requires hiring of unqualified persons.[3]

A hiring practice that has adverse impact on any legally protected group is prima facie evidence of discrimination. Therefore, the practice is illegal unless it

Uniform Guidelines
▪ guidelines developed by the EEOC that identify legally acceptable selection standards. They assist employers in determining the impact of their selection procedures based on the entire selection process.

disparate impact
▪ occurs when substantial underrepresentation of protected class members results from employment decisions that work to their disadvantage.

disparate treatment
▪ occurs when protected class members are treated differently from others. If disparate treatment has occurred, then most likely intentional discrimination exists.

adverse impact
▪ occurs when the selection rate for any race, sex, or ethnic group is less than four-fifths of the rate for the group with the highest selection rate.

can be justified by a business necessity. If a selection procedure has no adverse impact, it is generally assumed to be legal. The question is then: What constitutes adverse impact? According to the Uniform Guidelines, adverse impact is determined with a rule known as the "four-fifths" or "80 percent" rule, as follows:

> A selection rate for any race, sex, or ethnic group, which is less than four-fifths (4/5) (or eighty percent) of the rate for the group with the highest selection rate will generally be regarded by the Federal enforcement agencies as evidence of adverse impact, while a greater than four-fifths rate will generally not be regarded by Federal enforcement agencies as evidence of adverse impact.[4]

When the selection process denies employment opportunity to a disproportionately large number of a protected class, it is deemed to be in violation of Title VII. The burden of proof that adverse impact exists falls on the plaintiff. However, once adverse impact is shown, the employer is required to demonstrate the business necessity of its selection procedure and to establish the validity of its selection process. A "business necessity" through BFOQ can be used to explain disproportionate hiring practices even if the 80 percent rule is violated. However, the federal courts have interpreted BFOQ conservatively, making it a difficult case to prove. This concept has been further modified by *Wards Cove Packing Company v. Antonio*. In this case the Supreme Court held that a racial imbalance in one part of the workforce does not by itself establish a prima facie case of discrimination. Additionally, the plaintiff in a case must demonstrate that no business reason for the workforce imbalance was present.

Disparate treatment in selection standards is also prohibited. Disparate treatment is a practice that treats individuals differently because of race, color, religion, sex, national origin, age, physical or mental disability, or veteran status. This restriction requires that all applicants be treated alike in all aspects of the selection process. An example would be a firefighter interview panel spending an average of 45 minutes with each male candidate but an average of 10 minutes with each female candidate. Therefore, any selection standard or test must be judged by a standard of fairness, which means that all applicants are treated alike and provided with an equal opportunity to obtain employment or promotion.

Federal legislation through the Uniform Guidelines and the courts have required organizations to demonstrate that their selection standards are job related, valid, fair, and reliable. These requirements have led to major improvements in the selection and testing processes for firefighters and fire officers throughout this country. They have also improved the diversity of many departments because the selection decisions are equally applied and valid with respect to their content and application. The selection standards have caused managers more closely to evaluate what they do in the fire service and who qualifies to perform this type of work. An improved workforce that is more capable in serving the public is the result.

Selection Methods

After an effective recruitment program has identified sufficient job applicants, the fire department faces the task of choosing the best ones for hire or promotion. The selection process involves judging candidates on a variety of dimensions, ranging from the observable and measurable (i.e., years of experience) to the abstract and personal (i.e., leadership potential). In the process, an organization relies on a number of selection devices, all of which must satisfy strict legal requirements. Additionally, they must be accomplished within a specific period of time and within a certain financial budget. As one progresses through the selection process, the effects on the individual candidate and the department as a whole must be carefully considered at every step.

Because vacancies in the fire service attract such a large number of applicants in most departments, the sequence of the selection devices is also important. The sequence affects how the fire department manages large numbers in a timely and cost-effective manner. For this reason a written examination is usually the first step in qualifying for a position on the fire department as an efficient way to handle large numbers of candidates at one time. A significant number of applicants are rejected at this point for both entry-level positions and promotion within the department.

The next step in the process is normally the physical performance or physical ability test. This test takes more effort than the written examination, but is still able to handle a large number of candidates. Generally, the more costly and time-consuming steps are logically phased in later in the process. For example, a background investigation would be expensive to conduct early in the process for a number of applicants who may not be hired. For the most part, the sequence of steps in the hiring process will depend on the number of selection points in the process and the number of expected candidates. Many legal requirements come into play, such as the Americans With Disabilities Act (ADA), under which the fire department cannot conduct a medical physical until after the determination to hire has been made. Fire service personnel managers need to be aware of the legal requirements and then blend them into the selection process.

In order to develop a selection process that is efficient, legal, and selects the most appropriate candidates, the department will need a multistep process that consists of a number of selection devices sequenced to fit the needs of the particular department. The following devices are the most commonly used in the selection of firefighter applicants:

- Application form
- Written examination
- Physical ability test
- Interview
- Reference check
- Background investigation
- Medical examination

The actual sequence of these selection devices depends on local circumstances, the number of applicants, and the time constraints of the hiring process. The seven items listed would be considered the minimum number of steps in a fire department selection process. Here is a review of each step individually and in detail.

APPLICATION FORM

The application form is a document that seeks information about the applicant's background and present status. Normally, this information is used as an initial or preemployment screen to show whether the applicant meets the minimum job requirements. This selection device needs careful attention as the fire department develops its selection process. This type of form often asks for information regarding educational achievement, previous work experience, driver's license requirements, resident status, and other personal information. The questions asked on the application form must be clearly job related. Federal and state laws, court decisions, and administrative rulings have drastically modified the kinds of preemployment questions that may be asked through application forms. For example, inquiries about arrest records are not permissible, but employers may ask about convictions. Likewise, questions that have no relevance to the job should be eliminated, such as martial status.

The design of the application form needs to be a joint effort among the fire department, the personnel department, and the legal office of the jurisdiction. In this manner all of the pertinent issues and potential problems can be eliminated as one develops the form. If the fire department uses a general government application form, it will most likely have to include an addendum for information specific to the fire department. Fire departments should retain all applications and hiring-related documents for at least 3 years. The Immigration Reform and Control Act (IRCA) requires that within 72 hours of hiring, an employer must determine whether a job applicant is a U.S. citizen, registered alien, or illegal alien. Employers use the I-9 form to identify the status of potential employees and should submit it with the application form.

The accuracy of applicant-generated information is also a concern, even though in many cases falsifying an application form is an offense resulting in termination of employment. Studies have shown that approximately 30 percent of applications contain false information. The most common distortions relate to length of employment and previous salary.[5] In jurisdictions with residency requirements for firefighters, the department particularly has to be on guard for falsified information. As an applicant progresses in the selection process, additional verifications of the information submitted on the application form can be made.

APPLICATION FORM

Advantages	Disadvantages
■ Provides historical reference for review by employer	■ May include inaccurate or exaggerated information
■ Documents past achievements	■ Must ensure that information collected is job related

WRITTEN EXAMINATION

Written examinations administered in the selection of candidates for the fire department are usually cognitive ability tests designed to differentiate individuals based on their mental as opposed to physical capacities. These types of exams measure an individual's ability to learn and to perform a job and are appropriate for making a selection from a group of inexperienced candidates. The cognitive ability test can measure verbal comprehension or a person's ability to understand and use written and spoken language. It can measure quantitative ability concerning the solving of mathematical problems and also test for reasoning ability, which refers a person's capacity to develop solutions to diverse problems. The Wonderlic Personnel Test and the General Aptitude Test Battery (GATB) are two widely used cognitive ability tests.

One pertinent issue concerning the use of written exams is that the level of difficulty of the exam must relate to the job. If a fire department is using an exam to select firefighters that is then proven in court to be inappropriate for that purpose, many difficult issues follow. Again, the importance of a thorough job analysis is highlighted by this discussion. The cognitive ability level tested must pertain only to the requirements of a firefighter. The job analysis will be the basis for this determination; test items will have to be developed to this level of ability. In many ways written examinations have been the most controversial component of the selection process and the subject of court cases that have challenged these types of tests due to their adverse impact, particularly on African Americans. Two factors mitigated this problem in recent years. First, in many cases the tests that were thrown out usually were not proven to be job related via a valid job analysis. This discrepancy has been corrected by a more thorough and professional approach to the development of written exams. Second, more recent data suggest that race differences in scores on cognitive ability tests have decreased over the years.[6] Therefore, these types of examinations have less adverse impact than before.

Written examinations, if properly constructed and administered, do have positive results with respect to the validity and reliability sought in selection devices. The test results are objective and free from personal bias, with results expressed numerically, lending them to statistical analysis. They are also easy to administer to a large group of applicants and can be scored quickly using scanning devices. They are a necessary part of the selection process both for entry-level firefighters and for promotional purposes. However, the written exam should be only one factor in the hiring decision and so should be given appropriate weight in that process. Small differences in scores on written examinations are unlikely to indicate important differences in the candidates. One must remember that cognitive ability is just one necessary skill in a firefighter; many more, such as physical ability and psychomotor skills, also figure prominently.

Several fire departments use an entry-level written examination that includes a study guide for the applicant, which is provided to the applicant in advance of the examination. Questions are then asked on the examination that cover the study material, as well as more general questions. These exams are designed to test one's ability to read and comprehend the texts and other written materials typically found in the recruit school for firefighters of that jurisdiction. However, no requirements or questions included require previous ability or knowledge as a firefighter, only the potential to use the material at the training academy and in the job analysis. Many fire departments have found a good deal of success with this method.

WRITTEN EXAMINATION/COGNITIVE ABILITY TEST

Advantages

- Highly reliable
- Verbal reasoning and numerical tests have shown high validity for a wide range of jobs
- The validity rises with increasing complexity of the job
- May be administered in group settings where many applicants can be tested at the same time
- Scoring of the tests may be completed by computer scanning equipment
- Lower cost

Disadvantages

- Nonminorities typically score one standard deviation above minorities, which may result in adverse impact depending on how the scores are used in the selection process
- Written examinations are the subject of frequent litigation

PHYSICAL ABILITY TEST

Without question, it takes a high level of physical ability to be a firefighter or fire officer. Applicants' strength, reflexes, coordination, and other physical skills are important to their success on the job and their individual safety. In these cases the physical ability test is relevant not only in predicting successful job performance but also in predicting the occurrence of occupational injuries and disabilities.[7] The validity of physical ability tests for certain jobs is strong. However, these tests, especially strength tests, tend to have an adverse impact on women, certain races, and applicants with disabilities.

Two basic questions must be resolved with respect to the use of physical ability tests. First, is the physical ability essential to performing the job? The answer to this question in the fire service is yes, and it has been substantiated on countless occasions. Job analysis studies have consistently shown the need for certain levels of physical ability in order to be a firefighter. Second, is there a probability that failure to perform the physical ability results in some risk to the safety of the applicant, coworkers, or the client? If the answer is yes, then in some circumstances adverse impact against those with disabilities is warranted under the provisions of the ADA.

The key concerns and the real difficulty in developing these tests are how much strength, how much aerobic capacity, and how much agility are necessary? What are the minimum accepted levels, and does exceeding that level necessarily mean better performance on the job? Extraordinary feats of strength, stamina, or agility, even if occasionally done, are inappropriate as a test standard.[8] Once again, the job analysis will drive many of the decisions that need to be made in this phase.

Another consideration is that in many fire department physical ability tests, the setup for the test requires the candidate to perform the tasks in a specific way. Training in techniques and different equipment specifications can improve performance. Some people can get the job done, but do it in a different manner. Many fire departments help to mitigate this effect by allowing candidates to practice on the components of the physical agility tests. Other departments go even further and set up classes that train individuals to pass the physical ability tests. These efforts have shown great success, especially with those who may be the subject of adverse impact regarding the physical ability tests.

In many fire departments the physical ability test is structured to resemble actual fire department activities. Although it is in general a good and accepted practice, the test needs to reflect the actual job and not just look like it. The job analysis study, which should include data on how frequently certain tasks are done, will be the guiding document. Some departments have actually encountered problems because the tasks required on their physical ability test were contrary to their own training standards. For example, a candidate is made to lift a ladder by him- or herself, which the department's training manual clearly states is a task requiring two people. Do not be subject to these types of discrepancies.

Firefighter physical ability tests that use fire department equipment as identified by the job analysis have withstood many court challenges. The equipment used should be able to be operated safely by untrained individuals. In some cases the department may have to simulate actual fire department activities, such as a dummy drag to represent the rescue of a victim. In the case of *Zamlen v. City of Cleveland*, the city was upheld, even with significant adverse impact, because the physical ability test did parallel the actual tasks that firefighters perform on the job.

The International Association of Firefighters (IAFF) and the International Association of Fire Chiefs (IAFC) have worked together to address the issue of physical ability testing with the development of the Candidate Physical Ability Test (CPAT). The IAFC/IAFF Candidate Physical Ability Test is a validated, function-based physical ability test created for the fire service consisting of eight job task simulations and a time limit. The U.S. Department of Justice worked closely with the IAFC and IAFF during the comprehensive program development process.

This program covers all aspects of administering this test: developing recruiting and mentoring programs, preparing candidates to be successful, and setting up and administering the test. The preparation section in which potential candidates are mentored and prepared to take the CPAT has been especially advantageous to women candidates. The entire validation process is discussed in detail, as well as the legal issues that departments might face when implementing the actual testing. The CPAT is a comprehensive program that is to be implemented in its entirety; partial use of this document is prohibited.

The goals of the IAFC/IAFF Candidate Physical Ability Test are to improve the quality of life of all uniformed personnel and to provide for future physically qualified candidates. Hiring physically capable people will promote better service for the community, will improve firefighter performance, and will assist firefighters in experiencing healthy careers and retirement. Many fire departments in the United States have adopted the CPAT test with very good results. One of the major benefits is that it is a valid test and a comprehensive package, so fire departments that do not have the resources to develop such a procedure can use this process. Some fire departments are beginning to use a regional approach to the CPAT test, whereby a number of fire departments share in the results of the tests, thereby decreasing the individual cost to each department.

The physical ability tests are for the purpose of determining which candidates possess the minimum ability to perform the job of firefighter. They are not contests and should not be timed or graded as such. Time is a valid component of the test, but should be used to determine the maximum time permitted to perform certain tasks. Most fire department physical ability tests include a maximum time to perform the test and are scored on a pass–fail basis. The argument over whether a better time or higher degree of strength demonstrated on the tests is valid to the entry-level requirements for a firefighter continues.

PHYSICAL ABILITY TEST

Advantages

- Can identify individuals who are physically unable to perform the essential functions of a job without risking injury to themselves or others
- Can result in decreased costs related to disability/medical claims, insurance, and workers' compensation

Disadvantages

- Costly to administer
- Requirements must be shown to be job related through a thorough job analysis
- May have age-based disparate impact against older applicants

INTERVIEW

Interviews are frequently utilized as a component of fire department hiring or promotional processes. Common sense dictates that a manager must see and talk to the individual the department is considering for hire. Even though interviews are a common part of the selection process in many occupations, they have historically not been valid predictors of success on the job.[9] Many factors and personal bias can come into play during the interview process. The fire department personnel manager needs to be fully aware of these factors so the interview process can be as professional as possible.

Illegal Interview Questions

Employers conducting interviews should not ask about any of the following, because to not hire a candidate because of any one of them is discriminatory:

- Race or ethnicity
- Sex
- Religion
- National origin
- Birthplace
- Age
- Disability
- Marital/family status
- Pregnancy

One of the major problems with interviews is the lack of standardization, which can occur as each interviewer covers different information or interprets the same information differently. As a result, different interviewers reach different conclusions regarding the same candidate. A more structured approach to the selection interview, in which all interviewers cover the same basic content, can improve reliability and the success of the interview process. In addition, proper training of the interviewers in interview techniques and situations can improve the selection process.

An interviewer or interview panel making a selection recommendation must be able to recognize the factors that determined that decision. The following are recommended as a minimum to prevent EEOC concerns with the interview process:

1. Identify the objective criteria related to the job that is to be sought in the interview process.
2. Specify the decision-making criteria used in the process.
3. Identify the training provided to the interviewers.
4. Keep accurate records of the process.
5. Provide multiple levels of review for difficult or controversial decisions.
6. Use structured interviews to the extent possible.

The interview should be designed to create a proper environment with concern for the physical setting, such as a quiet, disturbance-free room, and also to create an atmosphere to help the candidate to feel comfortable. The focus of the interview should be to obtain from the candidate job-related information appropriate for the position being sought. Many seemingly innocent questions that may be asked in the interview process are illegal or discriminatory and must be avoided. The interview process also provides an opportunity for the candidate to ask questions regarding the department or position that will help the candidate determine whether he or she can work in and adapt to that particular organization. In fire department interviews, questions regarding shift work and possible station assignments are likely to arise, for example. The result of the interview process then is a two-way exchange of information that enables the department and the candidate to be better informed regarding suitability for the position.

Interviews span the range from highly structured and directive to free flowing with little structure. As part of the selection process, they therefore must be valid and reliable, especially in cases of adverse impact. The more structured the interview is, using job-related questions, the greater the validity and reliability of the process. The two basic types of interviews—unstructured and structured—are reviewed for your information.

In the unstructured or nondirective interview, the interviewer asks probing, open-ended questions. In this type of interview the candidate is urged to do the talking and to go into a great amount of detail. This approach can reveal information that would not come up in a more structured process. The unstructured interview is often time consuming per candidate and results in different information from each candidate, making comparisons difficult. Because this process encourages open discussion, potentially discriminatory or inappropriate information for job interview purposes may be discussed. Most fire departments interview a large number of applicants for firefighter positions, and the unstructured interview is not suited for this purpose. However, for promotional purposes, especially at higher ranks, this format may have many benefits, because it is a more in-depth process.

The structured interview consists of a series of job-related questions consistently asked of each applicant for the job. This method is conducive to uniformity from one interview to the next in that it treats all applicants in the same manner.

The use of structured interviews reduces much of the subjectivity and inconsistency found in unstructured interviews. This type of process can handle a large number of candidates in an efficient manner. In many cases the questions asked in this type of process are job knowledge questions to determine the degree of expertise of the candidate. Also, situational questions may be used in which a hypothetical job situation is posed and the appropriateness of the candidate's response is evaluated by the interviewer or a panel of interviewers.

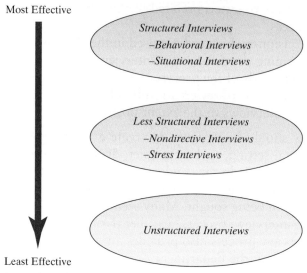

Source: Human Resource Management, Tenth Edition.

For the interview purposes of most fire departments, the structured interview using a panel is probably the best format. In this circumstance several individuals simultaneously interview one applicant. Because the interviewers all hear the same responses, panel interviews produce more consistent results. Panel interviews may also be less susceptible to the bias and prejudice of the interviewers, especially if the panel members come from diverse backgrounds and cultures.

A structured interview conducted by a panel can potentially intimidate or create stress in a candidate. The aspect of a firefighter applicant walking into a room and seeing four interviewers in full-dress uniforms staring at him or her from across the table can be a nerve-racking experience. The fire department will need to minimize the stress on the part of the candidate. If the interview is too rigid, the interviewers may neglect chances for follow-up questions, and the candidate is unlikely to provide any information spontaneously. One way that I have found helpful in reducing stress and getting the candidate going lies in the sequence in which the questions are asked. By asking an easy question first in the interview process, preferably one familiar to the candidates, such as one about themselves, the process goes much smoother.

Behavior-Based Interviews

Interview information at times is vague and can be difficult to quantify in a meaningful manner. One method to improve the effectiveness and validity of the interview process is to implement "behavior-based interviews." Behavioral based interviewing is based on discovering how the candidate has acted in specific employment-related situations. The logic is that how the candidate behaved in the past predicts how he or she will behave in the future. In a structured interview, the candidate will be asked a series of questions that typically have straightforward answers, such as "What are your strengths and weaknesses?" In a behavioral interview, an employer has determined via job analysis and other techniques what skills are needed in the person to be hired and will ask specific questions to find out whether the candidate has those skills. For example, in an interview for promotion within a fire department, a question may be "Describe how you handled a firefighter lateness issue and explain what positive techniques were employed to improve the situation."

Here are a number of reasons that make behavior-based interviews a better selection device as compared to other interview techniques:

- They involve identification of job requirements and job-related questions.
- Interviewers conduct similar interviews with every candidate for a position, enabling comparison of "apples to apples."
- The questions are behavior-based. This means that interviewers obtain information primarily from candidates, rather than from opinions.
- The evaluation process involves comparison of answers to defined hiring criteria. This promotes a decision based on the candidate's skills in doing the job rather than being based on interviewers' gut feel.

Although there is general acceptance of the effectiveness of behavior-based interviews, it is still useful to see quantitative evidence to that effect. The following examples show that behavior-based interviews help interviewers realistically determine who will be high performers:

Green, Alter, and Carr, 1993.   This study examined the validity of a behavior-based interview process. Results showed that interviewers' ratings positively correlated with job performance. The researchers found that behavior-based interviews help to "predict" who will be high performers based on the interview ratings.

Orpen, 1985.   This study analyzed the validity of a behavior-based interviewing process and an unstructured interview process. Results showed the behavior-based interview ratings to correlate more highly with performance measures than the unstructured process did. This suggests that behavior-based interviews successfully predict performance better than an unstructured process does.

Pulakos and Schmitt, 1995.   This study compared the validity of past-experience interviews and situational interviews. The past-experience interview ratings correlated more highly with performance ratings than the situational interview ratings did. This suggests that behavior-based interviews predict a candidate's future performance better than situational interviews do.[10]

As can be determined from this information on behavior-based interviews, they work best when the candidate has acquired appropriate job knowledge. For that reason, their use in entry-level firefighter selection may be limited. However, for promotional interviews or for any situation in which one is making a selection based on job knowledge, they are more effective than standard interview techniques.

In order for the interview to be successful for the candidate and the department, it must be well planned. The physical location of the interview should be pleasant and private with a minimum of disruptions. The interviewers must be trained to perform their duties. Some jurisdictions provide no training or orientation for the interviewers, which puts these interviewers in a precarious position if they are challenged legally and makes the interview process less effective. The proper training of interviewers can heighten their awareness to common interview problems such as central tendency, halo effect, premature judgments, and other situations that complicate the interview process. It can ensure that each person on the panel understands the questions and the most appropriate responses in order to improve the consistency within the process. Many departments identify possible responses to questions and evaluate them on a scale so that the interviewers understand what type and level of response is appropriate. Several departments allow the interview panels to practice by having a number of noncandidates go through the process first, which gives the interview panel experience working with each other prior to the first real candidate. These techniques improve the validity and the reliability of the interview process. The money spent in training the interview panel is easily justified when considered in light of the consequences of a poor selection or a legal challenge to the process.

INTERVIEW

Advantages

- Useful for determining whether the applicant has requisite communicative or social skills for the job
- Interviewer can obtain supplementary information
- Used to appraise candidates' verbal fluency
- Can assess the applicant's job knowledge
- Allows the applicant to ask questions that may reveal additional information useful for making a selection decision
- May be modified as needed to gather important information

Disadvantages

- Subjective evaluations are made
- Decisions tend to be made within the first few minutes of the interview with the remainder of the interview used to validate or justify the original decision
- Interviewers form stereotypes concerning the characteristics required for success on the job
- Negative information seems to be given more weight
- Not much evidence of validity of the selection procedure

REFERENCE CHECK

Personal reference checks are established to provide additional insight into the information submitted by the applicant. References may include previous employers and supervisors as well as friends and acquaintances. Personal references are likely to focus on the personal qualities of the applicant and may not include significant job-related information. In addition, because the references are supplied by the job applicant it would be unusual for them to be other than favorable reviews of the applicant. The credibility of some of these is suspect and most organizations do not place a high value on personal references.

Depending on the sensitivity of the job applied for, the personal references may extend beyond what was submitted by the applicant. For example, neighbors and others may be contacted to verify information submitted or to get an impression of the job applicant. This type of contact is used by many police departments, but few fire departments.

Outsourcing of reference checks is an expanding business. Employers contract with a third-party vendor to conduct reference checks and to access specialized databases. Often these firms have individuals with special training in conducting these types of investigations. In addition, the use of these types of companies can offer a degree of limited legal protection to the employer.

REFERENCE CHECK

Advantages	Disadvantages
■ Additional information beyond what applicant submitted	■ May be biased toward applicant
■ Employer references are more reliable	■ Some employers are reluctant to provide information for fear of legal action

BACKGROUND INVESTIGATION

A background investigation involves communicating with previous employers and others who can provide information on the applicant, including conducting a criminal history check of state and national records. The background investigation serves two purposes: It verifies the information the applicant gave to the department as being accurate and provides supplemental information that can be useful in the selection process. It helps detect cases in which applicants falsify or exaggerate information on application forms.

The fire department enjoys a right of entry to businesses and private residences that no other government agency has, including law enforcement. Fire departments also treat many people when they are disoriented or incapable of caring for their property and belongings. Firefighters drive large vehicles at high rates of speed through heavy traffic. For these reasons and many others, fire departments have a special duty to ensure that firefighters are honest and have exemplary driving and criminal records. This standard is necessary to maintain the character of and respect for the profession, and because the public deserves no less. Another reason for the increased emphasis on background checks is to prevent negligent hiring lawsuits.

If a reasonable background investigation in not conducted, or if negative information found in the investigation is not acted on, the department may be legally liable for **negligent hiring**. Negligent hiring is defined as a situation in which an employer is liable to a third party for injury because the employer knew or should have known of the employee's dangerous proclivities. For example, in *Ponticas v. K.M.S. Investments*, the employer hired a caretaker for an apartment building. This employee had been convicted of several crimes including armed robbery, burglary, and auto theft. Using his passkey to enter one of the

negligent hiring
■ a situation in which an employer is liable to a third party for injury because the employer knew or should have known of the employee's dangerous proclivities.

apartments, he raped a tenant. The court held that the employer had a duty to exercise reasonable care in hiring individuals who because of the nature of their employment might pose a threat to members of the public. The primary consideration in negligent hiring is whether the risk of harm could have been reasonably foreseen.[11]

In a number of court cases individuals have sued their former employers as a result of what the employers said to other potential employers that prevented the individuals from obtaining jobs. Because of such issues, organizations have been advised by their legal departments to give out only name, employment dates, and title. In many cases this has the effect of rendering employment checks worthless. Under the federal Privacy Act of 1974, a governmental employer must have a signed release from a person before it can provide information about that person to other potential employers. It is a good practice to obtain these releases in the event that information is given out. The best practice is to stick to the job-related facts and not embellish or add personal opinions that may be difficult to substantiate at a later date. Thirty-five states have laws that protect employers from civil liability when giving out reference information in good faith that is objective and factual in nature.

Many fire departments check an applicant's credit history. The logic is that an individual with a poor credit history may signal irresponsibility. This assumption may be questioned and employers that use credit checks must comply with the Fair Credit Reporting Act. This act requires disclosing that a credit check is being made, obtaining written consent from the person being checked, and furnishing the applicant a copy of the report. Credit history checking is most appropriate for applicants in jobs in which they use, access, or manage money as an essential job function. Fire departments may or may not fall into this category depending on their structure and function.

Procedures should be established for the department to conduct criminal background checks of applicants for the fire department. In many jurisdictions the fire department will have to coordinate this task with the local law enforcement agency. State and local criminal backgrounds are easier to obtain than national criminal background checks using the Federal Bureau of Investigation (FBI). A proper check requires fingerprints of the applicant and sufficient lead time to complete the national criminal background check. In addition, the driver's license of applicants should be investigated to ensure that their driving history meets the standards established for the department. Many states now notify an employer when someone in their employment has a driver's license violation. This notification needs to be prearranged with the state's Department of Motor Vehicles and should be utilized if available.

Departmental regulations often require that any firefighter arrested for or charged with a criminal act or a significant motor vehicle violation immediately report it to the department. This requirement would also include the final disposition of any criminal or traffic cases. Such reports allow the department to be aware of the criminal acts of those within its employment on a continuing basis and to take appropriate action where warranted to protect the reputation of the fire department and the public it serves. It also reduces liability for negligent retention, which is when an employer continues to employ someone whom it should not.

Many fire departments are finding it increasingly difficult to obtain information about applicants beyond the criminal and motor vehicle records. One of the main reasons is that employees have been successful in suing previous employers for defamation regarding information that they have released. Hence, many organizations are restricting the information they will give out to only verifying the dates of employment and job title. Some will provide information on a former employee only after they have obtained written consent from that employee.[12] However, the hiring agency should check with previous employers and obtain any information that they might provide. This area of job applications is often falsified, especially with respect to previous duties and responsibilities. Some "red flags" on applications require further investigation before conclusions are made, such as numerous employers in a short time frame and voids in one's employment history.

The use of polygraph tests to verify background information has been controversial at best. One use of the polygraph has been to confirm the information supplied to the organization on the application form and other questions regarding one's character and honesty. The Employee Polygraph Protection Act of 1988 severely limited the use of polygraphs in the private sector to those involved in or affecting interstate commerce. The employer is prohibited from requiring, requesting, or suggesting that an employee or prospective employee take any type of polygraph test. Although this act does not apply to government agencies, some have followed the provisions. For example, the state of Maryland prohibits the use of polygraphs in employment screening, except for law enforcement agencies. If an organization uses polygraph testing in its selection process, this testing should not be the sole disqualifying factor and should always be used in combination with other techniques.

The background investigation process should be planned in advance and approved by the legal office. The same procedures must be used for all applicants in order to establish fairness and reduce discrimination. A valid job-related reason is necessary to obtain the information in the background investigation, and the information once obtained should be maintained as confidential. The time and effort expended to perform a comprehensive background investigation is well worth it to the fire department and those it serves.

BACKGROUND INVESTIGATION

Advantages

- More thorough review of past actions by job candidate
- Reduces negligent hiring claims
- Verifies applicant information
- Proper legal review prior to conducting the investigation

Disadvantages

- May be costly and time consuming to properly conduct
- Polygraph use may be questionable
- Credit checks need to be job related

MEDICAL EXAMINATION

The main qualifying point about medical examinations is that they cannot be given until the job offer has been made. Under the Americans With Disabilities Act (ADA) of 1990, it is unlawful to test for or inquire whether an applicant has a disability before a job offer is extended. For this reason and because of the costs involved, a medical exam is normally the last step in the selection process for firefighters in most jurisdictions. After the department has ranked the candidates for selection and determined to whom job offers will be made, then the medical physical is administered.

Decision making for the fire service personnel manager is not a part of this aspect of the selection process, because the decisions are medical in nature and need to be made by a physician or a medical review board composed of a group of physicians. The difficulty is in the identification of the medical standard to which the applicant will be held. For example, is uncorrected vision approved up to 20/40 or to 20/100 or to 20/140? For medical standards many fire departments use the requirements of Chapter Two of NFPA 1001, Standard for Fire Fighter Professional Qualifications. This national consensus standard is quite extensive. Legal challenges under the provisions of ADA have greatly complicated this subject in the past few years and will continue to do so in the foreseeable future. However, once the medical standard or policy is established, then the decisions as to who qualifies are ones to be made by a physician.

It is important for the physicians making the medical decisions to understand the nature and rigors of firefighting work. A clear description as to the type of activities that firefighters perform can be drawn from the job analysis study that has been previously performed for this position and provided to the attending physician. This information will also be helpful in medical evaluations to return current firefighters to duty after an injury or illness. Many governments have specialty fire department medical clinics, in which physicians have a high degree of knowledge of the work and activity of firefighting.

A fire department can conduct medical evaluations that are job related and consistent with business necessity. A fire department does not have to hire an applicant with a disability when it can demonstrate that the work of firefighting would pose a direct threat to the public or coworkers. The department is required to reasonably accommodate a qualified individual with a disability seeking a firefighter position unless the accommodation creates an undue hardship. Reasonable accommodation in an occupation in which one may respond to virtually any type of emergency at any time is often difficult to provide, because of the lack of set working conditions and locations. However, where possible, it should be done.

The medical evaluation conducted in the selection process should include a drug screen for illegal and abused substances. Drug screening either may be a part of a medical exam or may be performed separately. Procedures for the handling and processing of urine and blood samples should be exact and in accordance with recommended practices. The accuracy of the drug test varies according to the type of test used, the item tested, and the quality of the laboratory where the test samples are sent. Because of the potential impact of prescription drugs on test results, applicants should complete a detailed questionnaire on this matter prior to the test. Whether urine, blood, saliva, or hair samples are used, the process of obtaining, labeling, and transferring the samples to the testing lab should be outlined clearly and definite policies and procedures established. If an individual tests positive for drug use, then a second confirmation test needs to be performed on the same sample to ensure accuracy. This confirmation test is very important and organizations should not initiate any action against the employee until the confirmation test is completed and the test results are verified. Testing for illegal drugs is not considered a medical examination under ADA and is permissible. During the recruitment process the substance abuse policy of the fire department should be made clear to all applicants.

MEDICAL EXAMINATION

Advantages	Disadvantages
▪ Determines previously undetected conditions that affect performance and health of the job candidate	▪ Expensive to administer
▪ Reduces potential disability and workers' compensation costs	▪ May present ADA issues
▪ Establishes a medical baseline for future claims	
▪ Decisions based on job-related medical criteria	

At this point we have reviewed many of the selection methods used by fire departments in the process of hiring new firefighters. These methods require a great deal of planning, need to be performed in the proper sequence, and must be applied in a fair and unbiased manner. These selection criteria are necessary to provide the public with the best-qualified applicants for the position of firefighter.

SELECTION METHODS

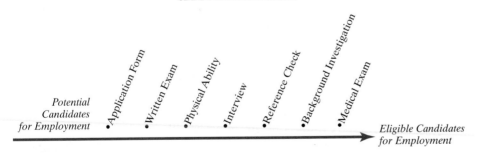

Assessment Centers

The existence of assessment centers can be traced back to World War II, when both the German and the British military utilized this technique. German military psychologists used the assessment center process to recommend and select German military officers because it was deemed better to select leaders based on a number of behaviors. In 1942, the British Military War Office Selection Boards utilized assessment centers. Their introduction stemmed from the fact that as a result of the existing system, a large proportion of those officers it had predicted would be successful were being returned to their unit as unsuitable. The assessment center approach subsequently adopted attempted to depict more accurately the types of behavior that a British military officer was required to display in order to be successful in his job. The tasks included leaderless group exercises, selection tests, and individual interviews by a senior officer, junior officer, and psychiatrist, respectively. This new system resulted in a substantial drop in the proportion of officers being returned to their unit as unfit for duty.

In the United States assessment centers were initially used by the Office of Strategic Studies (now the CIA) to select spies during World War II. Subsequently, the use of assessment centers was taken up by private sector companies such as American Telephone and Telegraph Company (AT&T), which began using them for management selection in 1956, as well as Standard Oil, IBM, Sears, and General Electric. The differences between the U.S. and U.K. approaches largely stemmed from their original background and their initial use. In the United Kingdom a greater emphasis was placed on group exercises with an appointed leader, group discussions, and long written exercises, whereas in the United States greater emphasis was placed on in-tray exercises, leaderless group exercises with assigned roles, and two-person role plays.[13]

An assessment center is composed of a series of evaluative exercises and tests used for selection and development. Information about an individual's potential for promotion is gathered systematically and analyzed as the candidate participates in a series of tests, interviews, simulations, and exercises. The assessment center concept is an effective, although expensive, approach to supervisory and managerial selection. Because of the expense and the inability to handle large numbers, the assessment center is not normally used in the selection process for entry-level employees. The assessment center is developed so that the candidate performs activities similar to those encountered in the actual job, so the simulated exercises that make up the assessment center must be based on thorough job analyses. It is crucial to any assessment center that the tests and exercises reflect the job content and types of situations faced by candidates within their work environment. The validity of the assessment center is high as is its reliability because it closely mirrors the actual work environment.[14]

The assessment center is an integrated approach to selection whereby a battery of tests and exercises is administered to a person or a group of people across a number of hours or days in some cases. Assessment centers give candidates opportunities to demonstrate behaviors and skills that are manifestly job related, and the results of these evaluations are more readily accepted by candidates and the organization. The results of assessment centers are increasingly being used to determine the type and sequence of developmental activities that candidates for managerial and executive positions exhibit on the job. Assessment centers are particularly useful in the following situations:

- Required skills are complex and cannot easily be assessed with interview or cognitive tests.
- Required skills include significant interpersonal elements such as those found in management positions.

- Assessment centers allow a wide range of criteria to be assessed, including group activity and aggregations of higher level, managerial competences.
- Multiple candidates are available and it is acceptable for them to interact with each other.

Assessment centers have grown in their use. In 1973 only about 7 percent of companies were using them. By the mid-1980s, this had grown to 20 percent, and by the beginning of the year 2000 it had increased to approximately 65 percent.[15]

One of the major advantages of the assessment center concept is that it allows the candidate the opportunity to be evaluated on a number of skills and attributes. A fire officer in today's world is called on to perform many tasks effectively. He or she may be in command of the first-arriving units at a dwelling fire in the morning, supervise inspections before lunch, counsel a firefighter later in the day on performance-related issues, and develop a written report for the battalion chief by the end of the shift. Because of the variety and complexity of a fire officer's job, the fire department needs an assessment tool that best replicates the work environment in this sense. Assessment centers are very effective for this purpose, and I believe they offer the flexibility and challenge necessary to fit the fire service work environment. Candidates in the assessment process have the opportunity to display many of their abilities, and a weakness in one ability will not disqualify them, just like in real life. Some fire departments still promote fire officers based on a written examination and perhaps an interview process. Written examinations certainly have their place in demonstrating job knowledge, but are much too limited to use as the sole selection criterion. It is a disservice to the candidates and the department to have such a one-dimensional system for promotion.

Not only do the candidates have the opportunity to display the full range of their qualifications in an assessment center, but they are also evaluated by a pool of assessors. No candidate in the process is subject to the rating of a single assessor. A group of assessors leads to the development of a consensus on the performance of the candidate in the assessment center process. A higher degree of observance and evaluation by the assessor panel tends to reduce the effect of any bias that one assessor may have as it relates to the candidate. Some departments use a mix of assessors from both inside and outside the department to increase objectivity.

Another advantage of the use of assessment centers is that they tend to demonstrate less adverse impact on minorities than do other selection devices, particularly written examinations. Both minorities and women acknowledge that assessment centers provide a fair opportunity to demonstrate their capabilities in managerial positions.[16] The assessment center process takes into account many more dimensions of one's ability to perform certain jobs. The process examines more of the whole person and total ability as opposed to a small segment of individual ability.

The training of the assessors is important to the success of the assessment center process. They will need to understand thoroughly the levels of performance they are evaluating. An agreement on the proper action of the candidate in the assessment center must be accomplished prior to the initiation of the evaluations, which will require training and background information as to the appropriate response. This response will be related to the job analysis and specific job functions at the level of the position being tested for. Many departments predetermine possible responses on a scale from unacceptable to outstanding to guide the assessors and maintain consistency. The assessors will then be trained to recognize these attributes and in this process discuss any issues that need clarification. The assessor training in many cases will take at least 2 days in order to cover the process and the use of the evaluation instruments completely.[17]

Assessors used in fire department assessment centers will most likely be fire department officers. It is a good practice to have officers from more than just one department participate. The use of fire department officers is recommended because they are thoroughly familiar with the jobs for which the candidates are being assessed. Their involvement in the assessment center process contributes to its acceptance by participants and others within the department. Additionally, it may help to have civilian personnel experts and managers participate because their perceptions and experiences will also improve the process.

Some progressive fire departments have developed a regional approach to the training of assessors. Several fire departments pool their resources and train a number of potential assessors to be used in future assessment centers. This ensures that the assessor training is consistent and that a pool of qualified assessors is available at all times. Sharing of resources in this manner tends to reduce the costs of conducting assessment centers.

In summary, here are the major advantages and main components of assessment centers:

- Assessment techniques based on job analysis
- Multiple measurements of performance
- Multiple assessors used in the process
- Assessors specifically trained for the assessments
- Systematic method of recording and evaluating behavior
- Behavior evaluated to a performance standard by pooling of assessments
- Potential adverse impact reduced

Many different possible individual assessments can be conducted in the process. The goal is to replicate actual important job-related functions and then evaluate the candidate's performance on a variety of different skills. Several of the more popular assessment exercises are listed as follows:

- *In-basket* exercises are designed to determine a candidate's administrative ability in making decisions and establishing priorities with written correspondence. Typically an in-basket has a number of items that must be acted on in a certain way and with a certain priority. This type of exercise is easy to evaluate because it is not dynamic and can be graded at a later date if necessary. More involved assessment centers actually include interruptions and emergencies that occur in the real world.
- *Performance counseling* exercises place the candidate in the position of an officer who must counsel a firefighter on a performance-related issue. The prospective officer is given information just prior to the assessment that describes the situation and the circumstances that led to the problem. He or she then performs the performance counseling session, which is evaluated by the assessors. Some fire departments use a professional actor to play the role

of the firefighter. In this way, the response and actions of the individual to be counseled are the same for each candidate.

- *Emergency scene simulation* exercises are designed to evaluate one's actions on the scene of an emergency. By the use of a tabletop, simulator, or actual evolutions, the decision-making ability of the candidate under emergency conditions is evaluated. Scenarios can be developed for structural fires, hazardous materials, and any other possible response by the fire department. Many of these simulations can be elaborate; the department then uses these systems for training officers as well as for evaluating them. The degree of anxiety and pressure present on the scene of an actual emergency is often present in these exercises.

- *Training presentation* exercises are used to evaluate the ability of the candidate to present a drill or community presentation. The candidate typically presents a 30- to 45-minute session and is evaluated on verbal ability, presentation skills, and techniques for holding the interest and attention of an audience.

- *Staff meeting* exercises simulate the conducting of a meeting to exchange information. The candidate receives background information and then creates an agenda and conducts the staff meeting. Different roles can be played by those in the staff meeting, and dynamic situations must be dealt with as the meeting progresses. In the situation in which groups are involved in the assessment, the department has to be careful to present the same basic scenario to each candidate. The more people in the process, the more difficult this can be.

- *Leaderless group* exercises are simulations of meetings in which the candidates react to the meeting format and information exchange. The person to be assessed is not leading the meeting but attending as a participant. This person will be evaluated on his or her interaction with the group and success in getting consensus within the group. Several candidates can be assessed at one time in this process. However, these types of exercises can be involved and may be difficult to evaluate.

As one can see, the types of assessment techniques that are used are limited only by one's imagination. However, the assessment techniques a fire department does use must be job related and appropriate for the level of position they are testing for. For example, the training presentation and performance counseling assessments may be right for a first-line supervisor, but the staff meeting exercise may not be. As in many other facets of personnel management, the job analysis guides technique selection.

Assessment centers are normally found in career fire departments; however, some volunteer fire departments also use them. The College Park Volunteer Fire Department, which protects the University of Maryland and surrounding areas, uses a full assessment center process in order to select the volunteer officers of that department. It has developed an outstanding system for promotion that has been in place for many years with excellent results.

Promotional Process

In many cases the assessment center will serve as the entire evaluation process for promotion, and in others it will be one component of the process. Much of this depends on the promotional process used and the number of potential candidates the department may have. This determines who is eligible, how they are evaluated, and how they are selected for promotion. Many decisions are involved in this process, which requires close coordination with the personnel department of the jurisdiction.

After firefighters become fully proficient at their rank, many naturally look toward career advancement within the organization. Some aspire to leadership positions so that they have more impact on the department; some want the higher status, pay, and benefits; and some think the job is easier. The latter are really surprised when they are actually promoted and must perform the duties of a fire officer. Whatever the motivation, which we will leave aside for a moment, a defined process needs to be administered in an equitable manner. Promotions in most fire departments are very competitive, and sometimes a fraction of a point in one's score can make a difference in when or if a firefighter is promoted. Therefore, the promotional system is subject to a great deal of interest and scrutiny.

The fire department must be careful to ensure that all elements of the promotional process are job related, valid, and fair. Sometimes the assumption is made that if an individual is a good firefighter, then he or she will be a good officer. This assumption is often true, but sometimes it is not. The job functions of a firefighter are distinctly different from those of a fire officer, and success at the firefighter level may not indicate one's ability to be an officer. The department has to be careful that all of its selection criteria come from the job analysis of the fire officer position. Some departments also give credit to candidates based on experience in the department. Experience is difficult to quantify and evaluate in the promotional process. The experience in this case is at the previous rank level and the testing is for the new rank level; they are dissimilar. It may be best not to make job experience a part of the system, knowing that it will become evident during the different assessments.

One of the first decisions is to determine who is eligible for promotion. Normally "time in grade" and the successful completion of certain elements of the current position are required before one can apply for promotion. My experience is that most fire departments require 3 to 6 years of experience as a firefighter before one can apply to be an officer. Some departments also require certain national professional certifications and/or college credits or degrees for promotional eligibility. The fire department will also need to determine how often to initiate the promotional process, depending on the size of the department and how many expected promotions are made each year. Some departments conduct the promotional process every 2 to 3 years in order to provide opportunity and to be in a position constantly to select the best people. Smaller departments may carry out the process only when there is a vacancy because their turnover is very low. Whichever policy is used, make sure that it is widely disseminated and understood by those eligible for the process.

After or sometimes even before the promotional process is announced, individuals eligible will want to prepare themselves. Most fire departments publish a list of the materials from which the examination items will be developed. It may include departmental orders and procedures, fire service textbooks and references, local government regulations, and the like. The test items, whether they are for a written examination or for an assessment center, have to be job related. The criteria for selection of textbooks and reference material have to include this. Candidates should be encouraged to develop study groups and to be prepared thoroughly for the promotional process. Many departments conduct classes that prepare personnel for the promotional process, which is a good idea. Just make sure that no one involved in the development of the test or assessment items is presenting these classes.

The promotional process that the fire department uses may be based on past practice or on consultation with the union, managed by the personnel department, or contracted out to a professional test development company. The test and assessment elements discussed in this chapter could be used as a part of this process in one form or another. The fire department personnel manager will have to make these decisions based on the particular circumstances of his or her department.

As an example, here is a promotional system that I am familiar with that has been demonstrated to be effective over a long period of time. The Recruitment and Examination Board developed the promotional policy and system for a department with nearly 800 career employees. The system was designed to make valid assessments about a candidate's ability to be an officer using a full range of knowledge, skills, and abilities for such a position. The process used for the first-line supervisory position of fire lieutenant follows the process for other promoted ranks within the department, adjusted for KSAs, of course.

PROMOTIONAL PROCESS FIRE DEPARTMENT

| Written Examination | Officer Candidate School | Assessment Center | Eligibility List |

These basic four steps comprised the system for promotion within the fire department. Because the department had a large number of personnel who desired promotion, a screening mechanism was needed. The written examination was used for this purpose and included test items to determine one's job-related knowledge of the abilities necessary to be a fire officer. The written examination score counted for 30 percent of the final score for the candidate. The top qualifiers on the written examination were then allowed to attend a 3-week officer candidate school. The officer candidate school consisted of a learning environment that prepared a firefighter to make the transition to fire officer. College professors taught classes on leadership and management; and senior fire officers instructed on emergency scene strategy and tactics, personnel procedures, and the many other abilities necessary to become a successful fire officer. The officer candidate school was an important part of the process, and many personnel frequently made favorable comments about how well it prepared them to be officers.

The assessment center consisted of four basic components. In an emergency scene assessment, the candidate was evaluated on three different simulated emergencies. Normally these included a single-unit response, a multiunit response, and a multiple-alarm incident. An in-basket exercise included items common to the rank of company officer, or higher depending on the officer level being tested for. The candidate also had to complete a 30-minute training presentation and a performance counseling session that the assessor panel evaluated. The assessment center score counted for 70 percent of the final score for the candidate.

At the conclusion of the assessment center, the scores were tabulated and the candidates were placed on an eligibility list that was good for 2 years. Promotions were made from this list as vacancies became available. By county personnel law, the fire chief could promote any one of the top three candidates for each position. However, the department promoted from the eligibility list in rank order. It was deemed appropriate that if the department developed the promotional criteria and system, then it was as equally appropriate to abide by the results of that system. Therefore, promotions were offered as the candidates appeared on the final list.

As can be seen from this example, a complete system was in place for the promotional opportunity of that department. Job-related promotional criteria were specified, a process was employed using appropriate techniques, and the result was a valid and fair system. The unique aspect of this system was the 3-week officer candidate school. This school was instrumental in the success of the process and allowed the department to have better trained and qualified leaders.

Summary

The selection process begins with a careful analysis of the job to be filled and the knowledge, skills, and abilities required for successful job performance. Applicants are assessed for their potential in a manner that is valid, reliable, and fair, and with relevance to the job. Hiring and promotional criteria are developed in a multistep system that takes into account the needs of both the department and the candidate. Relevant legal and administrative guidelines are followed in this process. The result is an efficient system of hiring and promoting, thereby ensuring the future success of the organization.

Review Questions

Use these questions to review the material in this chapter and for discussion purposes.

1. What is the importance of validity and reliability in the selection process for either hiring or promotion?
2. How do you know when adverse impact occurs in the testing process, and what are several ways in which it can be eliminated?
3. Discuss several ways in which written examinations and physical ability tests can be improved as selection devices for firefighters.
4. At what point can a medical physical be required of a candidate for the position of firefighter?
5. Discuss several advantages of the assessment center concept, and describe the assessment devices that you would recommend for the rank of lieutenant.

Case Study

You are the fire chief in a diverse community that is served by a fire department consisting of 575 employees. There is a local IAFF union that legally represents the firefighters of the department. In the past, promotional issues and the process have been agreed on by the department and the union. There is also a minority firefighter group that is organized as a member of the International Association of Black Professional Firefighters (IABPFF). Although the minority percentage of the general population of your jurisdiction is approximately 28 percent, the minority percentage of fire officers in your department is 4 percent.

The local IABPFF group has approached you and the mayor regarding the low number of minority fire department officers. It plans to file a discrimination lawsuit, contact the Department of Justice EEOC section, and conduct a press conference to announce this. The mayor requests that you report to him on the promotional process currently used within the fire department.

Develop the promotional process that you would like to see used in this circumstance. Provide a detailed plan as to the process and how you arrived at these conclusions. Identify the legal issues and how the selection devices were validated. Discuss how you would implement this plan given the current state of affairs in this fire department.

Fire Service Professional Qualifications

This chapter contains information that will assist fire officers in meeting the job performance requirements (JPRs) of NFPA 1021, Standard for Fire Officer Professional Qualifications:

4.2 Human Resource Management—Fire Officer I

4.2.5 Apply human resource policies and procedures given an administrative situation requiring action, so the policies and procedures are followed.

5.2 Human Resource Management—Fire Officer II

5.4.1 Develop a policy or procedure, given an assignment, so that the recommended policy or procedure identifies the problem and proposes a solution.

6.2 Human Resource Management—Fire Officer III

6.2.3 Develop procedures and programs for promoting members, given applicable policies and legal requirements, so that the process is valid and reliable, job related, and nondiscriminatory.

Endnotes

1. Stephen L. Guinn, Gain competitive advantage through employment testing, *HR Focus, 70,* September 1993.
2. Wayne R. Mondy and Robert M. Noe, *Human resource management* (Upper Saddle River, NJ: Prentice Hall, 1996).
3. Kenneth L. Sovereign, *Personnel law,* 4th ed. (Upper Saddle River, NJ: Prentice Hall, 1998).
4. Uniform Guidelines on Employee Selection Procedures, Sec. 3D, *Federal Register, 43,* 1978.
5. D. G. Lawrence et al., Design and use of weighted application blanks, *Personnel Administrator,* March 1992.
6. A. R. Jenson, White–black achievement differences: The narrowing gap, *American Psychologist, 39,* 1984.
7. J. R. Hollenbeck et al., Lower back disability in occupational settings: A human resource management view, *Personnel Psychology, 42,* 1992.
8. Robert E. Osby, Guidelines for effective fire service affirmative action, *Fire Chief,* September 1991.
9. T. L. Brink, A discouraging word improves your interviews, *Human Resources Magazine, 37,* 1992.
10. Novations Group, *Evidence of the effectiveness of behavior-based interviews,* retrieved September 20, 2004, from http://www.novations.com
11. Sovereign, *Personnel law.*
12. Research Institute of America, *Employment Coordinator,* April 1985.
13. Charles Woodruffe, *Assessment centres: Identifying and developing competence* (New York: State Mutual Book and Periodical Service, 1993).
14. B. J. Cohen, L. Moses, and J. Byham, *The validity of assessment centers: A literature review* (Pittsburgh: Development Dimensions Press, 1994).
15. Wayne F. Cascio, *Managing Human Resources,* 7th ed. (New York: McGraw-Hill, 2006).
16. *Assessment Centers,* retrieved October 10, 2007, from http://www.ChangingMinds.org
17. Randal Schuler and Susan Jackson, *Human resource management,* 6th ed. (New York: West Publishing, 1996).

Suggested Readings

Brink, T. L. 1992. A discouraging word improves your interviews. *Human Resources Magazine, 37.*

Cascio, Wayne F. 2006. *Managing human resources,* 7th ed. New York: McGraw-Hill.

Cohen, B. J., L. Moses, and J. Byham. 1994. *The validity of assessment centers: A literature review.* Pittsburgh: Development Dimensions Press.

Dessler, Gary. 2001. *Management: Leading people and organizations in the 21st century,* 2nd ed. Upper Saddle River, NJ: Prentice Hall.

French, Wendell. 1990. *Human resources management,* 2nd ed. Boston: Houghton Mifflin.

Guinn, Stephen L. 1993. Gain competitive advantage through employment testing. *HR Focus, 70,* September.

Hollenbeck, J. R., et al. 1992. Lower back disability in occupational settings: A human resource management view. *Personnel Psychology, 42.*

Jenson, A. R. 1984. White–black achievement differences: The narrowing gap. *American Psychologist, 39.*

Lawrence, D. G., et al. 1992. Design and use of weighted application blanks. *Personnel Administrator,* March.

Mathis, Robert L., and Juhn H. Jackson. 2003. *Human resource management,* 10th ed. Mason, OH: Thomson Learning.

Mondy, Wayne R., and Robert M. Noe. 1996. *Human resource management.* Upper Saddle River, NJ: Prentice Hall.

Osby, Robert E. 1991. Guidelines for effective fire service affirmative action. *Fire Chief,* September.

Schuler, Randal, and Susan Jackson. 1996. *Human resource management,* 6th ed. New York: West Publishing.

Sovereign, Kenneth L. 1998. *Personnel law,* 4th ed. Upper Saddle River, NJ: Prentice Hall.

Woodruffe, Charles. 1993. *Assessment centres: Identifying and developing competence.* New York: State Mutual Book and Periodical Service.

7 CHAPTER

Training and Development

Source: MF

KEY TERMS

accreditation, *p. 186*

active learning, *p. 163*

affective, *p. 165*

assessment, *p. 161*

certification, *p. 162*

cognitive, *p. 165*

instructor liability, *p. 180*

psychomotor, *p. 165*

taxonomy, *p. 165*

OBJECTIVES

After reading this chapter, you should be able to:

- Describe the process of training assessment.
- Understand adult learning principles.
- Describe how to use the concepts of curriculum development.
- Describe test item construction.
- Understand distance education.
- Describe the qualities of an effective instructor.
- Explain certification systems.
- Know how to improve training safety.

Training lies at the very foundation of the fire and rescue service. The fire service profession is fortunate in that its culture encourages and strongly supports training and development on a continuous basis throughout one's career. All firefighters and emergency medical services (EMS) personnel expect to be well trained to provide a high level of service to the public and for their personal protection as well. This high level of training is necessary in a public safety profession that may be called on to respond to any type of emergency at any time. Training is also very important in any occupation that has the potential for serious injury and death. Unfortunately, the history of the fire and rescue service is replete with the sacrifices made by firefighters in the line of duty. Proper training is the most effective way to increase safety and operational readiness.

No other public service that I am aware of spends as much of its on-duty time in training as does the fire service. From the firefighter's first day on the job or as a volunteer, he or she is constantly involved in training and developing the necessary skills to meet whatever challenge may lie in the next response. The specialization and the technology available to today's firefighter or paramedic require a high level of knowledge and ability to function properly in such an environment. Many highly developed training academies, schools, and institutions focus on the continuing professionalism of the fire and rescue service. The challenge lies in providing the proper training in a cost-effective manner to the right people at the right time.

Training Assessment

Training is a planned effort by an organization to facilitate the learning of job-related knowledge and skills by its employees to improve employee performance and further organizational goals. Notice that in this definition, training is planned by the organization. *Planning* implies a process of analysis, consideration of alternative methods, and decision-making ability regarding the training program. Therefore, a training plan should begin with an **assessment** of the need for training and a specific plan as to how it will be delivered and evaluated. The training of the most concern within the fire department involves the acquisition or improvement of job-related knowledge and skills. A broad-based general education is important, but in the fire service, the major concern is the improvement of job-related skills that can be directly applied toward increased performance and safety and the advancement of the goals of the organization.

Training is becoming more important as the business world changes due to heightened competition within the global economy and the changing demographics of the workforce. For example, in 2004, approximately 87 percent of employees in the United States worked in service-based industries as the shift from manufacturing to service continued. The assessment today that a manager would need to make regarding the development and implementation of training programs can be very broad based. Some of the considerations would include the following:

assessment
■ the process of documenting, usually in measurable terms, knowledge, skills, attitudes, and beliefs. An assessment of the need for training will include a specific plan as to how it will be delivered and evaluated.

- *Changes in the workforce*—Shifting demographics will necessitate training programs that take this into account. Trainers and educators will need bilingual skills and cultural awareness.
- *Changes in technology*—Keeping pace with technology will involve new training programs and retraining requirements on very short schedules. This also entails using technology to deliver training programs in a more flexible manner.
- *Competition*—Increasing competition for resources will impact the funding and time available for training programs.

- *Team-based approach*—More employee involvement in teams will necessitate more training programs that teach these organizational and team-building skills.
- *Employee empowerment*—Employees will need to be able to do more and think independently as well as have the discipline to work in an unsupervised environment such as telecommuting.[1]

In assessing the training needs of the organization, three distinct levels of analysis must be considered: organizational, operations, and individual analysis.[2] Each level is important in the development of a comprehensive training program.

Organizational analysis reviews the organization as a whole to determine where training is necessary. At this general level, it is important to analyze training needs as they relate to the organization's overall goals and objectives. For example, if a fire department decides to provide advanced life support service to the community, then a significant portion of this goal will be the training necessary to certify firefighters as paramedics. The alignment of organizational goals and training programs in this manner is necessary. Likewise, if the department has training programs that are not clearly connected to the goals of the organization, then perhaps the program could be eliminated. A central component of the analysis is a review of the resources available for training and their allocation within the department. When making the organizational analysis, an awareness of the external environment in which the department operates is required. Be sure to take advantage of opportunities in the environment to improve the training within the department and to make the programs more cost effective.

Operations analysis looks at the specific behaviors necessary to perform a job in a competent and safe manner. This analysis provides the fire department with the content of its training program. It requires the establishment of performance standards, descriptions of how the work is to be accomplished, and the knowledge and skills necessary for successful performance. The job analysis for various positions, performance appraisals, identification of undesirable events, and other pertinent information also fall within this assessment level. For example, if the safety records indicate that the department is experiencing a high number of backing accidents, then adjustments to the driver training program in handling vehicles in reverse may be appropriate.

The job performance requirements (JPRs) of the standards of the National Fire Protection Association (NFPA) can be useful in this stage of the assessment. A department can use the standards for firefighter (NFPA 1001) and fire officer (NFPA 1021) in the design of the training program to ensure that it fully meets or exceeds the national standards. The NFPA standards are national consensus standards that are widely accepted by almost all fire service organizations. These standards also serve as the basis by which performance is measured in firefighter and fire officer **certification** programs. A fire department can save a tremendous amount of time and effort by adopting these national standards into its training program.

Individual analysis determines how well and to what extent the employee is implementing the knowledge, skills, and abilities of a position. The gap between the desired performance and the actual performance defines the individual training needs. Here, training must be focused in order to best improve the performance of the individual. This need can be identified by performance that is evaluated by the organization, or the employee, through self-analysis, can determine where

certification
- the confirmation of certain characteristics of an object, person, or organization. The certification is usually issued by an organization to an individual after successfully testing to the standards approved by an accredited agency.

improvement is necessary. In certain circumstances, such as in the introduction of new equipment, all employees must have the training.[3]

The various levels of analysis provide the fire department with a great deal of information regarding the status of its organization and its training needs. Take the time and conduct this analysis in a forthright manner. Do not jump to conclusions and assume that certain training is necessary; it may not be. The department wastes valuable resources when personnel attend an unnecessary training program. When the fire department conducts the analysis, it is also helpful to use a group that includes a mix of individuals of different ranks and assignments within the department. They bring a number of perspectives and viewpoints to the process that increases the accuracy of the assessments.

The training assessment process also provides the fire department with a great deal of information regarding the training priorities, which guides decisions regarding the allocation of resources for training programs. Managers then determine whether the department has the budget, time, and expertise to provide the training or whether outside experts are needed. The fire department, with limited resources, must make decisions as to which training program is the priority.

Adult Learning Principles

In order for the transfer of knowledge to occur, it must be conveyed to the student in a manner that will be the most receptive. Adults have specific learning styles and needs that are quite different from those attending secondary schools and higher educational programs. Many of the students in fire service training programs have not been in a classroom for many years, and their levels of ability and experience vary. In addition, the motivation for being in the class differs among the various students. These factors create a unique blend of students for the instructor to work with and achieve the results desired within the training program.

Instructors must be cognizant of the characteristics and motivation of adult learners and incorporate this information into their lesson plans and programs. Five basic characteristics are associated with adult learners:

1. Adults tend to be self-motivated.
2. Adults bring many experiences and opinions to the training session. Using these experiences in the problem-solving process will contribute to their learning.
3. Adults learners desire relevant information that they can use to improve themselves, both personally and professionally.
4. Adult learners want to apply the knowledge and skills learned into their daily work immediately.
5. Most adult learners are motivated by an interpersonal desire to better themselves, not by getting a raise or promotion.[4]

Adults learn best by **active learning,** a process whereby one learns by doing, not by observing. When the student can use the knowledge in an actual setting, the subsequent retention of the material is high. Adults need to touch and do the activities, not watch slide shows and videotapes. With a higher degree of realism in the lesson, the participants will be more active and the retention of the material will be enhanced. The most recent curriculum for the training of emergency medical

active learning
- a process whereby one learns by doing, not by observing. This includes a higher degree of realism with active participants that will enhance retention.

technician–basic (EMT-B) incorporates the concepts of active learning. This new course is much different from the older EMT-A courses, which included a large amount of lecture and demonstration. The current EMT-B course involves the student in practical application and uses realistic scenarios for training purposes. This approach has improved the learning environment and the success rate of the students.

The following chart shows the percentage of learning from the various senses and the degree to which information is retained by the student, depending on the way the student received the information.[5]

Learning and Retention

LEARNING	RETENTION
83% by sight	10% by reading
11% by hearing	20% by hearing
3% by smell	30% by sight
2% by touch	50% by sight and hearing
1% by taste	70% by talking
	90% by doing

As this chart indicates, students learn primarily by seeing, but they retain the knowledge by doing. Reading alone is not a good method of retaining information. The more the instructor can involve the student through the use of different senses, the greater will be the retention of the information, hence the use of audiovisual materials and discussion. The more active and increased the participation of the student in actually experiencing and doing the activity, the higher is the level of retention. This information presents an interesting question regarding many of the distance learning concepts that are currently available. Distance learning has many advantages in flexibility for the student and in the economy of training program costs. For example, Internet-based training programs are available 24 hours a day for access by students. However, due to the delivery method of this instruction, it may be difficult to achieve high levels of retention. In order to involve the student, the program design needs to include a high level of interactivity. Some sophisticated computer programs are using these techniques whereby the student at certain prompts must do something. These types of programs are a great improvement over video broadcasts and other techniques by which the student only views the material.

The objective is to have the training material delivered in an effective manner to allow for the maximum retention by the student. The method of delivery and the training material must match for the best results. For this reason the fire service still has a great need to perform practical evolutions, because the student will have to perform the tasks based on instant recall in an emergency. Therefore, a large amount of training time is spent in doing practical activities to reinforce the learning and allow for the best possible retention. When reviewing training programs, make sure that the program is set up to stimulate the desired retention level and be very suspicious of briefcase- and videotape-only training programs.

Another important consideration is the "learning style" of adults. Adults tend over time to favor one particular learning style that fits their individual needs and that they have found to be successful in the past. Three basic learning styles can be described as auditory, visual, and kinesthetic. *Auditory* learners are good listeners and they concentrate on the spoken word, both what they hear from others and what they will repeat to themselves. They prefer to watch and listen to discussions between the instructor and students as well as student-to-student discussions. They also tend to be skilled at textbook learning as they may sound out the words to actually hear the information as they study. *Visual* learners do best by watching the instructor teach the course as in the process of explanation and writing on the board or by the use of overheads or other visual presentations. *Kinesthetic* learners prefer to engage in learning activities that involve their physical participation. They want activity and to be a part of what is being taught, not sitting in a class listening intently to the instructor for long periods of time. Practical training sessions in firefighter and emergency medical training programs are ideally suited for this learning style.

The difficulty for the instructor is that any one class will most likely have all three of the basic learning styles present at any given time. The key to instructional success is to ensure that presentations include elements that are appropriate for all three styles. Use of different techniques, presentation materials, and an instructor who knows how best to blend the different styles of learning to teach the course objectives are important. Incorporating training materials that relate to the different auditory, visual, and kinesthetic learning styles provides the student the best chance for retention and mastery of the course material.[6]

Taxonomy of Educational Objectives

A **taxonomy** provides a method for the classification of a system, in this case a system of educational objectives. It consists of a set of categories that encompasses the learning outcomes expected from instruction provided in training programs. This classification system was developed by psychologists, teachers, and test experts for use in curriculum development, teaching, and testing. Much of the early work in this field was conducted by Benjamin S. Bloom of the University of Chicago and others. The taxonomy is divided into three categories: (1) the **cognitive** domain, (2) the **affective** domain, and (3) the **psychomotor** domain. The Taxonomy of Educational Objectives provides for the classification of possible instructional outcomes and further defines these objectives from the simple to the more complex. This system can be used to identify and state training program objectives and the hierarchical relationship of the nature and level of learning outcomes.

COGNITIVE DOMAIN

The cognitive domain includes those objectives that emphasize intellectual outcomes, such as knowledge, understanding, and thinking skills. The cognitive domain is structured as follows:

Knowledge	Ability to remember previously learned material
Comprehension	Ability to grasp the meaning of material
Application	Ability to use learned material in situations
Analysis	Ability to understand component parts and structure

taxonomy
■ a method for the classification of a system; with regard to training, a system of educational objectives.

cognitive
■ includes those objectives that emphasize intellectual outcomes, such as knowledge, understanding, and thinking skills.

affective
■ includes those objectives that emphasize feeling and emotion, such as interests, attitudes, appreciation, and methods of adjustment.

psychomotor
■ includes those objectives that emphasize motor skills, such as handwriting, operating machinery, and speaking.

| Synthesis | Ability to put parts together to form a new whole |
| Evaluation | Ability to judge the value of material |

AFFECTIVE DOMAIN

The affective domain includes those objectives that emphasize feeling and emotion, such as interests, attitudes, appreciation, and methods of adjustment. The affective domain follows a hierarchical structure from the receipt of stimuli to the development of value systems that define one's lifestyle. The affective domain is structured as follows:

Receiving	Receipt of a particular stimuli
Responding	Willingness to actively participate and respond to stimuli
Valuing	Worth or value a student attaches to a particular object
Organization	Bringing together different values and resolving conflicts
Value complex	Value system that controls behavior

PSYCHOMOTOR DOMAIN

The psychomotor domain includes those objectives that emphasize motor skills, such as handwriting, operating machinery, and speaking. The psychomotor domain objectives play an important role in the performance of skills. The psychomotor domain is structured as follows:

Perception	Use of sense organs to guide motor activity
Set	Readiness to take a particular action
Guided response	Early stages of learning a complex skill, including imitation
Mechanism	Complex performance with learned responses that have become habitual
Overt response	Skillful performance and proficiency
Adaptation	Well-developed skills that the individual can modify
Origination	Creating of new movement patterns to fit special circumstances[7]

The Taxonomy of Educational Objectives provides a three-domain scheme for the classification of instructional objectives. This information can assist in developing the objectives for a fire department's training program and for identifying the different levels within the program. Appropriate instructional methods and testing procedures are available for each of the different domains and for the particular level within a domain. The three domains of the taxonomy are not mutually exclusive, and more than one may appear in an objective, although one should be dominant. This information is useful in thinking about how firefighters learn and how one should approach the development of fire service training programs.

Curriculum Development

The development of the training lessons and programs that result in the fire department's curriculum must be performed in a systematic and thoughtful manner. One of the key characteristics of an effective training system is that it is designed according to an instructional design process. Several processes can be used for this purpose. This section examines an established model that has proven successful in both the military and the fire service training programs. T. W. Good developed the instructional systems development (ISD) concept in the U.S. Air Force, and it has been used with positive results. Since the 1960s, the Air Force has been using ISD as a means to develop and provide appropriate training for space-age demands. In addition, the Maryland Fire and Rescue Institute (MFRI) of the University of Maryland has adopted this model for its curriculum development process. MFRI is well known for the development of training curriculum for the fire service and has many of its training programs published on a national basis.

The instructional systems development concept as a training model is a cyclical process designed to ensure the development of training programs that are educationally sound and present the student with the appropriate information for a particular subject. The model provides a proven scientific and systematic approach to training that leads the development team through an efficient and comprehensive process. At every step in the process, feedback and interaction are essential components of the system. This model provides the pathway for training staff to follow that will guide them to an effective product that meets the needs of the organization and the student.

MFRI INSTRUCTIONAL DEVELOPMENT PROCESS

The MFRI has utilized the ISD process for a number of years with resounding success. The process is initiated by a quality team assigned a certain subject area from which to develop a training program. The quality team consists of MFRI faculty and staff as well as officers from surrounding fire departments. The fire department officers serve as subject matter experts (SMEs) and bring to the team relevant and up-to-date operational experience. The team is provided with the resources it needs to function, including appropriate research, computers, and clerical services. The team then proceeds to implement the four phases of the MFRI Instructional Development Process.[8]

PHASE ONE—ASSESSMENT

In this phase the proper analysis and assessments are made at various levels. The following three assessments are conducted:

Needs analysis. Statement of subject, demonstrated training need, target audience, goals of the course, number of sessions, equipment requirements, and so forth

Occupational analysis. Identification of the job tasks as they relate to job performance requirements

Instructional requirements analysis. Development of logistical and fiscal impact statements and information for development and delivery costs, personnel, and equipment requirements

At this point the results of the assessment are reviewed in the context of the organizational analysis to determine whether the organization should proceed with the project. The costs in terms of budget, personnel, and equipment are judged in light of the need for this particular training program as indicated by the organizational analysis previously performed. If it is deemed organizationally feasible to continue, then the development team proceeds to Phase Two.

PHASE TWO—PROGRAM DEVELOPMENT

This phase consists of the actual program development process. Research and other work are conducted to assimilate the content of the training program. Phase Two consists of the following five steps:

Student performance objectives. Determining what performance outcomes the students of the course will be expected to achieve.

Enabling objectives. Assessing information and training material required to bridge the gap from the present state to the level required by the student performance objectives.

Student evaluation process. Establishing the type of evaluation techniques used to measure performance and achievement by the student. The development of the various written and practical examinations and other assessment criteria are developed at this step.

Develop lesson plans. Developing the actual lesson plans using a standard format. The MFRI process develops both student manuals and instructor guides.

Acquire instructional materials. The audiovisual and support requirements for the program are identified and a budget is established. Many programs are now supported by computer-generated graphics stored on CD-ROMs.

At this phase, much of the real work of developing the training program is accomplished. Depending on the available resources and the scope of the program to be developed, this phase may take several months or more to complete. The development of training programs is an expensive undertaking, which is why one needs a structured and efficient instructional development plan. The quality team approach works well because the development work can be assigned and the writing of specific chapters or lessons can occur simultaneously. It is very important that the work be well coordinated so that the style and layout of the finished product are similar and flow well. The quality team also allows for a high degree of feedback and discussion, which improves the final product.

PHASE THREE—PILOT TEST

The pilot test phase ensures that the program is readily used by instructors and is received well by the students. In some cases the development team, usually composed of people who have mastered the subject, may have overlooked an item because of their skill and familiarity with the subject material. In addition, after actually teaching a full course, the instructor may have very valuable feedback for the development team. In some cases it is a good idea to have several pilots of the course before it is finalized.

Prior to the course being released, the MFRI program conducts a number of "Train-the-Trainer" sessions. Instructors who are assigned to teach the course attend these sessions to obtain the teaching philosophy of the course and to review new material. A member of the training program development team usually leads these sessions. This orientation affects the future success and instruction of the program in the field.

PHASE FOUR—EVALUATION

Continuing evaluation and feedback are components of each phase of the instructional development process. Every time the course is given, instructor and student evaluations are completed. After the course has been taught for a period of 1 year, the development team is reassembled to review the comments from the instructors and students and then to make any necessary revisions to the program.

Perhaps the most relevant evaluation of one's training program is the documented outcomes of firefighters' behaviors when they return to duty in their respective departments. Can they actually do what they have been taught? Feedback from firefighters and supervisors is constantly sought to determine whether the training programs are effective in terms of results on the job. In evaluating training programs, one should look for and measure the change in four basic areas[9]:

Reaction	How do the participants feel about the training program?
Learning	To what extent have the participants learned what was taught?
Behavior	What on-the-job changes in behavior have occurred?
Results	To what extent has training improved results in terms of productivity, accident reduction, quality improvements, and so on?

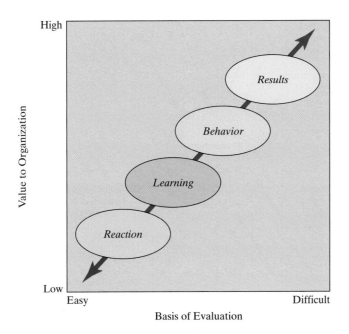

All training programs should be completely reviewed at 5-year intervals to ensure that the programs stay current with relevant information and that they are fully compliant with national standards. The process just described is one example of curriculum development of a state training agency that has been in place for many years with successful results. The following chart is a graphic representation depicting the relationship and feedback among the four phases of the curriculum development process.

Curriculum development is not an easy process; it takes a considerable commitment of resources and time. Most fire department training academies will not have the time and budget resources to devote to large development projects. Many struggle to teach the classes that are necessary, much less develop a curriculum. Actually, it would be inefficient for each fire department to develop its own programs. Training curriculum can be purchased to allow one to instruct most fire service and emergency medical subjects, and include the appropriate lesson plans and support materials that meet the national standards. In addition, many programs are available from the National Fire Academy (NFA) that are very well done and are in the public domain. It is economical and practical to take advantage of as many of these programs as possible, and then develop programs in-house only for special circumstances or equipment unique to one's department.

MFRI INSTRUCTIONAL DEVELOPMENT FLOWCHART

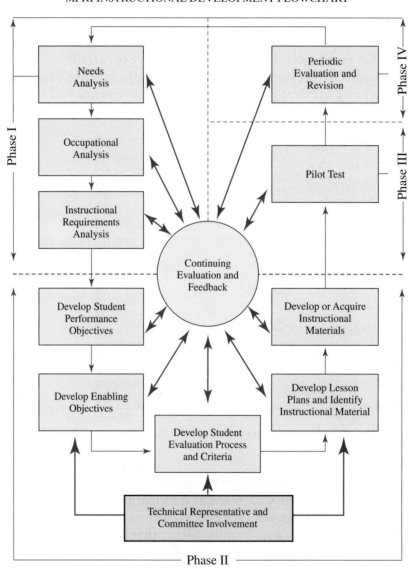

MFRI Instructional Development Flowchart.
Source: MFRI Archives.

Test Item Construction

A critical component of the development of training programs is the method by which the performance will be measured. It is necessary to demonstrate that the student has learned and can apply the lessons. The purpose of testing is to support teaching and to integrate performance measurement more fully into the educational process. Proper evaluation has many advantages: It contributes to program accountability and improves future learning. For the student, it measures success, indicates areas of improvement, and also provides feedback to the instructor, enabling him or her either to adjust the course to a faster pace or to review sessions that have been covered but not mastered.

The previous discussions in Chapter 6 involving test validity, reliability, job relatedness, and fairness are all appropriate regarding the development of test questions. These subjects will not be repeated here as this section concentrates on how to actually develop the test questions.

The process of developing questions for evaluation proposes will be greatly enhanced by having knowledgeable and competent people involved in and overseeing the process. It should be at least a two-step process that entails the development of the test items and then a review by a separate group or committee. This second step enhances the quality of the evaluation process and assists in its defense if challenged. It is good practice to ensure that individuals on the test item construction committee are recognized technical experts on the subject matter and have the following attributes:

- Knowledge of practices used in test item writing
- Knowledge of preferred test item formats
- Knowledge of rules for writing each of the various test item types
- Knowledge of communication skills
- Excellent command of the subject's technical terms and procedures

The appointment of a technical review committee to evaluate the work of the test writing group is a good idea. This group has the following responsibilities in the test item review process:

- Members will be subject matter experts.
- Reach consensus on knowledge measured and agree on relevance to and requirement for the job and course content.
- Review correctness of technical terms and test item construction.
- Review and agree on the results of revisions.
- Date and document rejected test items.[10]

The test construction group and the technical review committee work together to ensure that the test items and process are appropriate and well done. One task is to verify that the test items are of the proper difficulty in relationship to the content of the training program. The test construction committee must ensure that the questions be developed to the learning level of the lesson. In reference to the preceding discussion on the Taxonomy of Educational Objectives, if a cognitive lesson is written and presented at the Comprehension level, then the test item should also be from that level. For example, you would not use a test item from the Application level, because it would be from a higher level than the one in which the lesson was presented. The test construction committee judges the level of difficulty when creating test items but determines the actual difficulty level after the test is administered and scored. A number of statistical measurements can be employed to determine the difficulty and reliability of test questions. The point bi-serial calculation is a statistical method to discriminate among students who perform well on test items and those who do not. Reliability can be measured by evaluating the measures of central tendency (mean, median, and mode) and measures of dispersion such as standard deviation.

WRITTEN TEST ITEMS

A common method used for evaluating students is written test items. They are objective, are easy to administer and score, and can be given to large groups in an efficient manner. The two basic types of written examinations are the selected response item and the constructed response item.[11]

Selected response items present tasks or objectives to students through various situations and require a student to choose from the alternatives provided. Examples of selected response items are

- Multiple-choice test items
- Matching test items
- True or false test items

Constructed response items or *open-ended questions* require the student to create a response rather than choose a response already provided. Examples of constructed response items are

- Completion (fill-in-the-blank) items
- Short-answer items
- Essay items

Either the selected response or the constructed response items are appropriate as test items. They must match the level of instruction and be properly formatted and written. Because students vary in their ability to take tests, it is good practice to vary the test items, which may help students perform better in the examination process. Essay items are used in many tests, but are more difficult to grade than are the other test items mentioned in this section.

Many test bank services are also currently available. These companies employ test construction professionals and SMEs to develop examinations that relate to NFPA professional qualification standards or to certain training programs. These services may save a lot of time and also allow one to have access to a larger test bank with stronger statistical measurements.

PRACTICAL TEST ITEMS

Practical test items require the student to demonstrate proficiency in the application of psychomotor skills. The application of skills and knowledge is demonstrated according to established criteria and evaluated by someone with proven ability in that skill. A skills checklist is a popular method of evaluating individual or group performance. The skills checklist presents an objective list of the skills that must be demonstrated with proficiency to show mastery of a behavior. Many fire departments are requiring 100 percent compliance with the listed skills in order to pass the course or evaluation successfully. The student stays at the current level until able to demonstrate, with no exceptions, a mastery of a given skill before moving on to higher-order skills.

These skill test items can be used for discrete or individual skills and for a series of skills that must be completed in a proper sequence in order to be successful. Many of the practical or skill items are graded on a pass–fail basis.

Source: MFRI Archives.

Distance Education

Many colleges and universities are answering the challenge of rapid technological change and enhanced educational systems by developing distance education programs. At its most basic level, distance education takes place when a teacher and student(s) are separated by physical distance, and technology (i.e., voice, video, data, and print)—often in concert with face-to-face communication—is used to bridge the instructional gap. These types of programs can provide students with a college education; reach those disadvantaged by limited time, distance, or physical disability; and update the knowledge base of workers at their places of employment. Fire departments, with their shift-work schedules and potential interruptions due to emergency responses, are well suited to engage the distance education environment.

Rapid technological advances have made distance education or online courses increasingly more accessible, friendlier, and much more social. Since the early 1990s, these advances and the exploding need for lifelong learning have resulted in a quantum growth in the number of online courses available and the number of students taking these courses. Over 23 million students were enrolled in distance learning courses in 2002, up from 710,000 in 1998. Fifteen percent of all higher-education students are taking distance learning courses, up from 5 percent in 1998.[12]

Many educators ask whether distance students learn as much as students receiving traditional face-to-face instruction. Research comparing the two methods of instruction indicates that teaching and studying at a distance can be as effective as traditional instruction when the method and technologies used are appropriate to the instructional tasks, when there is student-to-student interaction, and when there is timely teacher-to-student feedback.

In March 2006 the U.S. Department of Education's Office of Postsecondary Education produced a report titled *Evidence of Quality in Distance Education Programs Drawn from Interviews with the Accreditation Community.* That report identified several factors that distance education programs should have, such as the following:

- Logically sequenced course development schedule
- Faculty training sessions aligned with the development schedule
- Use of a common platform for online courses
- Consistency in course formats
- Multiyear schedule of course offerings

The accomplishment of these objectives will create a coherent curriculum rather than a collection of courses. This requires a systematic and coordinated approach to distance education planning.

Historically, the fire service has relied predominantly on two methods of training: knowledge delivery and skills demonstration. Traditionally, knowledge-based training has been delivered through instructor-led classroom training followed by skills demonstrations and practice by the students. Although these training methods have had their success, today's fire service is vastly different from its predecessors. Fire service students have demonstrated a capacity to adapt quickly to the new learning methods via distance education. Much of the resistance to distance education has been that of the instructors who do not want to change their styles and fear losing control of the classroom. If a successful distance education program is to be implemented in one's jurisdiction, this needs to be dealt with in a proactive manner. The first group that will need to be converted is not the students, but the current cadre of fire service instructors.

HOW IS DISTANCE EDUCATION DELIVERED?

A wide range of technological options is available to the distance educator. These technologies fall into four major categories:

Voice.   Instructional audio tools include the interactive technologies of telephone, audio conferencing, and shortwave radio. Passive (i.e., one-way) audio tools include tapes and radio.

Video.   Instructional video tools include still images such as slides, preproduced moving images (e.g., film, videotape), and real-time moving images combined with audio conferencing (one- or two-way video with two-way audio).

Data.   Computers send and receive information electronically. For this reason, the term *data* is used to describe this broad category of instructional tools. Computer applications for distance education are varied and include the following:

- *Computer-assisted instruction (CAI)* uses the computer as a self-contained teaching machine to present individual lessons.
- *Computer-managed instruction (CMI)* uses the computer to organize instruction and track student records and progress. The instruction itself need not be delivered via a computer, although CAI is often combined with CMI.
- *Computer-mediated education (CME)* describes computer applications that facilitate the delivery of instruction. Examples include e-mail, fax, real-time computer conferencing, and World Wide Web applications.

Print.   Print is a foundational element of distance education programs and the basis from which all other delivery systems have evolved. Various print formats are available, including textbooks, study guides, workbooks, course syllabi, and case studies.

EFFECTIVE DISTANCE EDUCATION

Without exception, effective distance education programs begin with careful planning and a focused understanding of course requirements and student needs. Appropriate technology can only be selected once these elements are understood in detail. There is no mystery to the way effective distance education programs develop. They do not happen spontaneously but evolve through the hard work and dedicated efforts of many individuals and organizations. In fact, successful distance education programs rely on the consistent and integrated efforts of students, faculty, facilitators, support staff, and administrators.

Online degrees and educational programs are very competitive if they are earned at quality, accredited, reputable programs. Much of the research in distance education has consistently produced a finding called the *no significant difference phenomenon*. There seems to be no difference in grades or learning levels between distance and traditional students. It really depends on the quality of the program, just as it does in traditional educational programs. I have been a faculty member at University Maryland University College teaching a distance education program for 10 years. The diversity of the students is much greater because they are from all over the United States instead of from only one geographic area. The program is taught in an interactive manner using the Internet, and the work that the students must complete exceeds many traditional lecture-based courses. I was skeptical at first, but have come to appreciate fully what distance education programs have to offer.

For many fire service students, distance education programs have made degree opportunities available to them that were previously denied. Few universities offer bachelor degree programs in Fire Service Management. The Degrees at a Distance Program established by the National Fire Academy (NFA) has been instrumental in developing degree opportunities for fire service students. Regardless of where firefighters live or what their shift schedule is, they can take college classes toward a degree. The benefit that this will have in the continued professionalism of the fire service is profound.

As with any education or training program, there are distinct advantages and disadvantages to the various methods of program delivery. Before you engage in distance education programs, it is important to understand and appreciate these differences fully.

Advantages of Distance Learning

- *Convenience*—One can take courses in the comfort of one's home or office.
- *Flexibility*—Students often do not want to or cannot go to school full-time. They can take courses when they want to and set their own pace.
- *Availability*—The number of courses and range of student support services have exploded with the use of the Internet.
- *Time savings*—No time is wasted commuting to classes or the library.
- *No interruption of the job or career*—Working students do not have to relocate, reschedule their jobs, or take leaves of absence to go to school.
- *Rich diversity*—Classes are made up of a wide diversity of generally older, highly motivated students from many walks of life in different parts of the world.

Disadvantages of Distance Learning

- *Commitment*—There will be periods when demands on time from employers, family, and friends make it difficult to find time to study.

- *Time management*—The student cannot procrastinate. Work, family, and social commitments must be constantly time managed.
- *Information management*—Students must be able to actively manage their course information. Online learners receive a lot of information at once.
- *No instructor coaching*—If a student's learning style is one in which he or she likes personalized attention from teachers, this will be more difficult.
- *Independence and isolation*—Students should be prepared to work alone with limited access to lecturers, tutors, and fellow students. Students often are expected to find their own resources for completing assignments.
- *Technology savvy*—Students must be comfortable with constant change and using technologies because most online courses use technology for teaching and communication.
- *Acceptability*—Although perceptions are changing, there is still a stigma attached to distance-delivered degrees.

WHICH TECHNOLOGY IS BEST?

Although technology plays a key role in the delivery of distance education, educators must remain focused on instructional outcomes, not on the technology of delivery. The key to effective distance education is focusing on the needs of the learners, the requirements of the content, and the constraints faced by the teacher before selecting a delivery system. Typically, this systematic approach results in a mix of media, each serving a specific purpose. For example:

- A strong print component can provide much of the basic instructional content in the form of a course text, as well as readings, the syllabus, and day-to-day schedule.
- Interactive audio- or videoconferencing can provide real-time face-to-face (or voice-to-voice) interaction. This is also an excellent and cost-effective way to incorporate guest speakers and content experts.
- Computer conferencing or e-mail can be used to send messages, assignment feedback, and other targeted communication to one or more class members. It can also be used to increase interaction among students.
- Prerecorded videotapes can be used to present class lectures and visually oriented content.
- Fax or e-mail can be used to distribute assignments, make last-minute announcements, receive student assignments, and provide timely feedback.

Using this integrated approach, the educator carefully selects among the technological options. The goal is to build a mix of instructional media, meeting the needs of the learner in a manner that is instructionally effective and economically prudent. One excellent example of the use of distance education technology is the Fire and Emergency Television Network (FETN). This system uses satellite broadcasts directly to the fire station to deliver fire and EMS training course content. This is supplemented by manuals, workbooks, and examinations that can be administered at the work location. Many fire departments broadcast this to all fire stations within their jurisdiction to supplement their training programs with timely and interesting content. The FETN broadcasts are repeated on a regular basis so that all shifts can take advantage of the programming. This option has worked very well for many fire departments.

Every year it seems as though training requirements increase while volunteer firefighter recruitment declines. The trend of increased training demands causes volunteer and rural firefighters to spend an ever-increasing number of nights and weekends training at the fire station. Many of these fire departments are substantial distances from fire training academies. Consequently, in order to recruit and train volunteers, it is imperative for fire departments to modernize training programs and support distance education delivery systems that will provide the most effective, time-efficient learning environment without sacrificing quality or skills-based training. The ability to receive training programs at home at any hour of the day will substantially improve the level of training for volunteer firefighters. Even if the entire course cannot be conducted online, and many cannot, a portion can be accomplished in this manner. The student would then have to report to a training session to complete skills proficiency in order to be competent. This would greatly reduce the amount of time spent in a classroom environment. Many blended learning programs like the preceding example are becoming very popular and are successful with fire service and EMS training programs. If a community needs to have a viable volunteer firefighter program, then distance education programs will support this in a positive way.

Distance education is here to stay; it is a rapidly advancing component of virtually any educational system. The advantages of using distance education technology within fire departments far outweigh any disadvantages. The use of simulation within distance education programs is also growing. This will be a major step forward for fire service training, especially with respect to company and command officer training programs. All fire service training academies should be incorporating distance education programs and technology into their curriculums.

The Instructor

Many important factors are involved in the successful training of an individual firefighter or multiple firefighters. It is my opinion that the most important factor is the instructor, the person most responsible for the success or failure of the training process. I have witnessed poor training sessions in multimillion dollar training facilities and have seen outstanding training provided from the back step of a pumper in an open field. The difference in these examples is the instructors and how they go about meeting the challenges that they face. Properly qualified, competent, and enthusiastic instructors are often the difference that means the most to the students and the lessons to be learned.

Fire and emergency service instructors are frequently the first personal experience that firefighters encounter in their department. Every firefighter vividly remembers the first day of recruit school in the fire department: a day filled with anxiety, fear of failure, and the promise of a new career. The instructor was the person who largely determined the firefighter's success, not only from the course material but also as a role model. Instructors have to possess a very large array of technical expertise, instructional skills, as well as the proper attitude and concern for the students under their supervision. Instructors are lifelong students themselves because they are constantly learning new concepts and techniques that they will then pass on to their students. It is a difficult and dynamic job, but one that is essential for a fire department to be able to prepare its members properly for emergency response.

Instructors need to be good communicators in that they both listen and speak well and they understand the communication and learning process. Good leadership abilities are important for instructors as is a high regard for safety of their students. The sixth edition of the IFSTA *Fire and Emergency Services Instructor* text has developed a list of the desired qualities of a good instructor, which are discussed next.

QUALITIES OF A GOOD INSTRUCTOR

- *Ability to understand and work well with people*—Know the needs of students and be able to inspire confidence as well as impart knowledge. Be open minded, fair, and willing to communicate in an effective manner, building mutual respect and pride of accomplishment.
- *Desire to teach*—Teaching is not an assignment; it is an opportunity. Through professional development and preparation, the desire to improve others is evidenced by the actions of the instructor.
- *Competence*—One cannot teach what one does not know or understand. Technical competence, experience, and the ability to follow regulations demonstrate the ability to be an instructor. This includes the appropriate course and instructor evaluations that provide valuable feedback to the instructor.
- *Enthusiasm*—Competent instructors generate interest and enthusiasm in the subject material, thereby increasing the participation and desire of the students to learn.
- *Motivation*—Learning is goal oriented and the instructor has to instill the proper amount of desire and motivation for the students to want to do well. Communicate the end result of the training activity and how to get there in a safe manner.
- *Ingenuity and creativity*—One style of instruction does not fit all subjects and students. Instructors need to be able to vary their techniques based on the subject matter and resources at hand. Innovation and the use of new technology will be received in a positive manner.
- *Empathy*—Remember what it is like to be a student. Provide assistance and understand the feelings, abilities, and difficulties of those learning new material.

Source: MFRI Archives.

The International Fire Service Training Association (IFSTA) is an association of fire service personnel dedicated to upgrading firefighting techniques and safety through training. IFSTA's mission is to identify areas of need for training materials and to foster the development and validation of training materials for the fire service and related areas. An important aspect of IFSTA is the validation process of its training materials. Committees of technical experts meet and work at the annual validation conference addressing the current standards of the National Fire Protection Association (NFPA) and other standards-making groups as they apply to IFSTA-validated manuals. These technical experts review all manual drafts and verify that the contents are valid for use in the fire and emergency service fields.

IFSTA has also developed ethical standards for instructors that are very important and appropriate, which are reprinted here with its permission.

CANON OF ETHICS FOR THE FIRE AND EMERGENCY SERVICES INSTRUCTOR

The preservation of human life, the freedom from injury, and the protection of valuable property depend upon the careful and thorough execution of instructor duties. Instructors recognize and value their work, whether performed on a full- or part-time basis, as a volunteer or as a career. In recognizing the importance of performing correctly, instructors pledge themselves to the highest standards of professional and ethical conduct, which includes a commitment to perform the following:

- Achieve and maintain professional competency.
- Advance instructor professional competency through networking and mentoring.
- Teach only those subjects that you are qualified to teach.
- Prepare for each presentation because a life depends on it.
- Evaluate program results with honesty.
- Work to continually improve programs and presentations.
- Use only current and accurate material, information, and statistics.
- Perform instructor duties with integrity.
- Respect the work of other instructors through the courtesies of crediting their ideas and materials appropriately and through compliance with copyright laws.
- Recognize when your conduct does not meet this canon of ethics and resolve to improve.

INSTRUCTOR QUALIFICATIONS

Qualified and competent instructors are essential to the success of any training or educational endeavor. Fire departments have a unique advantage in training and certifying fire service instructors via the NFPA Professional Qualifications Standards. The standard that applies to fire service instructors is NFPA 1041, Standard for Fire Service Instructor Professional Qualifications. NFPA 1041 establishes the minimum qualifications and competence required for three levels of fire service instructor. Although it is not required unless enacted at the state or local level, NFPA 1041 serves as the national minimum standard for instructors.

NFPA professional qualification standards are written in the job performance requirement (JPR) format. JPRs are derived from the minimum job performance required for a specific job based on the standard. These are grouped according to job duties and functional areas of responsibility within a job. The JPR for each duty defines what an individual should be able to do in order to perform that duty successfully. The professional qualification standards are useful in designing and evaluating training programs, measuring job performance, as well as forming the basis for certification systems. Fire service instructors are eligible to be certified to the NFPA 1041 standard using these criteria.

Before fire service instructors are allowed to teach a class on their own, they need to meet certain requirements. This is for the safety of the students as well as having a method to ensure proper instruction in accordance with the lesson plan. Technical competence is important, but is not the only consideration. We have all seen people who have outstanding technical ability yet cannot teach or impart this knowledge to others. The technical competence needs to be paired with the ability to teach and be an instructor that leads the class toward the objective of acquired knowledge. Minimum standards need to be met, and then a program of student and instructor evaluations to ensure that the job is being performed properly is important. Instructors should also have to acquire regular professional development to ensure that they stay current within their field.

The state of Maryland has had a well-established emergency services instructor certification system since 1984. The Maryland Instructor Certification Review Board (MICRB), which is a part of the Maryland Higher Education Commission, determines minimal instructor qualifications and recertification criteria. By Maryland state law, any fire, rescue, or emergency medical instructor has to meet the requirements of NFPA 1041 via a prescribed training course. The instructor then has to attend a skills program to determine that he or she has the necessary practical training skills. This is followed by the new instructor teaching an interim class under the supervision of an experienced instructor and two successful evaluations by instructor evaluators. The certification is for a 3-year period and the instructor has to be recertified at 3-year intervals. To be recertified the instructor must complete continuing education in instructional methodology and be successfully evaluated by an independent evaluator. This system has developed a very highly qualified group of instructors that have served the interests of the student and training safety very well. To control and improve the instructional process it starts by controlling and establishing stringent qualifications for instructors.

INSTRUCTOR LEGAL LIABILITY

instructor liability
■ a situation in which an instructor is financially and legally responsible, such as when an act is committed that causes injury to another; therefore, the instructor may be responsible for any damage incurred. Liability may be civil or criminal.

Instructor liability is a situation in which an instructor is financially and legally responsible, such as when an act is committed that causes injury to another;

therefore, the instructor may be responsible for any damage incurred. Liability may be civil or criminal.

Civil liability should be a concern of all instructors. Civil law is defined as that body of law that addresses rights between individuals and in general the burden of proof is "fair preponderance of the evidence." The general charge in a civil suit is a "tort," or a civil wrong. A tort is an act committed by one or more parties that causes injury to another, for which the law allows a remedy of monetary damages. In many training situations in which an injury occurs to a student, the instructor faces an accusation of negligence. Negligence is defined as "the failure to exercise the care that the reasonable prudent person would have exercised under the circumstances which causes damages to another." Generally, in order for an action of negligence to exist, three elements have to be present:

1. An act or omission—An act has to occur, similar to a crime has to be committed, for a criminal charge to be sustained. Often the question is whether there was a "duty" to act, and in most cases it is clear that an instructor has a "duty" to protect students from undue harm.
2. Damages to the plaintiff—The victim has to suffer actual damages and the negligence has to be from the proximate cause of the damage. Damages can be property, personal injury, pain and suffering, loss of income, and so on.
3. Breach of standard of care—If an instructor fails to provide the level of care that a reasonable and prudent person would, the injured party can sue for negligence.[13]

You can reasonably assume as an instructor that if you are not following the policies of NFPA 1403, Standard on Live Fire Training Evolutions, and an injury occurs that you will possibly be held liable for that injury. National standards, laws and regulations, and industry-wide standards form the basis of the standard of care required in these legal actions. This also why it is important that instructors be properly trained (NFPA 1041) and be able to prove that in a court of law if necessary. Other legal definitions that are important to instructors follow:

- *Gross negligence*—An aggravated form of negligence that involves an extreme departure from the ordinary standard of care. Instructors may lose immunity in cases where gross negligence is proven.
- *Recklessness*—An aggravated form of negligence involving willful, wanton conduct wherein the instructor had knowledge that harm was likely and chose to act despite the risk.
- *Respondeat superior*—Legal doctrine that holds employers liable for the acts of employees committed within the scope of the worker's employment.
- *Torts Claim Act*—Establishes government liability and procedures by which an individual can file a claim against the government. Governments can limit their liability in many cases.

In limited cases instructors have been charged with criminal violations as well as civil actions. There is no government immunity for tort claims if the instructor is determined to be grossly negligent in his or her actions and no immunity at all for criminal charges. As an instructor one does not have to be a lawyer, but does have to be aware of the potential for legal issues such as just described. If an instructor is operating outside of departmental policy and not following prescribed regulations, his or her liability potential dramatically increases. The message here is very clear: By following policy and procedure and doing what is proper, both the instructor and the students will benefit.

Determining who is staffing the training academy and their qualifications and desire to be an instructor is critical to the success of a department's training program. It is not an assignment to be taken lightly. Supervision of the training environment at the academy is very important and requires a real conscientious effort. Many of the training tragedies and failures that have occurred within the fire service could have been prevented by more competent and qualified instructors. Instructors who are overzealous, disregard safety procedures, or do not follow approved lesson plans are the making of an unfortunate accident. The sad note of this is that it is the student that often bears the brunt of this by injury or death. There is no question that unsafe acts taught at the academy tend to stay with the department for a very long time. Always ensure that only the best and most qualified individuals are assigned to the training academy positions.

Fire Training Academies

Numerous types and levels of training and educational programs are available to firefighters throughout the United States. Each has a specific role in the training process at the local, state, or national level. Fire training programs can range from 6 hours to several hundred hours depending upon the course content. Fire departments that provide emergency medical services have additional training requirements, which vary depending upon the level of service provided. Local fire training academies operate at the city, county, or municipal level and generally provide training services to one fire department or to a specific geographical region.

Most large city fire departments have their own fire training academy designed to create situations that represent the emergencies the firefighters of that jurisdiction will encounter. These training academies are staffed by experienced and professional fire service instructors who guide and manage the training programs to best serve their community. Each state operates a state fire training program designed to provide training to a large number of firefighters in a larger geographical area. Many state fire training programs or state fire training academies are associated with universities or colleges. Others are a part of state government or are under the auspices of a separate state-level board or commission. Each year the state fire training programs in the United States cumulatively train over 900,000 firefighters. To accommodate these large numbers of students, many state fire training academies have large training facilities and regional training systems.

Source: MFRI Archives.

In order to conduct realistic fire training programs, training centers and academies must have the appropriate training props and equipment. These specialized items allow for the simulation and replication of the actual emergencies to which the firefighter must respond. In essence, these training centers must construct a specially designed building that can be set on fire and burned several times each day. Structural burn buildings are constructed with materials that have been tested and proven to withstand the constant heating and rapid cooling that occur during live fire training evolutions. Special fire bricks, coatings, and high temperature tiles are used to protect the structural elements of the building and make these buildings very expensive to construct. Due to environmental concerns, these buildings can also be difficult to locate. To reduce some of the environmental concerns of burning petroleum-based fuels and Class A combustibles such as wood products, many new fire training props use propane gas as the fuel. Firefighter trainees need a place to learn how to lift and place ground ladders, use aerial ladder trucks that can reach up to 135 feet, rappel out of upper floors, practice high-angle rescues, and advance hose lines. To accommodate these needs, fire training centers must have tall structures suitable for these exercises. These buildings, generally five to ten stories high, are designed to facilitate a large number of training scenarios. Buildings such as these are in constant use at any fire training academy.

Firefighters are often called upon to perform specialized rescues. To accommodate this need for training many fire training centers have elaborate props that can be used to simulate these situations. For example, numerous centers have confined space rescue props that are used to simulate rescues from below ground or tight spaces, props to simulate hazardous materials spills from tank trucks, and props that allow the simulation of building collapses. The grounds of the training academy can be used to simulate mass casualty incidents requiring a coordinated response and a host of other activities that are regularly created for training purposes. These innovative props and the ability to create a realistic training environment are critical to the success of the firefighter training program.

In addition to all of the specialized structures and training props, fire training academies must have adequate classroom space to conduct didactic training sessions. While much of what firefighter trainees receive is hands-on training, significant classroom work is required as well. Firefighters, for example, must learn hydraulic theory to operate pumps and move water, the different phases of combustion, the proper method of assessing and managing an emergency medical patient, and survival techniques. In general, before any training prop is used on the training ground, the firefighters have already attended classroom sessions to learn the procedures and techniques required to handle such dangers in a safe manner.[14]

Source: MFRI Archives.

Fire service training is becoming more and more technologically advanced. Much of the course work previously taught at the fire training academy is now available online in computer-assisted or distance learning programs. Many departments use fire simulators that employ computer graphics and programs to train command-level fire officers. Human patient simulators driven by computers and pneumatic systems are used in emergency medical training for firefighters and paramedics. Systems are being developed that will use virtual reality techniques to simulate fires so realistically that an actual fire will not even be necessary. As technology advances, so does the world of fire service training and education.

Certification Systems

A great deal of training and education occurs in the fire service on a daily basis throughout the United States. How the fire department and training agencies measure, compare, and quantify this training is important to the profession. If the training is accomplished in accordance with national standards and the results are certified by an independent body or organization, then the system is productive. The public expects qualified, competent, and professional personnel, be they volunteer or career, to respond to emergencies. There must be a certification system to ensure this and to provide standards that are recognized as being professional in terms of intellect, competence, and responsibility.

THE PROCESS OF CERTIFICATION

The first step in this process is to have national standards against which evaluations can be made. In the fire service profession the undisputed leader in this field is the National Fire Protection Association (NFPA). The mission of the NFPA, which was organized in 1896, is to reduce the burden of fire on the quality of life by advocating scientifically based consensus codes and standards, research, and education for fire and safety-related issues. The NFPA consensus process relies on personnel from a wide range of professional backgrounds who serve on more than 200 technical committees, each reflecting a balance of affected interests. The NFPA Standards Council, which oversees these technical committees, administers the codes and standard-making activities and regulations. If the council determines the need for a proposal project, it either assigns it to an existing committee or establishes a new committee. If the council feels the project is big enough to span the scope of more than one committee, it may appoint a technical correlating committee to direct the work of the technical committees.

In the fire and rescue service, the document that guides training and certification systems is the NFPA 1000, Standard for Fire Service Professional Qualifications Accreditation and Certification Systems. The minimum criteria for certification as well as the assessment and validation requirements of the certifying agencies are established in this standard. The professional standards that have been developed in this series of standards are widely used and acknowledged within North America. Many fire service training programs have been designed to meet the minimum requirements of the professional qualification standards. The NFPA professional qualification system standards that are eligible for certification in most states are as follows:

NFPA 1001, Standard for Fire Fighter Professional Qualifications
NFPA 1002, Standard for Fire Apparatus Driver/Operator Professional Qualifications
NFPA 1003, Standard for Airport Fire Fighter Professional Qualifications
NFPA 1006, Standard for Technical Rescuer Professional Qualifications
NFPA 1021, Standard for Fire Officer Professional Qualifications
NFPA 1031, Standard for Professional Qualifications for Fire Inspector and Plan Examiner
NFPA 1033, Standard for Professional Qualifications for Fire Investigator
NFPA 1035, Standard for Professional Qualifications for Public Fire and Life Safety Educator
NFPA 1041, Standard for Fire Service Instructor Professional Qualifications

The professional qualification system of the NFPA has established national standards to which individual training programs and performance can be evaluated. The NFPA professional qualification standards are now in the job performance requirement (JPR) format, which makes evaluation of performance easier as it relates to the knowledge, skills, and abilities of a particular standard. In addition, the standards are revised every 5 years so that they remain current and reflect the true status of the fire and rescue services. When individuals within the fire service have been certified as meeting the national standards, they may be recognized at the local, state, and national levels for demonstrated proficiency within the standard to which they have been certified. This enhances their credibility on the individual and departmental levels. Certification is a statement of accomplishment that improves one's worth as well as transferability to other departments. Being certified to national standards also improves the general professionalism of the fire and rescue service.

Source: MFRI Archives.

accreditation

■ given to entities that have been recognized or approved by a higher institution for meeting a set of standards. Organizations that issue credentials or certify third parties against official standards are themselves formally accredited by accreditation bodies.

CERTIFICATION AGENCIES

For this discussion, it is important to understand the difference between certification and accreditation. **Accreditation** is given to entities that have been recognized or approved by a higher institution for meeting a set of standards. Therefore, agencies and entities are accredited. Firefighters and officers do not become accredited; they become *certified*. In general, training agencies receive accreditation and individuals are certified for having completed an accredited program.

In 1970, a number of interested national fire service groups created the Joint Council of National Fire Service Organizations, and two of its seven original organizational goals were as follows:

- To develop nationally recognized standards for competency and achievement of skills development, technical proficiency, and academic knowledge appropriate to every level of the fire service career ladder
- To establish realistic standards of educational achievement and provide every member of the fire service with equal opportunities commensurate with professional requirements[15]

A committee was established under the auspices of the Joint Council and in 1972 created the National Professional Qualifications Board, commonly referred to as the pro board, to establish a national firefighter certification system. This board is now called the National Board on Fire Service Professional Qualifications (NBFSPQ), and its board of directors consists of representatives of the following organizations: International Association of Fire Chiefs, National Fire Protection Association, North American Fire Training Directors, National Association of State Fire Marshals, and the International Association of Arson Investigators. This board establishes the criteria for the certification of fire personnel to the national professional qualification standards. The board also establishes the accreditation process for state and local certification systems. Generally, the NBFSPQ accredits only one entity in each state as the certifying body of that state. The NBFSPQ has a well-defined process for accreditation that includes an extensive site visit by the accreditation team, and then a recommendation is made to the board. Once a state is accredited, it must then meet requirements for reaccreditation at specific intervals.

The International Fire Service Accreditation Congress (IFSAC) was established in 1991 as a national accrediting body for the certification of fire personnel and for the accreditation of college-level fire service educational programs. IFSAC grew out of a national meeting hosted by the National Association of State Directors of Fire Training and Education in 1990 and is located at and supported by Oklahoma State University. The IFSAC system consists of an Accreditation Congress, which is composed of one representative of each state or entity participating in the program, and a board of governors consisting of seven members from the Accreditation Congress. The Accreditation Congress establishes the policy for the IFSAC system. In addition, a degree assembly coordinates the accreditation of college-level fire service degree programs. The IFSAC system uses the NFPA professional qualifications standards for certification purposes.

Another excellent way to obtain recognition for training programs and increase their value to students is to have them evaluated for equivalent college credit. The American Council on Education (ACE), an independent, nonprofit organization founded in 1918, serves as an umbrella organization for colleges and universities. ACE provides a forum for discussion and decision making on higher-education issues of national importance and seeks to coordinate the interests of all segments of the higher-education community. The ACE College Credit Recommendation Service is a project of the ACE Center for Adult Learning and Educational Credentials. Founded in 1942, the center pioneered the evaluation of education and training attained outside of the formal classroom. Many major corporations such as Xerox, Ford Motor, Microsoft, and others have had their training courses evaluated for equivalent college credit.

Upon request, the ACE College Credit Recommendation Service will conduct a site visit by education experts of the training facilities, curriculum, record keeping, and evaluation and testing procedures of that jurisdiction. If approved, the training courses of the jurisdiction will be listed in the *National Guide to Educational Credit for Training Programs,* which is available to academic advisers for reference while counseling students and awarding equivalent college credit. In order to utilize the ACE listing, the student must request that the college or university academic counselors consult the national guide for the specific course review and college credit recommendation. The awarding of the appropriate transfer credit is then approved and, at many institutions, at no cost to the student. The ACE listing does not imply automatic college credit; it must be accepted by the individual school, college, or university.

The approval and listing of a fire department's training programs in the ACE guide provide a step toward *seamless education,* wherein an individual may progress toward higher education without the restriction of system boundaries. This recognition of fire service training programs validates and emphasizes a department's commitment to high-quality education and provides another positive incentive for firefighters to complete training. It enhances and brings recognition to the entire training effort, and I highly recommend pursuit of this, if possible.

Career Development

A career is a general course that a person chooses to pursue throughout his or her working life. Individual and organizational careers are not separate and distinct, and an individual whose career path cannot be achieved within the organization may eventually leave. Or, if opportunities are not available elsewhere, an employee may "leave" the department but still continue to draw a paycheck. Sometimes this practice is known as *retiring in place.* It is to the advantage of both the fire department and the employee to have career options and paths that keep employees interested in and dedicated to their careers. The department utilizes the formal approach of career development to ensure that firefighters and officers with the proper qualifications and experience are available when needed and that their careers are productive until their retirement date.

People change and view their careers differently at various stages of their lives. Some of this change is due to the aging process and from opportunities for growth and status. What motivates and challenges a recruit firefighter is not the same as that for a senior battalion officer. For some people, the job is just a job, and they receive their growth and status from other activities in life. This scenario is somewhat unusual in the fire service because the demands and the rewards intrinsic to the job are high compared to those of many other professions. It is difficult to be a firefighter and just get by; firefighters are generally very involved and passionate about what they do. For many people in the fire service, the difficulty is in letting go of the job for a while and pursuing other interests.

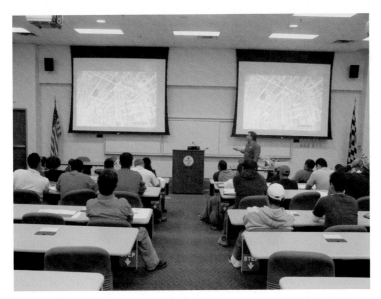

Source: MFRI Archives.

Even within the same profession, people have different aspirations, backgrounds, and experiences. They are molded by the constant interaction with the environment. However, some general categories have been identified in research conducted by Edgar Schein, which accounts for the way people prepare for and develop a career. He called these *career anchors.*[16] One of the implications of career anchors is that the personnel management of the fire department must be flexible enough to provide for opportunity and alternative paths to satisfy an individual's varying needs within the goals of the organization. The five major categories are as follows:

Managerial Competence. Individuals in this category want to manage people and be involved in the future of the organization. The promotional system within the fire department is important to them. The difficulty lies in people who desire this path but are not promoted. These employees need ways to achieve status and involvement without necessarily being promoted to a higher rank.

Technical Competence. These people seek the continuous development of technical talent. Individuals in this category seek positions as driver/operators and serve well on speciality teams, such as haz-mat response, high-angle rescue, and confined space.

Security. These employees want stability in their careers. They often see themselves tied to a particular organization or geographic location. This is where one finds the firefighters who have served their entire career in one station or bureau.

Creativity. Entrepreneurial individuals want to create something that is their own. They work well on special projects that they can identify with and are dedicated and hardworking.

Autonomy. Independent people desire to be free from organizational constraints. These types can be a challenge in professions that require a high degree of teamwork, such as firefighting. They can be useful on special projects and can perhaps be well suited as fire department representatives to intergovernmental agency projects. They may be the only fire department person there and may gain some degree of independence from this role.

Learning involves creating a change in one's behavior; when enough individual change occurs, the behavior of the department changes. In order to foster learning and continually move toward a higher level of knowledge and ability in each person, a positive environment needs to be created. Some fire departments continually encourage their firefighters to seek new skills and knowledge. These departments are frequently the subject of articles in professional journals; they innovate and contribute to the profession in a significant way. Their programs and procedures are sought after, and not by accident many of their members move on to become chiefs of other departments. A closer look would reveal an organization that has dedicated itself from top to bottom to being a learning organization. It builds an environment of knowledge and encourages all members of the department to live up to their full potential.

A key aspect of this process is to help firefighters know why they should learn. People learn best when they understand the purpose and expected outcome of the training activity. Clear objectives and a valid needs assessment provide much of this information. Relevant training allows the firefighters to use their own experiences in the training process. This experience creates a bond to the training program and heightens attention to the material. The department needs to make the training opportunities available and encourage participation. Often when a budget crisis erupts, the first cut is training opportunities, especially if they involve travel. Although some reductions may be inevitable, it is best to take a long-term view. If training is the first thing that is cut, it sends a strong message as to the real priorities of the department.

Learning organizations provide frequent feedback and information on the performance of their members. The feedback focuses on the specific behaviors involved and not on the individual. Failure represents an opportunity to learn and try again. Mentoring programs guide and encourage firefighters toward higher levels of performance. Learning organizations provide the support and coordination of training and educational programs, and they allow flexibility in scheduling so that the maximum number of people may take advantage of the opportunity. This commitment to learning includes having the fixed facilities to train properly in a safe environment specifically created for that purpose. These types of organizations respect the value of a general education as well as the detailed technical knowledge necessary in the fire service. They are likely to require completion of college degrees and reward this achievement within the department.

Many attributes make a learning organization a reality. An essential attribute, in my opinion, would be the attitude of the department: the strong desire to acquire knowledge and strive for excellence at each opportunity. The involvement of everyone in this process, working constantly to be better and provide better service to the community, is not a matter of budget, because some departments have successful training programs on minuscule budgets. They rely on resourcefulness and the can-do attitude of their personnel to propel them forward. When visiting departments with this attitude, one can almost instantly sense it by their behavior and the way they approach challenges. The goal of a fire service personnel manager is to instill this attitude in his or her training program.

Training Safety

It is a sad reality for the firefighting profession, but each year in the United States approximately 100 firefighters are killed while on duty and tens of thousands are injured while performing their duties. In 2007, the National Fire Protection Association (NFPA) released a study that analyzed the last 30 years of firefighter fatalities. It reported that the average number of on-duty firefighter deaths to occur annually has dropped by one-third over the past 30 years. In the late 1970s, on average 151 firefighters were killed on duty annually. So far in this first decade of the twenty-first century, the annual average is 99 deaths per year. In 2006, 89 firefighters died as a result of on-duty injuries. This shows the improvements made in firefighter health and safety over the past three decades, but the number of firefighter deaths has stabilized the past several years in the 90 to 100 per year range.

There are two major forces driving this decrease in deaths. First is the drop in the number of on-duty fatalities annually attributed to sudden cardiac death. In the first 5 years of this study, an average of 65 such deaths occurred. In the most recent 5 years, the annual average dropped to 41. Better physical fitness, cardiac health, and monitoring programs have been attributed to this decrease. The other significant drop is in the number of deaths at the scene of fires. In the first 5 years, an average of 79 deaths occurred annually on the fireground; fifty-nine of those 79 deaths were at fires involving structures. In the most recent 5 years, that number has dropped to an average of 34 deaths annually on the fireground, with 22 of those deaths at structure fires. During this same period the number of structure fires has decreased also. The World Trade Center disaster on September 11, 2001, resulted in the largest loss of firefighters from a single incident in the United States.[17]

What I find most disturbing is that, according to the NFPA report *U.S. Firefighter Deaths Related to Training* (June 2006), between 1996 and 2005, 100 firefighters in the United States died while engaged in training-related activities; this represents 10 percent of all on-duty firefighter fatalities. The deaths occurred during a broad range of activities, including apparatus and equipment drills; physical fitness; live fire training; underwater/dive training; and while attending classes or seminars. This means that in any given year approximately 10 firefighters die while being trained to be a firefighter; and more importantly as other firefighter fatalities are decreasing, the training-related firefighter fatalities have increased over the past 10 years. I am not aware of any occupation in the United States that has such a dismal record in the training of its employees. This cannot continue; fire departments and fire officers must be held accountable for their actions both criminally and civilly.

The majority of deaths during training occur while the firefighter is engaged in physical fitness activities. Findings for firefighter fatalities overall indicate

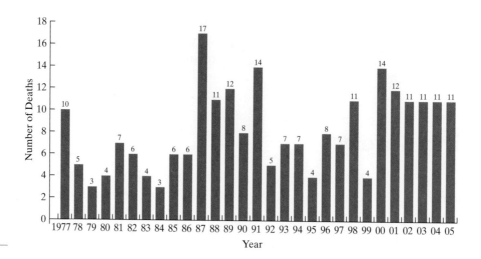

Deaths of U.S. Fire-fighters During Training, 1977–2005.
Source: NFPA, 2006.

consistently that the number-one cause of on-duty firefighter fatalities is sudden cardiac death, and that holds true, also, during training activities. In the training situations, as in other firefighter cardiac death studies, many of the firefighters who died in training activities had known preexisting heart conditions. To prevent deaths related to physical fitness and fire training activities, firefighters should engage in such activities only after being evaluated by a physician and meeting the appropriate medical standards. Physical fitness training programs should be overseen by certified fitness trainers and be gradual in the level of increased physical activity. Better initial medical screening and better physical fitness program management will assist in preventing these types of training deaths. The NFPA recommends several steps that should be taken to reduce the risk of heart attacks among firefighters:

- Conducting annual medical evaluations
- Screening for coronary artery disease (CAD) risk factors
- Conducting exercise stress tests (ESTs) for those with multiple CAD risk factors
- Providing appropriate treatment for those risk factors and restrictions for firefighters with positive stress tests[18]

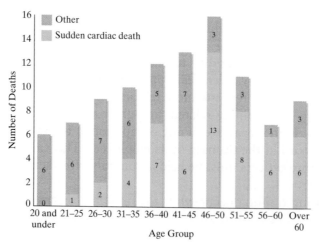

Training Deaths by Age and Cause of Death.
Source: NFPA, 2006.

PHYSICAL DEMANDS OF FIREFIGHTER TRAINING

Firefighting activities and firefighter training are strenuous activities and often require that firefighters work at or near maximal heart rates, in many cases for extended periods. Firefighters are sometimes called upon to push themselves to the limit. Efforts during training by firefighters often require extreme physical exertion that will adversely affect those who are not prepared for such environments. The major physical demands of firefighting and firefighter training can be grouped under the following major categories.

Cardiac Failure

There is a nearly linear relationship between heart rate and work rate. Heart rate increases from about 70 beats per minute at rest to a maximum value during exertion. There are several contributing factors to heart attacks. The most important factor is the inability to deliver oxygen to the myocardium. It is more likely that a heart attack will occur during episodes of high oxygen demand during strenuous exercise. Cardiac failure is also more likely in people who are not conditioned to physical exercise or who have other predisposing factors.

Thermal Stress

Heat stroke is a very serious health problem associated with working in hot environments. Its onset takes place when the thermoregulatory system fails, sweating becomes inadequate, and the body's only effective means of removing excess heat is compromised with little warning to the victim. Heat stroke develops progressively and subjects may experience some of the intermediate conditions prior to the onset of heat stroke. Heat exhaustion includes several clinical disorders and is typically considered as the direct precursor to heat stroke. It can have symptoms that may resemble the early symptoms of heat stroke.

Firefighters working in personal protective equipment (PPE) and during periods of heavy exertion in training activities must be monitored for these conditions as well as the monitoring of weather conditions and proper hydration, with frequent rehabilitation periods.

Inhalation of Contaminants

Although firefighters are usually protected against inhalation of airborne contaminants by their self-contained breathing apparatus (SCBA), on occasion breathing of toxic substances does occur. During intervals when the SCBA is not used for one reason or another, gas, vapor, or smoke particulates present in many fires can cause injury or death.

Thermal Burn

Training fires in structural burn buildings generally approach 800 to 1,000°F at the ceiling and 200 to 300°F at the 2- to 3-foot level within the burn room. The amount of fuel used in the training evolutions is a critical factor in the room temperature and the duration of the heat. Thermal burns can occur to firefighters who are not wearing their PPE properly. Acquired structures are very problematic with regard to thermal burns; it is difficult to control the fuel load because the building is combustible.

Disorientation and Panic

Psychological factors can be important determinants of appropriate responses during severe exertion when firefighting. Several studies have found that subjects who exhibited high anxiety were predisposed to respiratory distress while wearing a respirator. Firefighters can find themselves in unsafe situations if they make incorrect or unsound decisions. Several incident case histories appearing on www.firefighternearmiss.com involve confusion, incoherence, and poor judgment during training or live fires. Heat can cause degradations in cognitive ability, dexterity, and motor skills. At some point, it would be expected that the firefighter could not act on a decision even if capable of making it due to extreme disorientation and panic.

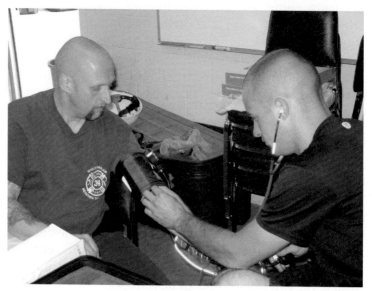

Source: MFRI Archives.

Physical Injury

Fire training activities often involve the use of power and hydraulic tools, climbing ladders, carrying and using hand tools, lifting heavy equipment, and so forth. Injuries due to slips and falls, lacerations, crushing injuries, and trauma can occur. Strains and sprains are the leading cause of injury in firefighter training as they are in firefighting during emergency response.[19]

Fire training deaths often occur during practical training exercises, most notably during interior structural firefighting training. Because the purpose of practical training is to familiarize personnel with the conditions they will face in an actual incident, full protective clothing, including self-contained breathing apparatus (SCBA), personal alert safety systems (PASS) devices, and accountability systems should be used where appropriate. All live fire evolutions at a minimum should conform to NFPA 1403, Standard on Live Fire Training Evolutions. Only those personnel who are qualified should be allowed to participate directly in training activities. Allowing unqualified personnel to take part in training or actual incidents could expose a fire department or fire officers to serious liability in the event of an injury or fatality.

Here are some unfortunate examples of the types of training activities that have resulted in firefighter fatalities:

- A fire department officer was charged with second-degree manslaughter for setting a fire in an acquired structure training exercise that burned out of control and killed a 19-year-old trainee and critically injured several other firefighters. The fire officer was subsequently found guilty of criminally negligent homicide.

- Two firefighters died while participating in live fire training at an acquired vacant structure. To produce a larger fire, urethane foam mattresses were added to the fire. As firefighters were conducting simulated search and rescue operations, the fire intensified and a flashover occurred, killing both firefighters.

- One firefighter trainee died while engaged in firefighting in a structure constructed of metal marine shipping containers that had been assembled to represent a shipboard scenario. Poor planning, use of facility and fuels that were not approved, lack of monitoring of trainees, and no RIT or EMS plan contributed to this situation. In addition, the recruits reported being physically and mentally abused by the academy instructors.

- Firefighter recruit died from heat stroke on first day of recruit school from overexertion. This incident occurred during a "Code Red" heat alert day due to extreme hot weather. No fitness training plan being in place, unqualified supervision of the physical training activities, and inadequate medical response all were factors.

- Firefighter died from a fall from the top of the training tower at a fire academy. Training evolutions were taking place during heavy rains at a time during which the state was under emergency operations for flooding and storm conditions. The firefighter fell while trying to descend from the roof via a ground ladder.

- Fire training academy conducted an event called "Hell Night," which consisted of 8 hours of nonstop aggressive training activities. Seven firefighters were transported to hospitals suffering from heat exhaustion and one firefighter was comatose for several weeks resulting in permanent injury and disability. Poor supervision and intimidation of recruits were major factors.

- Recruit firefighter training exercise was conducted in vacant row-house structure. Firefighter recruit became trapped on upper level due to the advance of seven separate fires set by instructors. Inadequate water supply, multiple fires, lack of supervision and planning, as well as noncompliance with NFPA 1403 were all contributing factors.

To be honest about it, every time I see a list of fatalities and the events that have caused them in the training environment such as those noted, it makes me sad. I feel this way because of the unnecessary death and injury that have been caused in what should be the safest environment that a firefighter is engaged with— the training environment. The cases just cited are not accidents; they are caused by a disregard of procedure and a "cowboy "mentality that really degrades and hurts our profession. What makes me even sadder is that these situations are tolerated until the point of a firefighter fatality, and then all are surprised that it happened within their department. We as a profession must not allow these circumstances to occur and must have the fortitude to speak up and intervene when we see events like these that are ongoing. The record of the fire service with respect to training-related fatality and injuries needs major improvement. Losing a firefighter during an emergency incident is hard enough; losing a firefighter during a training evolution is unforgivable.

One method to seek this improvement is to ensure that all training activities have a defined purpose and plan. In my opinion any training activity must include three basic elements: safety, educational purpose, and realistic activity.

Safety. Safety is of the highest priority and must be adhered to at all times. The most basic method to ensure this is to fully follow the NFPA 1403, Standard on Live Fire Training Evolutions; NFPA 1582, Standard on Comprehensive Occupational Medical Program for Fire Departments; and NFPA 1041, Standard for Fire Service Instructor Professional Qualifications. Knowledge of these standards is vital, but of higher importance is the discipline and accountability within the fire training environment to follow the standards in practice without exception to prevent injury and death.

Educational Purpose. Training has to be related to a valid educational purpose and objective. This means that an approved lesson plan is developed and followed and that training is not just made up as one goes along. Any training activity has to be directed toward a valid educational outcome and not involve a test of courage or gamesmanship. Following a lesson plan helps to ensure safety and prevents freelancing by instructors and students, which is a very common problem. This also helps to protect the instructors and the organization if an injury were to occur.

Realistic Activity. To be of real value to the student and the fire departments they serve, the training must be realistic. Firefighters have to be well-trained for their own protection as well as to be able to deliver high-quality service to the public. Some groups have advocated that live fire training no longer be used to train firefighters and they would recommend simulation only. I think this is an overreaction to a legitimate safety concern. For the best interest of all, the training has to have a degree of realism within the confines of adequate safety. This means that recruit firefighters are exposed only to fire situations that are appropriate for their level of training and experience and not more than that. Realism is important, but only to the level of the student's ability and with safety as the paramount objective.

HYDRATION AND FIRE TRAINING

Fluid replacement is the single most important component of an effective hydration rehabilitation program. Water is vital to every body system and comprises 60 percent of our total body mass. The results of a study conducted at the UMCP Center for Firefighter Safety Research and Development in 2006 indicated that dehydration has a significant impact on the physiological response to the stresses of firefighting. The study noted that training academies should provide instructions to participants prior to and during firefighter emergency training to encourage proper hydration.

The best practice is for the training academy to assume that students arrive dehydrated and actively take measures to avoid severe dehydration once training begins. Based on research, it is estimated that firefighters regularly lose 2 or more liters of sweat during strenuous firefighting activities. When considering the amount of sweat lost during these activities, it is important to note that under most circumstances firefighters begin sweating as soon as they don personal protective gear and continue to sweat even after the training evolution is completed. Adequate water should be available on site and students should begin hydrating upon arrival and throughout the training period.

The Maryland Fire and Rescue Institute has developed a hydration policy to deal with this issue. This includes urine charts in each bathroom of all of its training facilities and increased instructor and student awareness of this issue. Urine color can be a good indicator of possible dehydration and is easy to monitor. The general procedures of the hydration policy are noted here.

Prehydration

- The goal of prehydrating is to start the training session euhydrated (properly hydrated) and with normal plasma electrolyte levels. Prehydrating should begin at least several hours before the training session to enable fluid absorption and allow urine output to maintain normal levels.
- Prior to the training session, students should slowly drink 1 ounce of water for every 10 pounds of body weight at least 4 hours before the training session. If the student does not produce urine, or if the urine is dark or highly concentrated, the student should slowly drink an additional 1 ounce of water for every 20 pounds of body weight about 2 hours before the training session.
- Do not substitute beverages with alcohol or caffeine for water. Caffeine and alcohol act as diuretics and can exacerbate dehydration.
- Students should not attempt to hyperhydrate prior to a training session as it has been shown to provide no clear physiologic or performance advantage and can increase the risk of hyponatremia, a potentially lethal condition.

Preventing Dehydration

- The goal of drinking during the training session is to prevent excessive dehydration and excessive changes in electrolyte balance. The specific amount and rate of fluid replacement is highly variable depending on individual sweat rate, session intensity and duration, and environmental conditions.
- Ideally, students should create a customized fluid replacement plan based on pre- and posttraining session weight with the goal to prevent loss of more than 2 percent of baseline body weight during activity.
- In the absence of an individualized fluid replacement plan, students should drink water slowly and continuously during the breaks provided during the training session. Electrolyte replacement beverages may be beneficial in the most extreme training conditions, but the primary goal should be volume replacement, which is best accomplished with water.
- Students should continue fluid replacement even if they do not feel thirsty. By the time thirst is detected, the student is already dehydrated, which results in decreased performance and increased health and safety risk.

Rehydration

- The goal of rehydration is to fully replace any fluid and electrolyte deficit.
- Individuals should drink 20 ounces of fluid for every pound of body weight lost during the training session. If the total body weight lost during training is unknown, students should drink slowly and continuously until urine is no longer dark or highly concentrated.
- Consuming beverages and snacks with sodium will help expedite rapid and complete recovery by stimulating thirst and fluid retention.[20]

The strenuous nature of firefighter training, especially in high heat climates, needs the constant attention of instructors and students. By following the above noted procedures regarding hydration, students will be healthier and better able to perform their training functions.

MANAGEMENT OF FIRE TRAINING PROGRAMS

With the dismal safety record that the fire service has with respect to training, it is clear that significant management action needs to be directed at this issue. Existing behaviors and attitudes must change because these actions are creating the issues of noncompliance followed by tragedy. The fire chief has the prime responsibility to establish the proper safety culture within the department. Individual preferences and attitudes have to be overcome and officers held accountable for not following departmental policy and the appropriate NFPA standards. It is not difficult to spot training academy officers who have an agenda of making it unreasonably tough on the recruits or having to prove how dangerous the job of firefighter is. In an article titled "No Excuses" in the May 2005 issue of *Fire Chief Magazine*, Chief Gary Morris made the following very astute comments regarding training:

> Training academies and recruit training programs have a history of creating their own cultures. Left without oversight and guidance by organizational policies and proper procedures, the recruit training program can easily involve into rites of passage, unsafe practices, and abuses of new firefighters. Fire Chiefs must keep an eye on training officers who manage recruits with an autocratic, military style, as that can lead to verbal and physical abuse of trainees, along with unsafe training practices. While strong and direct supervision of recruits is essential for situations that involve known hazards, such as live fire exercises, it must not be abusive. The objective is to teach recruits how to work safely, not push them to the point of injury.

These comments by Chief Morris are very appropriate and every fire chief should take note. It is also clear that the United States military used to have a reputation for abusive training practices that has been corrected through effective management. I also firmly believe that the goal of a safety culture begins at the fire training academy. When instructors disregard procedure and do not follow safety regulations, it sends a clear message to all of the members of the department. Conversely, when instructors follow safe practices and set the example, it is emulated by others in the department. So if your goal is to have a culture of safety in the fire department, the training academy is the place to start.

It is clear that the responsibility for the safety of the training program is with the officers in charge of such activities. Recruits are hesitant to look weak and to report situations before they become critical. In some cases they do not even know how much danger they may be in. Therefore, close supervision and proper student-to-instructor ratios that lend themselves to close supervision are important. Generally, this is one instructor to five students. Clearly, the students are responsible for their own safety as well if they have been trained to do so and need to report situations of health and safety as soon as possible without fear of retribution.

If a serious injury or fatality of a firefighter occurs in training, it is a devastating circumstance for the department and everyone involved. Because the firefighter has been entrusted to the care of the department, these are very emotional events. Legal action will soon follow and many fire chiefs and instructors have lost their jobs because of these types of circumstances. As with almost anything in life, a safe, proactive approach means that the opportunity for this to occur in the department will be remote.

Summary

Good public service and outstanding organizations are not accidents; they are the result of effective training and development. Training is important in the fire and rescue service because lives depend on well-trained firefighters. Leaders in the fire service have a solemn responsibility to ensure that personnel are ready to respond to any emergency and to do so in a safe manner. This difficult task requires a great deal of planning and resources.

To ensure training effectiveness, the need for training must be established through the careful analysis of the organization and its personnel. The use of adult learning principles allows the students to receive and retain the training curriculum. Proper testing, certification, and evaluation verify that the student has mastered the training material and is ready to perform safely in emergencies. Career development opportunities available throughout one's career present challenges and maintain a high degree of interest.

Review Questions

Use these questions to review the material in this chapter and for discussion purposes.

1. Identify and explain the three levels of the training assessment process.
2. What are the five basic characteristics associated with adult learners, and how can you adjust your training programs to accommodate these needs?
3. Describe the four phases of the instructional systems development concept, and compare them to your department's current curriculum development process.
4. Why is it important for firefighters and fire officers to be certified? How can these certifications be obtained?
5. What steps can your fire department take to create a learning environment?

Case Study

You are the newly appointed fire chief in a career department consisting of 400 employees operating out of 18 fire stations. In your orientation to the department you have come to the conclusion that training needs major improvement. Your observations on the emergency scene include a lack of SOPs, poor safety compliance, and the use of tactics that are inappropriate for the fires encountered. There is a training academy that was built in the 1960s; the structures and facilities are in need of repair. The four uniformed personnel and one administrative person assigned to the training academy have been there for over 6 years. The fire department's budget is constrained and there is no opportunity for a major increase in funding in the near future.

How do you go about developing a plan to improve the training and education of this fire department? What steps do you take and in what priority? Explain in detail the methods and types of systems that you would employ to improve this situation.

Fire Service Professional Qualifications

This chapter contains information that will assist fire officers in meeting the job performance requirements (JPRs) of NFPA 1021, Standard for Fire Officer Professional Qualifications:

4.2 Human Resource Management— Fire Officer I

4.2.3 Direct unit members during a training evolution, given a company training evolution and training policies and procedures, so that the evolution is performed in accordance with safety plans, efficiently, and as directed.

5.2 Human Resource Management— Fire Officer II

5.2.1 Initiate actions to maximize member performance and/or to correct unacceptable performance, given human resource policies and procedures, so that member and/or unit performance improves or the issue is referred to the next level of supervision.

6.2 Human Resource Management— Fire Officer III

6.2.4 Describe methods to facilitate and encourage members to participate in professional development to achieve their full potential.

6.4.8 Research, evaluate, and project training requirements, facilities, and buildings needs, given appropriate data that reflect community needs and resources, to meet departmental training goals.

7.2 Human Resource Management— Fire Officer IV

7.2.3 Establish and evaluate a list of education and in-service training goals, given a summary of the job requirements for all positions within the department, so that all members can achieve and maintain required proficiencies.

7.4.2 Evaluate and project training requirements, facilities, and buildings needs, given data that reflect community needs and resources, to meet departmental training goals.

Endnotes

1. Wayne F. Cascio, *Managing human resources,* 7th ed. (New York: McGraw-Hill, 2006).
2. I. L. Goldstein, *Training and development in organizations* (San Francisco: Jossey-Bass, 1989).
3. Cascio, *Managing human resources.*
4. William D. McClincy, *Instructional methods in emergency services,* 2nd ed. (Upper Saddle River, NJ: Pearson Education, 2002).
5. Pennsylvania Department of Health, *Rescue instructor curriculum* (Harrisburg, PA: Pennsylvania Department of Health, 1983).
6. McClincy, *Instructional methods in emergency services.*
7. Norman E. Gronlund, *Stating objectives for classroom instruction,* 3rd ed. (New York: Macmillan, 1985).
8. Maryland Fire and Rescue Institute, University of Maryland, *Curriculum development process and style guide* (College Park, MD: MFRI, 1998).
9. D. L. Kirkpatrick, Four steps to measuring training effectiveness, *Personnel Administrator,* November, 1983.
10. Maryland Fire and Rescue Institute, University of Maryland, *Test development and LXR test guidelines* (College Park, MD: MFRI, 1997).

11. Ibid.
12. Internet home page of The American Center for the Study of Distance Education, Penn State University, State College, PA, retrieved November 2, 2008, from http://www.ed.psu.edu/acsde
13. J. Curtis Varone, *Legal considerations for fire and emergency services* (Clifton Park, NY: Thomson Delmar Learning, 2007).
14. Angela Bennett et al., *Health and safety guidelines for firefighter safety* (College Park, MD: University of Maryland Center for Firefighting Safety Research and Development, 2006).
15. U.S. Fire Administration, *Personnel management for the fire service course guide*, April 1998.
16. Edgar Schein, How career anchors hold executives to their career paths, *Personnel*, May 1975.
17. Rita F. Fahy et al., *What's changed over the past 30 years?* (Quincy, MA: National Fire Protection Association, June 2007).
18. Rita F. Fahy, *U.S. firefighter deaths related to training, 1996–2005* (Quincy, MA: National Fire Protection Association, June 2006).
19. Bennett et al., *Health and safety guidelines for firefighter safety.*
20. Maryland Fire and Rescue Institute, *MFRI hydration policy* (College Park, MD: Author, 2007).

Suggested Readings

Bachtler, Joseph R., and Thomas F. Brennan. 1995. *The fire chief's handbook*, 5th ed. Saddle Brook, NJ: PennWell Publishing.

Cascio, Wayne F. 2006. *Managing human resources*, 7th ed. New York: McGraw-Hill.

Fahy, Rita F. 2006. *U.S. firefighter deaths related to training, 1996–2005*. Quincy, MA: National Fire Protection Association.

Fahy, Rita F., et al. 2007. *What's changed over the past 30 years?* Quincy, MA: National Fire Protection Association.

Goldstein, I. L. 1989. *Training and development in organizations*. San Francisco: Jossey-Bass.

Gronlund, Norman E. 1985. *Stating objectives for classroom instruction*, 3rd ed. New York: Macmillan.

Hall, F. S., and M. H. Albrecht. 1979. *The management of affirmative action*. Santa Monica, CA: Goodyear.

Kirkpatrick, D. L. November, 1983. Four steps to measuring training effectiveness. *Personnel Administrator*.

Kram, K. E. 1984. *Mentoring at work*. Glenview, IL: Scott Foresman.

Maryland Fire and Rescue Institute, University of Maryland. 1997. *Test development and LXR test guidelines*. College Park, MD: MFRI.

Maryland Fire and Rescue Institute, University of Maryland. 1998. *Curriculum development process and style guide*. College Park, MD: MFRI.

Maryland Fire and Rescue Institute, University of Maryland. 2006. *Health and safety guidelines for firefighter training*. College Park, MD: MFRI.

McClincy, William D. 2002. *Instructional methods in emergency services*, 2nd ed. Upper Saddle River, NJ: Pearson Education.

Pennsylvania Department of Health. 1983. *Rescue instructor curriculum*. Harrisburg, PA: Pennsylvania Department of Health.

Schein, Edgar. May, 1975. How career anchors hold executives to their career paths. *Personnel*.

Schuler, Randall S., and Susan E. Jackson. 1996. *Human resource management*, 6th ed. New York: West Publishing.

Varone, J. Curtis. 2007. *Legal considerations for fire and emergency services*. New York: Thomson Delmar Learning.

Performance Appraisal

Source: MFRI

OBJECTIVES

After reading this chapter, you should be able to:

- Describe the need for performance appraisal systems.
- Describe performance assessment techniques.
- Describe performance appraisal errors.
- Understand the elements of a performance appraisal system.
- Know how to prepare for the performance review meeting.
- Understand how to conduct the performance review.
- Describe 360-degree feedback.
- Understand legal implications of performance appraisals.

Performance appraisal is the formal, systematic assessment of how well employees are performing their jobs in relation to established standards and the communication of that assessment to the employee and the organization. How appraisal systems are used and the manner in which the results are communicated affect the organization in significant ways. The results of the appraisal

performance appraisal
- the formal, systematic assessment of how well employees are performing their jobs in relation to established standards and the communication of that assessment to the employee and the organization.

process can have major impacts on training and development, compensation, promotional decisions, and many other aspects of an organization. It is an important function of personnel management, but one that some supervisors unfortunately seem to avoid.

The goal of the performance appraisal process is to improve the quality of work and the individual employees involved in that work. The process of performance appraisal has the potential to affect strongly how employees feel about their department and themselves. In organizations such as fire departments, it is critical for both safety and service delivery reasons that performance be monitored and maintained at high levels and include individual and unit performance. Much can be gained from having a comprehensive and fair system of evaluating performance in the fire and rescue service.

The Need for Performance Appraisal Systems

One of the fundamental reasons for a performance appraisal system is to make decisions about employees. Who should be commended, receive merit pay, be transferred, or be terminated? The fire department must be able to make these decisions in a fair and equitable manner in the best interests of both the organization and the employee. An assessment of the performance of the individual employee is essential to this process. Therefore, a system designed to assess performance and to improve performance where lacking is necessary. The appraisal methods used and the manner in which results are communicated are important to an individual's productivity and morale. Employees tend to be apprehensive in systems in which either the criteria used to make the assessments are ambiguous or those assessments tend to be overly critical rather than helpful. The more the individual understands the appraisal system and the more the appraisals are used as developmental opportunities, the greater is the acceptance.

Performance appraisal systems are designed to determine work effectiveness using objective criteria. The uses of these systems assist in achieving higher productivity outcomes by determining and documenting how employees meet job specifications. A major challenge for performance appraisal systems is to define performance standards while maintaining objectivity. The focus of performance appraisal systems is on **evaluation** and **development.** The intent of the evaluative purpose is to inform employees of their collective performance over a determined period of time. The collected performance data are frequently used to reward high performance and contributions to the success of the organization. The intent of the developmental purpose is to identify problems in employees performing the assigned tasks and/or responsibilities. The collected performance data are used to provide necessary skill training or professional development.

Effective performance appraisals can provide accurate assessments of employee productivity and quality of work and can motivate employees to higher levels of performance by giving the employees helpful feedback. In order to do this successfully, the performance appraisal system needs to contain certain qualities

evaluation
■ is intended to inform employees of their collective performance over a determined period of time. The evaluations are used to reward high performance and contributions to the success of the organization.

development
■ is intended to identify problems in employee performance and then develop a plan of improvement to provide the necessary skill training or professional development.

that will lead to the desired level of success. Enumerated here is a list of the desired qualities to be found in an effective performance appraisal system.

- Management must actively support the performance appraisal system. Experience indicates that such support is perhaps the most important requirement for success. Without strong top management backing, the system, no matter how well designed, is doomed to fail.
- The performance appraisal system should be designed to fit the specific requirements of the organization. A fire department should not be using the same appraisal system as general government.
- Rating factors should be valid, as objective as possible, and based on a job analysis for the position to be evaluated. A degree of judgment is required in all appraisal systems just as it is in other management areas. Nevertheless, organizations should still seek to maximize the degree of objectivity in performance appraisals and to eliminate to the extent possible the use of subjective factors.
- Appraisals should be free of bias and guard against discriminatory practices. Performance ratings should not be influenced by race, sex, age, or other irrelevant factors.
- Procedures and administration should be uniform. A system's procedures and its administration should be standardized and uniform in their application.
- The system should include a review or appeals process. To help ensure fairness in the appraisal process, some type of review and/or appeal mechanism should be established as this will strengthen employee confidence in the system.
- Performance ratings should be documented. The reasons for appraisals and specific instances of positive or inadequate performance should be recorded in writing. Careful written documentation of performance may prove important should employees file a grievance alleging that the employer discriminated against them on the basis of race, sex, or age in making promotions or awarding merit pay increases.
- Supervisors should be properly trained and qualified. For performance appraisal to succeed, raters must be trained in the philosophy of the system, how it fits in with the organization's goals and strategies, how it will help managers, and the actual mechanics of the system, including how to use the evaluation forms. Supervisors need training to develop skills for observing and recording employees' behavior and for providing them with appropriate feedback.

Fire service organizations generally use performance appraisal for two basic purposes. One purpose is to measure performance in order to make salary and other administrative decisions regarding employees. Promotions, terminations, merit pay increases, and other decisions may hinge on these evaluations. This at times tends to put stress on supervisors performing the appraisals because of the consequences to the employee. The second purpose focuses on the development of the employee toward maximizing his or her performance for the benefit of the organization. In this role, the supervisor acts more like a coach or counselor than as a judge, which changes the atmosphere of the appraisal process. The developmental type of performance appraisal focusing on growth potential and direction tends to be easier for supervisors to accept and implement.

Performance appraisal is very important to the organization in general. Many personnel management functions are affected by the performance appraisal system, which can provide useful information regarding the effectiveness of training programs, the recruitment and selection process, career development, and other programs. For example, if many performance appraisals indicate that firefighters are having difficulty in operating self-contained breathing apparatus (SCBA), then more training is indicated in this subject area. If recruit firefighters are having difficulty performing their duties, it may be a sign that the selection process must be evaluated. Performance appraisal also provides information for other personnel functions, serving as the basis for pay increases, release from probationary status, performance awards, and many other actions. Let us define some of the terms that we will be using to discuss performance appraisal:

Performance refers to an employee's accomplishment of assigned tasks.

Performance appraisal is the systematic description of the job-related strengths and weaknesses of an individual or group.

Appraisal period is the length of time during which an employee's job performance is observed in order to make a formal report of it.

Performance management is the total process of observing an employee's performance in relation to job requirements over a specified period of time, setting goals, documenting performance during the period, and then providing feedback in the form of an appraisal.

PERFORMANCE MANAGEMENT

Job Analysis *Describes work and personnel requirements of a particular job*	➤	**Performance Standards** *Translates job requirements into identifiable levels of performance*	➤	**Performance Appraisal** *Documents the job-related strengths and weaknesses of individual employees*

Performance Management.

Performance appraisal, if done properly, can strengthen an organization as it prepares and develops the personnel of that organization. After all, the sum total of the individual performances is the performance of the organization. Effective appraisal requires a system that is properly designed and well communicated to the members of the department. It will provide the following benefits[1]:

Feedback and Recognition.   Most employees want to know how well they are doing on the job and be recognized for their contributions. Performance appraisal communicates the status of the work effort and provides recognition where warranted.

Personal Development.   Evaluation can reveal the causes of good and poor performance in an employee, which can then be discussed and a plan of improvement developed between the supervisor and the employee. Plans can be individually developed for each employee as the circumstances require.

Goal Setting.   After the cause is determined, then specific plans can be developed to assist the employee in reaching goals and making the desired improvements in performance. The goals may have timetables and milestones that must be met within a certain time frame.

Career Development. Long-term planning regarding the direction of one's career can occur with performance appraisal. It may involve a plan for promotion or the realization that this career is not right for the individual.

Documentation. Performance appraisal may be used to justify performance awards and accolades for an employee. In addition, it is necessary in documenting adverse employment actions such as suspension or termination, which may result in legal challenges.

A properly designed and implemented performance appraisal system is unquestionably essential for the success of a fire department. The role of the fire service personnel manager is to develop a system that meets the needs of the employees and the department and includes the necessary personnel and legal components.

Performance Assessment Techniques

The development of performance assessment techniques follows a thorough job analysis to determine the knowledge, skills, and abilities (KSAs) that the employee needs to be successful in the job. Some jobs allow for direct measures of performance, such as in a factory, where a certain degree of output and quality is expected during each shift. Other jobs such as the fire service require much more judgment on the behalf of the evaluator, because the work environment is more complex.

The challenge of monitoring performance is to obtain a fair and representative assessment of the overall work performed by the employee. Of the many ways to monitor performance, the most frequently used method is direct observation by the supervisor. Other methods include, but are not limited to, feedback from coworkers, customer complaints or compliments, review of finished product, review of paperwork submitted, and accident and injury data. The performance of the employee should be consistently monitored and reviewed throughout the evaluation period. Failure to monitor and document employee performance consistently can result in inaccurate performance information. In addition, employees need frequent and regular feedback so that they have the opportunity to adjust performance on an ongoing basis. A supervisor will need a number of sources of information from which to conduct a fair appraisal. The goal is to obtain a representative number of unbiased work examples that accurately reflect the total performance of the employee.

A number of performance appraisal techniques can be employed, but in general they focus on the general categories of employee **attributes, behaviors,** or **results.**[2] The attribute approach to performance appraisal concentrates on the extent to which individuals demonstrate certain desirable characteristics and traits within the job being evaluated. These attributes are evaluated as they relate to traits such as leadership, teamwork, judgment, and others that have been established as important to the position being reviewed. The difficulty with this type of evaluation arises because attributes such as leadership can be hard to define and consistently apply to a number of employees, especially when the evaluations are used for comparative purposes. The validity of these types of evaluations can be low, because the attributes cannot be directly observed but must be inferred from behavior.

attributes
- an approach to performance appraisal that concentrates on the extent to which individuals demonstrate certain desirable characteristics and traits within the job being evaluated.

behaviors
- an approach to performance appraisal that attempts to define the actual behaviors an employee must exhibit to be successful on the job.

results
- an approach to performance appraisal that focuses on the attainment of objective and measurable results of the individual or group.

The behavioral approach to performance appraisal attempts to define the actual behaviors an employee must exhibit to be successful on the job. The behaviors are job-related occurrences that are observed and recorded for evaluation purposes. Various performance evaluation techniques allow supervisors to assess the extent to which the employees are performing the required behaviors. These types of rating systems use actual and observable job performance and not traits or attributes of performance, and as such are more meaningful to the person being evaluated. Most of the techniques rely on in-depth job analysis, so the behaviors that are identified are measurable and valid. In the fire service, many of the most critical job behaviors may be infrequently performed depending on the assignment. The department in this case may have to set up training scenarios and then evaluate the performance of the firefighters as opposed to waiting for an actual event to occur.

The results approach to performance appraisal focuses on the attainment of objective and measurable results of the individual or a group. This approach assumes that subjectivity can be eliminated from the process and that results are the best measurement of what the employee is contributing to the organization. It tends to link the performance of the employee closely to that of the organization. Problems occur with this technique because the results in most organizations are the accumulation of the output of many people and processes. It is difficult to measure at the individual level and is often affected by other factors outside of the immediate work environment such as general economic conditions and the work of subcontractors. Another disadvantage is that this evaluation may inform you of the lack of results but may be weak in addressing individual behaviors that can improve the results.

The following section identifies and discusses several of the most frequently used performance assessment techniques. In reviewing these, think about how they could be applied to different positions within the fire and rescue service.

GRAPHIC RATING SCALES

One of the oldest and most widely used assessment techniques is the graphic rating scale. The rating scale appears as a line or series of boxes along which performance levels are recorded. The evaluator marks a box or area that best describes the performance of the individual being evaluated on different attributes of his or her performance. In many of the graphic rating scales, each attribute is divided into levels of performance, such as Outstanding = 5 and Unsatisfactory = 1. In some of the systems used, the levels of performance are numerically based, and others present raters with a choice of adjectives that describe the behavior of the employee. In terms of the amount of structure in the development of the graphic rating scales, they may differ in the following ways:

1. The degree to which the meaning of the response categories is defined
2. The degree to which the individual who is interpreting the ratings can tell clearly what response was intended
3. The degree to which the performance dimensions are defined for the rater

Graphic rating scales do present several challenges regarding their implementation within an organization. In some systems, several traits or factors are grouped together and the rater is given only one box to check. Often, when descriptive words are used in the scales, they have different meanings to different raters. Terms such as *initiative, dependability,* and *cooperation* are subject to many interpretations, especially when they are combined with performance levels such as *outstanding* and *satisfactory*. Clear definitions must be presented to reduce the effect of misinterpretations.

Contemporary versions of graphic rating scales are more likely to use only characteristics that are closely related to actual job performance and to exclude such characteristics as loyalty, attitude, and so on. In addition, systems use descriptive statements for performance levels for each attribute being evaluated. However, even with these improvements the graphic rating scale method in many cases is restrictive, is arbitrary, and lacks the essential communication opportunities to be of great value in today's work environment. Much of the popularity of graphic rating scales is due to their simplicity and ease of use by supervisors.

CRITICAL INCIDENT

The critical incident method requires that written records be maintained of highly favorable and highly unfavorable performance. At the end of the appraisal period, the evaluator uses these records along with other data to evaluate the employee performance. It requires the supervisor to be aware of and to interpret the critical incident properly. This method may have some advantages to the fire and rescue service as an adjunct to an ongoing appraisal system. Firefighters truly respond to critical incidents in the emergency services, and thus fire officers have the opportunity to gauge performance in the work setting with very important consequences. The critical incident appraisal system does tend to capture the highs and lows but may miss some of the important everyday events that make for a good employee.

RANKING

In the ranking appraisal systems, the supervisor places all employees from a group in rank order of overall performance. The obvious difficulty here is whether the employees are being ranked to a standard of performance or against each other. One major drawback is that the size of the differences between the rankings is not well defined. Ranking also means that someone must be last, which ignores the possibility that the last-ranked person in one group may equal the top-ranked person in another group. Therefore, it is difficult to make comparisons of performance on an organization-wide basis. Paired comparison is a variation of the ranking method in which the performance of each employee is compared to that of every other employee in the group. The comparison is often based on a single criterion, such as overall performance. These ranking systems may not foster cooperation in the workplace because employees will be ranked against each other for performance purposes.

BEHAVIORALLY ANCHORED RATING SCALES

BARS
■ behaviorally anchored rating scales; various performance levels shown along a scale and described in terms of an employee's specific job behavior.

In behaviorally anchored rating scales **(BARS),** various performance levels are shown along a scale and described in terms of an employee's specific job behavior. Based on a complete job analysis, important job behaviors are described on a scale. The description includes information as to the level of performance expected for that point on the scale. For example, on the performance evaluation for a firefighter, the use of SCBA would be evaluated. On the low end of the evaluation it may read, "Even after repeated instruction, the firefighter cannot properly don the SCBA in 45 seconds and does not completely understand when the unit is required to be used." On the upper end it may read, "Firefighter is well versed in the use and operation of SCBA in fire and other emergency situations. Can properly don the unit in under 30 seconds and completely understands emergency breathing operations."

Behaviorally anchored rating scales focus on specific job behaviors rather than on traits or characteristics. This type of performance appraisal system reduces the amount of subjectivity and relies more on the observation of important job functions. It also makes it easier to compare performance across an organization where many different supervisors are conducting the evaluations. They will have a standard description of varying levels of performance against which to rate employees. This approach also improves communication with the employee because specific job behaviors can be discussed and plans for improvement developed.

MANAGEMENT BY OBJECTIVES

Management by objectives (MBO) is a system that features an agreement between the supervisor and the employee on the objectives for a particular period and a review of how well the objectives were achieved. MBO systems do tend to create a good deal of communication between the supervisor and employee regarding the objectives and the process to achieve them. They also call for periodic reviews and ongoing communication throughout the entire evaluation period. MBO systems can give individual employees more control of their work, motivate them, improve their growth and development, and provide them with clear evaluation criteria. One review of the research on MBO has revealed an important finding regarding its effectiveness. Of seventy studies examined, sixty-eight showed productivity gains, suggesting that MBO usually increases productivity.[3]

management by objectives
■ a system that features an agreement between the supervisor and the employee on the objectives for a particular period and a review of how well the objectives were achieved.

Three key assumptions are critical to the success of an MBO appraisal system. First, an employee who is involved in planning, establishing objectives, and determining performance levels tends to demonstrate a higher level of commitment and performance. Second, clearly identified objectives encourage employees to work effectively toward achieving the desired results. Third, performance objectives are measurable and they define the results. This system avoids vague generalities, such as "initiative" and "cooperation," which are common in many appraisals. The results speak for themselves at the end of the evaluation period.

The MBO process can serve as a valuable component of a performance appraisal system and in general is implemented using the following four-step process:

1. Job review and agreement—The employee and supervisor meet to review the activities and expected results to be accomplished. The goal is to have a good understanding and agreement of what the expectations will be for the appraisal period.
2. Performance standards—Specific standards of performance need to be mutually developed and articulated. The standards must be detailed and measurable in a quantifiable manner. Standards that are too general or vague will cause problems later in the appraisal process.

3. Objective setting—Objectives to be achieved are established by mutual agreement of the employee and supervisor. These need to be attainable with a reasonable degree of effort on behalf of the employee. Objectives that are too low or too high will not serve the intended purpose.
4. Performance discussions—The objectives are used as a basis for continuous discussions throughout the evaluation period to provide feedback and direction. If justified, the objectives may have to be modified based on conditions not foreseen at the time of their development.[4]

MBO programs may place too much emphasis on measurable quantitative objectives and can lead to the neglect of other important responsibilities. These systems can get to be real "number games" if not managed properly. The tendency is to adopt only those objectives that are important to the supervisor, so this must be guarded against. In service organizations, especially emergency response agencies, it may be difficult to quantify the objectives realistically in some cases.

TOTAL QUALITY MANAGEMENT

total quality management
■ recommends that the major focus of performance appraisal should be to provide employees with feedback in the areas that they can improve. The focus of TQM is on teams and groups of people achieving results as opposed to on individuals.

Total quality management (TQM) suggests that the major focus of performance appraisal should be to provide employees with feedback about areas in which they can improve. One type of necessary feedback is subjective and comes from managers, peers, customers, and others about the personal qualities of the employee. Performance appraisal in this regard deals with such dimensions as cooperation, attitude, initiative, and communications skills. Another type of feedback is objective and is based on the work process itself using statistical quality control methods. These techniques provide employees with an objective tool to identify causes of problems and potential solutions, such as process-flow analysis, cause and effect diagrams, Pareto charts, and other similar evaluations.

The focus of TQM is on teams and groups of people achieving results as opposed to on individuals. Therefore, many of the evaluations will be on the achievements of the group as opposed to individuals within the group. The quality of the product or service is paramount and takes precedence over the quantitative aspects of performance appraisal. When performance expectations focus on process improvements as well as on the behavioral skills needed to provide a product or service, total quality, excellent customer service, and appraisal of individuals and teams become the standard for how business is done.

In summary, the type of performance assessment techniques used will be based on the circumstances within the fire department and the desired results. In my opinion, it is best to use a combination of techniques within the appraisal system to offer the best opportunity to capture the true performance of the employee and to provide for the finest supervisor and employee communication process. In many regards the critical factor is not which technique is used but how it is administered. If the supervisor is conscientious and concerned about employee performance and conveys this concern with a high level of communication about performance, the techniques will be successful. The goal is clearly to improve the performance of individual employees and the organization as a whole. *How* you get there is important; *whether* you get there is more important.

Appraisal Errors

The validity and reliability of the appraisal system are important to the overall success of the program. The more clearly defined the measurements are, the better the system of appraisal. In general, the behaviorally anchored rating scales are more valid and reliable than the trait-based systems. The overall validity and reliability of the performance appraisal system can be increased by the following factors:

1. Individuals are rated on observable dimensions of performance, rather than on traits.
2. The rating criteria are clearly defined and understood by supervisors and employees.
3. Rating scales that define levels of performance are job related, supported by job analysis, and carefully developed.
4. Managers are well trained in the application of the appraisal system.[5]

Supervisors are the most frequently used source of performance appraisal. Unfortunately, supervisors can make several kinds of errors when conducting appraisals that cause the process to be less reliable or less valid than it might otherwise be. The reality that performance appraisal often includes a degree of subjectivity has sparked a great deal of research on errors and bias. When criteria are not clearly specified, a variety of errors can occur during the rating process. Favoritism should be eliminated, but as long as humans are involved in the process it will be present to some degree. The following section identifies the types of rating errors and ways to eliminate them from the process.

HALO AND HORN EFFECT

Halo and horn errors occur with a tendency to think of an individual in general as good or bad, which is reflected in judgments of specific performance as good or bad. The supervisor allows the employee's performance on one aspect of the job to influence an overall evaluation of other aspects of the job. The rater does not make meaningful distinctions when evaluating specific dimensions of performance. An example would be when a firefighter who is skilled at fire operations is also evaluated as very good in fire prevention and EMS, even if he or she is not proficient in these areas, or vice versa.

CENTRAL TENDENCY

Rather than use extremes in ratings, some evaluators tend to rate all people as average, even when the performance actually varies. Evaluators with large spans of control and little opportunity to observe behavior are likely to rate most people in the middle of the scale. Some evaluation systems require the supervisor to justify in writing extremely high or low ratings. The rater may avoid controversy and extra work by giving only average ratings.

LENIENCY OR SEVERITY ERROR

Leniency occurs when, to avoid conflict, an evaluator rates all employees higher than they deserve to be rated on an objective basis. This error may occur when highly subjective performance criteria are used and the rater is required to discuss evaluation results with the employee. Severity errors occur when the supervisor rates all employees as "below expectations" because of unrealistic standards or selective perception. This behavior is due to a lack of understanding of the various evaluation factors. Upper-level supervisors in the organization need to be aware of these tendencies and correct them if found.

CONTRAST EFFECT

The contrast effect happens when the supervisor compares employees to one another rather than to an established standard based on the performance expectations for the job. It occurs frequently with ranking systems or if the criteria for performance are not clear. To correct for this problem, specific performance criteria and systems are needed to help all supervisors understand the standards and their application. Ratings should reflect performance relative to job-related criteria, not to other people.

SAMPLING ERROR

In the case in which a supervisor has seen only a small sample of an employee's work, the appraisal may be subject to sampling error. This tends to occur when the employee's work is largely out of the supervisor's view and the sample that the supervisor sees may weigh disproportionately in the overall appraisal. Ideally, the work being evaluated should be a good representative sample of all the work performed.

RECENCY ERROR

Recency error occurs when the supervisor evaluates performance based on recent incidents rather than on those occurring throughout the evaluation period. Because employees know when their scheduled evaluation is to occur, their job performance commonly improves just prior to this time. Individual performance should be considered for the entire evaluation period. The opposite of recency error is primacy effect. In this case the information that is received first in the evaluation period receives the most weight, as opposed to information received over the entire period.

FRAME OF REFERENCE ERROR

The supervisor compares an employee's performance to the rater's own personal standards for that job, rather than to those identified by the organization. Another aspect of this error is when a supervisor has an initial favorable or unfavorable judgment about an employee and then uses it to distort the evaluation of job performance in the appraisal process.

PERSONAL BIAS

Supervisors display their personal bias in the performance appraisal process due to race, religion, gender, disability, or age. Although an illegal activity prohibited by federal legislation, discrimination continues to be a problem in the appraisal process. Organizations should be particularly aware of its potential occurrence and not tolerate it under any circumstances.

COMMON APPRAISAL ERRORS

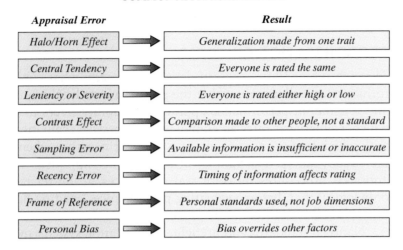

Appraisal Error		Result
Halo/Horn Effect	⇒	Generalization made from one trait
Central Tendency	⇒	Everyone is rated the same
Leniency or Severity	⇒	Everyone is rated either high or low
Contrast Effect	⇒	Comparison made to other people, not a standard
Sampling Error	⇒	Available information is insufficient or inaccurate
Recency Error	⇒	Timing of information affects rating
Frame of Reference	⇒	Personal standards used, not job dimensions
Personal Bias	⇒	Bias overrides other factors

Elements of a Performance Appraisal System

Performance appraisal is a very important aspect of fire service personnel management. It allows managers to link employee activity and performance to the goals of the organization in a meaningful way. Expectations of performance are established and evaluated at specific intervals, which assist employees with their development, allow the organization continually to improve, and maintain performance at a high level. Information derived from the performance appraisal is also helpful for many organizational administrative purposes, such as salary administration, promotion, retention, and many other personnel-related decisions.

The overriding purposes of an effective performance appraisal system are to link employee performance to the goals of the organization, provide information for administrative decisions, and offer useful developmental feedback to the employees. Fulfilling these three purposes is central to maintaining a competitive advantage through enlightened personnel management. A number of elements or components must be in place in order for the appraisal system to function at a high level. A fire department may have a great deal of flexibility in these decisions, or it may be tied to an existing system within the government or personnel office. Whatever the case may be, fire service personnel managers should attempt to ensure that all elements are in place and performed well. The following chart displays the major elements of the performance appraisal system.

Each of these seven elements is interrelated and depends on the others to make the system work effectively. This section reviews each element in detail.

PERFORMANCE CRITERIA

Through job analysis, one can determine exactly what constitutes effective performance as well as ensure that the process meets legal requirements and is valid and reliable. In performance appraisal, reliability refers to the consistency with which a supervisor rates an employee in successive evaluations or the consistency with which two or more supervisors rate performance when they have comparable circumstances. The validity of performance appraisal is the extent to which appraisal procedures measure real differences in job-related performance. It is only logical that, before evaluating performance, one first has to identify and define it for all concerned in the process.

National standards and firefighter certification systems may also be used in the performance appraisal process as performance criteria to be met by the employee. These standards must be demonstrated to be job related for the specific position.

TRAINING

In order to have an effective performance appraisal program, it is essential that the evaluators and those being evaluated be well aware of the different elements of the system and how they are to be implemented in the context of the department. Everyone involved in the appraisal process should clearly understand his or her role and responsibilities. The program should also include training for supervisors on how to evaluate performance, avoid appraisal errors, and utilize the process for performance development. Time and adequate resources need to be set aside to accomplish this training. Unfortunately, a fine appraisal process can be minimized by a lack of training, which results in improper implementation.

EXPECTATION-SETTING MEETING

At the beginning of the appraisal period, the supervisor conducts an expectation-setting meeting to review expectations of performance with each employee. In the expectation-setting or goal-setting meeting, the supervisor and the employee discuss and document the priority duties and the most important aspects of the employee's job. During this meeting, the level of performance that would meet or exceed expectations is established. The person who is to be evaluated should have carefully considered input into this process.

At the conclusion of the expectation-setting meeting, there should be a clear understanding of what is expected during the appraisal period and how it will be evaluated. Performance development plans should be agreed on and put in writing to prevent future misunderstandings. The supervisor should ensure that the employee understand not only what is to be accomplished during the review period and how it will be evaluated, but also the entire appraisal process including the employee's rights and appeals.

ONGOING FEEDBACK

The purpose of the performance appraisal system is to improve the work and performance of the employees of the organization. Appraisals can happen whenever appropriate or needed. Why would a supervisor wait for a formal performance review session in 6 months to discuss a performance issue that just occurred? The goal is constant improvement of personnel and operations. Therefore, supervisors need to deal with performance issues, both good and bad, as they arise. In this manner, the issues will be fresh in everyone's mind and attention can be focused on improvement.

Discussions regarding performance are more likely to occur throughout the year if the supervisor is also open to performance issues that employees may raise. On occasion, supervisor performance needs adjustment, which employees may bring to the supervisor's attention. It is also possible that expectations or goals will have to be adjusted during the evaluation period because of issues beyond individuals' control. Often these ongoing feedback sessions can be informal and conducted in a coaching-type format. Managers find that if these sessions are handled in a proper and considerate way, employees will be receptive to the feedback. They will appreciate the opportunity to improve their performance and in most cases respond positively.

SELF-ASSESSMENT

In preparation for the midway and year-end performance review meetings, employees should conduct a self-assessment of their performance using the evaluation form as a guide. First of all, this process gets employees to think about and consider their performance in a formal way. It helps prepare them for the meeting with their supervisor and enable them to be more active in these discussions, which will increase the communication and understanding essential to the performance review process.

MIDWAY FEEDBACK SESSION

At the halfway point of the performance review cycle, the supervisor should meet with each employee to conduct a midway feedback session. The purpose of this meeting is to go over employee performance through the initial phase of the review period. It provides an opportunity to review the expectations established at the beginning of the process and discuss the accomplishments to date. The timetable for completion of the expectations can be assessed and adjusted if good reasons exist. The supervisor should be open and understanding with the conversation focused on how to achieve what was planned.

Even though it is not the formal evaluation, midway feedback is the opportunity for the supervisor to review performance with the employee. The performance evaluation form could be used even if it were not submitted at this point. The key point is that the employee clearly understand the purpose of this evaluation of his or her performance. The supervisor should ensure that this understanding takes place by asking follow-up questions and engaging the employee in a discussion. It is a good idea to note the date and time of the meeting and any pertinent information discussed as a reference, especially if any major performance discrepancies were covered that may require further action at a future date.

PERFORMANCE REVIEW

At the end of the performance appraisal cycle, the supervisor and employee discuss the employee's performance throughout the entire evaluation period. If the preceding elements of the process were followed, this meeting should have no surprises. The employee would be well aware of the supervisor's opinion of his or her performance and would have been given ample opportunity and counseling to make any adjustments. This session results in the documentation of the performance on the prescribed forms and then forwarding it up the organization.

After completing discussion of each expectation of the performance review, the supervisor usually assigns an overall rating that represents the employee's cumulative performance for the appraisal cycle. The supervisor and employee indicate that they are in agreement with the individual ratings and overall review. Major areas of disagreement should be noted and acknowledged on the form. A process of review can be important to resolve key disagreements with respect to the evaluation. Normally an appeals procedure to the next highest level of supervision is in place.

The supervisor's appraisals of his or her employees should be reviewed by his or her supervisors. This system of checks and balances provides a level of assurance that the process is being completed properly and in a timely way. It also allows for a review to help prevent bias or other activities that are unjustly unfavorable to the employee without adequate documentation. Such a situation is unfortunate, but it does happen and it needs to be guarded against. After the completion of the performance review, the cycle starts again with the expectation-setting meeting for the next review cycle.

Preparing for the Performance Review Meeting

Perhaps the most prevalent single problem in the performance review process is the failure of the supervisor and the employee to communicate effectively. Many supervisors avoid it because they fear a confrontation, are not prepared, resist being judgmental, and for a host of other reasons. However, the performance review meetings that occur as a part of the appraisal process are essential to its success. To be an effective supervisor who has the ability to improve performance in those personnel assigned to one's company, the supervisor must be proficient in conducting these types of feedback sessions, which require substantial effort and proper preparation.

If the necessary coaching and feedback have occurred throughout the evaluation period, then the performance review meeting is much easier to conduct. It will include no surprises, and an established relationship will have been developed during this time. Essentially the miniconferences or meetings throughout the year build toward the formal performance review meeting as a natural extension of the previous efforts. The feelings of trust and understanding between the supervisor and the employee will have been well established at this point. If these meetings have been neglected throughout the year, then the performance review may be a traumatic event for the supervisor and the employee. A neglected process often starts with the supervisor saying, "I am doing this only because the department requires it, so here's your evaluation; go ahead and sign it."

The key to the success of the performance review meeting is in the preparation. One of the first things that a supervisor can do is gather the information recorded throughout the evaluation period, which may include notes in the personnel file, commendations, disciplinary actions, training, and evaluation materials. This information helps review the past year and focuses on the employee's performance. Nothing collected to this point will be new information to the employee, because he or she will have discussed it with the supervisor at a previous meeting regarding performance.

Supervisors use various methods to collect information on employee performance throughout the performance appraisal cycle. This activity can be time consuming depending on how many employees one oversees, unless the supervisor has a system. It can be further complicated if all appraisals are due during the same time frame, such as at the end of a calendar or fiscal year. A better system is to have the employee appraisals due at various times throughout the year to spread the workload and allow the supervisor to concentrate and do a better job. Two methods of collecting performance data are used as examples: the performance log and the T-account.[6] Both methods collect employee performance data as the review cycle progresses. The manager can modify these examples to suit his or her own personal style. The objective is to have a method to collect the data on a continual basis, which is more important than the exact method used to do so.

EXAMPLE OF A PERFORMANCE LOG

Use this form to record observations of employee's performance throughout the performance period.

Name:
Job Title:
Assignment:

Date	Brief Description of Performance Observed	Level of Performance

Several helpful steps regarding the performance review meeting are identified here:

- Identify relevant information that documents the performance of the employee regarding significant factors acknowledged in the performance review process, which will focus on the elements of the expectation-setting meeting that occurred at the beginning of the process.
- Examine how performance meets or exceeds the expectations agreed upon. Be sure to focus the feedback on the behavior or results and not on the person. At this point a plan should be forming in the supervisor's mind as to how to approach the performance review meeting.

- Prepare a draft of the meeting agenda, indicating points to discuss further with the employee. This plan will guide the meeting and make it easier to stay on track and not omit important information. The supervisor can also prepare a draft performance evaluation; but make sure that it is viewed as a "draft." What is being prepared is a draft of what will be covered in the meeting, not the final evaluation.
- Do not become so entrenched in opinions that you block out points and explanations that the employee may raise. A supervisor needs to be flexible and remain open to information from a different perspective.
- Prior to the meeting, have the employee conduct his or her own self-assessment. This assignment gets employees to think about performance and how they have conducted themselves during the evaluation period. This preparation will add to the productivity of the meeting. The self-assessment should be done well in advance of the date for the performance review meeting.

Any uncertainty on behalf of the supervisor about how to handle a certain situation that may occur during the meeting is best addressed prior to the meeting. The evaluator may need to contact his or her superior as one who may have more experience and insight into the problem. Also, any recommendation for a performance award based on the excellence of the employee's work needs prior approval. Informing an employee he or she is receiving a merit award and then not having it occur for some reason creates a negative situation and may damage the future relationship with the employee.

EXAMPLE OF A T-ACCOUNT

Record When, What, and How Well Throughout the Year

+	−
3/12 Attended training session on own initiative	1/18 Did not check apparatus at start of the shift
5/14 Good job in conducting fire safety talk at elementary school	2/19 Involved in vehicle accident with Engine 52 while responding to house fire. Cause was speed too great for road conditions
6/15 Rescue of child at fire incident #99-01235	

Once a supervisor feels confident and ready for the performance review session, then it can be scheduled with the employee. If the supervisor has prepared well throughout the year, the performance review can be a positive experience and facilitate improved performance of employees.

Conducting the Performance Review

Unprepared and poor supervisors dread this meeting. They should, because it is a clear indication that they are not capable of performing their job in a satisfactory manner. Often, the employee is more prepared and is ready to refute a hastily prepared performance review. The opportunity to improve performance and enhance communications is lost. This negative situation will be readily apparent to others within the department. Throughout my career, I have been constantly amazed at the times a supervisor would be discussing a firefighter's performance with me in a negative way, and when I checked the most recent performance evaluation of the employee, it exceeded standards in all categories. It was quite easy to determine who was not doing his or her job in these cases.

A number of positive steps ensure the success of conducting the performance review and are more likely to accomplish all of the objectives for the meeting. Preparation for and setup of the meeting take place in advance, so that both the supervisor and the employee are aware of what is going to occur and when. The meeting requires a time frame free from interruptions. A fire station hardly guarantees an interruption-free meeting, but certain time frames during a shift are historically less active. For example, if you work in a station next to a major highway, you would not schedule the meeting during rush hour, which is normally a busy response time for the company. The meeting should be conducted in private, with telephone calls and other activities on hold so that everyone can concentrate on the business at hand. A momentum to the communications builds during the meeting, which is inhibited by interruptions.

Opening the meeting in a manner that puts the employee at ease makes him or her comfortable with the process. Often, it is good to take some time and explain the overall purpose of the performance review and how it improves the department in general. Then the agenda moves on to review the expectations established at the beginning of the performance appraisal cycle. This review should provide a clear understanding of what was to be accomplished along with any modifications made at the midpoint review session.

Employee participation in the performance review session has already been encouraged by proper respect, trust, and communication that have been built throughout the year. However, some employees are more introverted than are others, and other techniques may be needed to provide for a good two-way exchange of information. The participation of the employee in the performance review session is very important. Studies have shown that when employees actively participate in the performance review, they are consistently satisfied with the process. Other than satisfaction with a supervisor, participation was the single most important predictor of satisfaction with the feedback session.[7] The importance of active communication and feedback within the performance appraisal process cannot be overstated.

Recognition of effective performance through praise and positive feedback lets employees know how much the department appreciates their performance and contributions. The purpose of the meeting is to give accurate performance feedback, which involves recognizing both effective and poor performance. Recognizing effective performance provides reinforcement for that behavior. Such recognition may have already taken place during the year in previous discussions with the employee, but during a review it becomes a permanent part of the personnel record.

When performance problems exist, focus must be on solving the problem, not on criticism of the individual. Discussion should address the employee's behavior or results, as opposed to personal attributes. Again, a review meeting should not be the first time that the employee hears of the issue, but the formality of the evaluation carries more weight. Having been confronted with the performance issues, employees often agree that a change is in order and will work in this direction. However, if the supervisor continues to come up with more and more examples of low performance, the employee may get defensive. The best approach is to make the point and then move on as to how it can be resolved.

Often a major disagreement at the time of the performance review session is the result of not discussing performance throughout the year. The employee has a different opinion of personal contributions and has been aware of nothing to the contrary throughout the review cycle. An employee may feel ambushed at the review session and act accordingly, often in a defensive and perhaps hostile manner. If defensiveness occurs, it is probably best to stop the meeting and reschedule the session. To improve poor performance, the supervisor must address the performance-related issues that are causing the inadequate results. It entails working closely with the employee to determine the actual cause and a plan of improvement. The supervisor will need to reengage this process and do the performance and communication steps that should have been done during the appraisal cycle. The formal appraisal may have to be delayed until this step is accomplished. This is not the preferred method but may be necessary in order to be fair to the employee being evaluated.

Supervisors must be aware that in many cases employees may know that they are not performing up to the required levels. But because the supervisor has not confronted or worked with them on these issues, they may assume that they are no real problem. When employees see it documented on their formal performance appraisal, it creates a major concern for them.

The formal performance appraisals are also opportunities to think about one's performance as a supervisor. Constant disagreements in this process or other negative issues are a signal to supervisors to examine closely their role and how they are implementing the appraisals. This introspection may be very valuable in improving the supervisors' performance and thus that of employees for whom they are responsible.

In cases of a performance deficiency, it is best to get employees to offer specific action plans that they feel will address the issues. This input allows them to identify closely with the improvement plan and helps them develop skills and abilities important for future career development. Some examples of what the employee may include are increased training opportunities, different job assignments, more time to perform certain tasks, closer supervision, and so forth. By working with the employee in this manner, a supervisor can fashion a plan that has a high degree of success.

The conclusion to the performance appraisal meeting should be positive and directed toward the future. Areas of agreement and disagreement must be documented, and expectations for the future should be clear. The supervisor and the employee should clarify the actions that each is responsible for in completing improvement plans and make a point to include a date to meet again to review issues and continue the communications process. Regardless of what was discussed in the performance review, the ending to the meeting needs to be on a positive note and with the anticipation of improvement or the continuation of good performance. A firm handshake and a commitment to work together are good ways to end the meeting.

PERFORMANCE REVIEW MEETING SUGGESTIONS

Positive Aspects	Negative Aspects
■ Prepare in advance.	■ Dominate the conversation.
■ Focus on actual performance.	■ Lecture the employee.
■ Be specific and provide examples.	■ Be overly negative.
■ Discuss specific actions to be taken for improvement.	■ Focus on only one aspect of performance.
■ Reinforce desired behaviors.	■ Compare the employee with other employees.
■ Discuss past, but focus on future.	■ Force agreement in all areas.[8]
■ End on a positive note.	

360-Degree Feedback

Many organizations use multirater or 360-degree feedback, which utilizes performance input from all directions (360 degrees). The performance review data are received from managers, subordinates, customers, peers, and others in this process.[9] The comments from the evaluators are summarized and then fed back to the individual to be evaluated. The power of the 360-degree feedback is that it provides clear and important performance information from a variety of sources. When several colleagues and customers state unambiguously that an employee needs to change, the message is clear.

If employees receive feedback only from their managers, they are acting on limited information. If employees also get feedback from other people with whom they work, then they can act on a more complete picture of perceptions. As 360-degree feedback has evolved, the range of people who give feedback has diversified. Many organizations that use 360-degree feedback now include feedback from suppliers and customers, in addition to the traditional sources within the organization such as subordinates, peers, managers, and team members. Fire departments have always received opinions and letters from citizens commenting on their performance. A survey to gather this information from the public in a specific manner would be valuable 360-degree feedback information.

Originally, 360-degree feedback was used to accelerate the growth and development of a talented employee. The belief was that understanding discrepancies between how we see ourselves and how others see us would enhance self-awareness and lead to personal development. Over the years, the uses of 360-degree feedback have expanded significantly to include not only executive development, feedback, individual development planning, and coaching but also career development, training, performance appraisal, compensation, succession planning, and team building.

A study in 2004 by Robert Eichinger on the patterns of rater accuracy showed that how long the rater has known the person has the most effect on the accuracy of a 360-degree feedback.[10] This study demonstrated that subjects in the group "known for one to three years" are the most accurate, followed by "known for less than one year," followed by "known for three to five years," with the least accurate being "known for more than five years." The study concluded that the most accurate ratings come from knowing the person long enough to get past first impressions, but not so long as to begin to generalize favorably.

The 360-degree feedback is fundamentally about improving job performance, and the key to effective change is convincing individuals that they personally need to change. Often, those higher in the organization find the need to change more difficult. Managers often believe that they would not have gotten to where they are if it were not for good performance, so why change now? This system provides clear information on the strengths and weaknesses of individuals on an objective basis and from several other sources. Its impact can be profound with regard to recognizing that a change is necessary.

The 360-degree feedback can provide valuable insight into one's behavior and performance. It can be used to supplement and improve the process of performance appraisal but is not the entire process. The respondents can judge the performance based only on what they observe, and in many cases they may be unaware of what is causing the performance. The feedback may be good feedback, but it will have to be supported with a plan for improved performance. This plan needs to include the support, direction, coaching, and other aspects that manage the individual through the change process.

The 360-degree feedback appears to be a good system, especially for upper-level positions within the fire department. Often the higher one is in the organization, the more difficult it is to get valid feedback on performance, and even more so in paramilitary organizations like fire departments. People may be hesitant to provide valid feedback because they may believe it will hinder their opportunities in the future or cause negative consequences in the departmental politics. Often, the feedback one really needs as a ranking officer in the department is from subordinates, who frequently are most reluctant to provide it. The 360-degree feedback system presents a way to elicit this valuable feedback and at the same time ensure some anonymity to the providers. This system may reveal valuable insights regarding the management of the fire department.

Advantages of 360-Degree Feedback

- Combined opinions provide an accurate, objective, and well-rounded view.
- It is a system in which many employees participate, thereby improving their feelings of inclusiveness and relevance.
- Some skills, such as leadership, may be best judged by subordinates and peers rather than superiors.
- Comments are difficult to ignore when expressed by a number of colleagues.
- It can lead to positive behavior changes, driven by peer review.
- It can be motivating for people who undervalue themselves.

Disadvantages of 360-Degree Feedback

- It is time consuming and costly, so the technique is often restricted to management levels.
- If too many appraisers are used, the results can be difficult to interpret.
- It has the potential to be destructive unless handled carefully and sensitively.
- It can generate an environment of suspicion, unless managed openly and honestly.

Legal Implications of Performance Appraisal

A cursory review of court cases indicates that perfect appraisals are not expected from employers. However, certain expectations dictate that the appraisal process be fair, objective, valid, and meet the legal requirements commonly applied to hiring and promotion. Since the mid-1990s, employees have won 64 percent of the wrongful termination suits decided by a jury trial. In the landmark case of *Albemarle Paper Company v. Moody*, the U.S. Supreme Court affirmed the EEOC's interpretation that performance ratings were "tests" and criticized the paper company's performance evaluation system as being too vague and subject to each supervisor's interpretation.

The 1978 Uniform Guidelines on Employee Selection Procedures clearly include performance appraisal procedures, which are subject to similar standards of reliability, validity, and job relatedness as all other selection techniques. The Uniform Guidelines indicate that the following aspects be present in a performance appraisal system:

1. A job analysis has been conducted to determine characteristics necessary for successful job performance.
2. These characteristics are incorporated into the rating instruments, which provide written standards for the evaluator.
3. Supervisors are trained in the use of the evaluation instruments, and standards are applied uniformly throughout the workforce.
4. A formal appeal mechanism or review of ratings by upper management is in place.
5. Evaluations and termination decisions are backed up with documentation of substandard performance.
6. Performance counseling is used to assist below-standard performers in improving their performance.[11]

Given that the results of a performance appraisal are often used to support a promotion, termination, salary increase, or other benefits, they are looked at very closely in employee discrimination suits. These legal implications have to be dealt with in a manner that does not adversely affect the entire performance appraisal program. Some legal advice would have the supervisor be so conservative and circumspect in what is said to the employee that the focus changes from reality to one of legalese with limited opportunity for improving either the employee or the organization. It is always important to have a good and valid system of performance appraisal that will withstand legal challenge. In many cases it is not the system that is the problem, but its implementation by various supervisors, so training continues to be a very important aspect. The general principles of equity and fairness should be a major aspect of any good employee evaluation process. The following points will improve the legal defense of performance appraisal systems.

- Performance appraisals should not be used in a merely punitive or retaliatory fashion. It is unprofessional for a supervisor to use the appraisal process to "get even" with an employee who has displeased or upset the supervisor in some way.
- Appraisals should not be used to discriminate against employees on the basis of race, religion, age, gender, disability, marital status, pregnancy, or sexual preference.
- Performance appraisal results should be fair, accurate, and supported by evidence and examples. The nature and effects of the employee's behavior should be clearly and accurately documented.
- An employee should have the opportunity to comment on his or her appraisal result, to express agreement or otherwise, and to appeal the result or at least request a review by a higher level of supervision.
- Appraisal results should not be used as the sole basis for promotion, remuneration, or termination decisions. A broad range of information should be considered, in which the employee's appraisal results may be significant but not necessarily the only consideration.
- Provide timely feedback, especially to marginal or poor performers. It is not fair or legally sufficient to offer no feedback to a poor performer for 12 months and then present that person with a bad appraisal.
- Appraisals should avoid inflammatory and emotive language. Ensure that criticisms relate to actual job requirements and are not based on mere personal or other irrelevant issues that have little or no connection with actual job requirements.
- Managers and supervisors required to conduct performance appraisals should be trained in appraisal principles and techniques. Conducting performance appraisals is one of the most demanding of all supervisory activities. It is a sensitive and sometimes controversial task, which, if mishandled, can lead to a legal challenge.
- Appraisal results should be treated as private and confidential information. Only personnel with an approved need to know should have access to an employee's performance appraisal information.

Although no performance appraisal system is likely to be totally immune to legal challenge, systems that possess the characteristics previously noted will be much more legally defensible. In addition, the legal requirements make for a better and more equitable system of evaluating performance, which all organizations should desire.

Summary

Performance appraisal is the formal, systematic assessment of how well employees are performing their jobs in relation to established standards and the communication of that assessment to employees and the organization. The manner in which the performance appraisal is conducted has significant impact on the organization's effectiveness. A number of assessment techniques can be used for this purpose, and certain appraisal errors should be avoided in this process.

Effective performance appraisal requires a comprehensive system to be in place to manage this process. Adequate preparation for performance appraisal meetings and reviews greatly enhances their effectiveness and prevents personnel problems. Additionally, a host of legal issues and concerns must be satisfied in order to have an acceptable performance appraisal system. Effective performance appraisal systems enhance the organization and ensure that employees are progressing toward continuous improvement of their duties and responsibilities.

Review Questions

Use these questions to review the material in this chapter and for discussion purposes.

1. Why is performance appraisal important to a fire department?
2. Identify three performance assessment techniques, and describe the positive and negative aspects of each technique. Which technique do you think is the most appropriate for use in your department?
3. Identify and describe five appraisal errors commonly found in appraisal systems.
4. Describe in detail how you as a supervisor would prepare for a performance review meeting with a firefighter who has a problem arriving for work on time.
5. Do you think that the 360-degree feedback would be appropriate for officers at the level of chief in your fire department?

Case Study

You are the lieutenant in charge of an engine company consisting of four firefighters. One is a recruit firefighter, two have less than 10 years on the department, and one is a 24-year veteran. They all have various levels of ability and commitment to the job. Each firefighter has his or her own individual strengths and weaknesses that tend to conflict with one another. The result is that the company does not function well as a team, even though on an individual level all four are satisfactory employees.

As the company officer you want to develop a performance appraisal system that communicates and facilitates the desired performance. How would you do this, and what structure and techniques should you employ? How could you improve the team performance and not negate the individual performance at the same time? Explain the system and criteria that you would use in this circumstance.

Fire Service Professional Qualifications

This chapter contains information that will assist fire officers in meeting the job performance requirements (JPRs) of NFPA 1021, Standard for Fire Officer Professional Qualifications:

4.2 Human Resource Management— Fire Officer I

    4.2.5 Apply human resource policies and procedures given an administrative situation requiring action, so the policies and procedures are followed.

5.2 Human Resource Management— Fire Officer II

    5.2.1 Initiate actions to maximize member performance and/or to correct unacceptable performance, given human resource policies and procedures, so that member and/or unit performance improves or the issue is referred to the next level of supervision.

    5.2.2 Evaluate the job performance of assigned members, given personnel records and evaluation forms, so that each member's performance is evaluated accurately and reported according to human resource policies and procedures.

Endnotes

1. Douglas T. Hall and James G. Goodale, *Human resource management* (Glenview, IL: Scott Foresman, 1986).
2. John R. Hollenbeck et al., *Human resource management: Gaining a competitive advantage,* 2nd ed. (New York: McGraw-Hill, 1997).
3. R. Rodgers and J. Hunter, Impact of management by objectives on organizational productivity, *Journal of Applied Psychology, 76,* 1991.
4. Robert L. Mathis and John H. Jackson, *Human resource management,* 10th ed. (Mason, OH: Thomson Learning, 2003).
5. Hall and Goodale, *Human resource management.*
6. University of Maryland Personnel Services Department, *Performance review and development training workbook for supervisors* (College Park, MD: University of Maryland, 1995).
7. W. Giles and K. Mossholder, Employee reactions to contextual and session components of performance appraisal, *Journal of Applied Psychology, 75,* 1990.
8. Mathis and Jackson, *Human resource management.*
9. Wayne F. Cascio, *Managing human resources,* 7th ed. (New York: McGraw-Hill, 2006).
10. Robert W. Eichinger and Michael M. Lombardo, *FYI: For your improvement, a guide for development and coaching,* 4th ed. (Minneapolis, MN: Lominger International Inc., 2004).
11. Robert W. Goddard, Is your appraisal system headed for court? *Personnel Journal, 60,* 1989.

Suggested Readings

Cascio, Wayne F. 2006. *Managing human resources,* 7th ed. New York: McGraw-Hill.
Dessler, Gary. 2001. *Management: Leading people and organizations in the 21st century,* 2nd ed. Upper Saddle River, NJ: Prentice Hall.
Giles, W., and K. Mossholder. 1990. Employee reactions to contextual and session components

of performance appraisal. *Journal of Applied Psychology, 75.*

Goddard, Robert W. 1989. Is your appraisal system headed for court? *Personnel Journal, 60.*

Hall, Douglas T., and James G. Goodale. 1986. *Human resource management.* Glenview, IL: Scott Foresman.

Hollenbeck, John R., et al. 1997. *Human resource management: Gaining a competitive advantage,* 2nd ed. New York: McGraw-Hill.

Klingner, Donald E., and John Nalbandian. 1985. *Public personnel management,* 2nd ed. Upper Saddle River, NJ: Prentice Hall.

Mathis, Robert L., and John H. Jackson. 2003. *Human resource management,* 10th ed. Mason, OH: Thomson Learning.

Matthes, Karen. 1992. Will your performance appraisal system stand up in court? *HR Focus, 69.*

Milkovich, George T., and John W. Boundreau. 1991. *Human resource management.* New York: R. R. Donnelley & Sons.

Rodgers, R., and J. Hunter. 1991. Impact of management by objectives on organizational productivity. *Journal of Applied Psychology, 76.*

Schuler, Randall S., and Susan E. Jackson. 1996. *Human resource management,* 6th ed. New York: West Publishing.

University of Maryland Personnel Services Department. 1995. *Performance review and development training workbook for supervisors.* College Park, MD: University of Maryland.

9

CHAPTER

Discipline

Source: MFRI

KEY TERMS

control of results, *p. 230*

discipline, *p. 229*

due process, *p. 237*

illegal acts, *p. 228*

personnel control, *p. 230*

specific action control, *p. 230*

Weingarten rights, *p. 238*

OBJECTIVES

After reading this chapter, you should be able to:

- Describe organizational control.
- Describe disciplinary actions.
- Understand progressive discipline.
- Know how to conduct disciplinary proceedings.
- Describe discipline in the fire service environment.
- Describe exit interviews.

There is no question that most employees conduct themselves in a proper and lawful manner within the organizational setting. They want to be a productive part of the organization and its success. However, on occasion, employees violate organizational rules and policy and possibly commit **illegal acts** that must be addressed in order to maintain the trust and viability of the organization as a whole. No one really likes to be involved in the disciplinary process, on either the receiving end or the administration end. Effective disciplinary action addresses

illegal acts
■ criminal offenses committed by an employee either on or off duty, as opposed to a rules violation.

228

inappropriate employee actions and behavior, not the individual employee as a person. Disciplinary action is also at the end of a spectrum of employee motivational and performance improvement techniques.

In general, **discipline** refers to the actions imposed by an organization on its employees for failure to follow the organization's rules, standards, or policies. Traditional approaches to discipline, based on punishment, are known to promote adversarial relationships within the organization. Many organizations now use a more effective approach that recognizes good performance and encourages employee commitment to the organization and its goals in a positive manner. A first step in the disciplinary procedure is to establish work rules consistent with the organization's goals and objectives. These work rules become the basis for disciplinary actions when the rules are broken. They are generally established jointly by management, the organization's human resources personnel, and employees, who should have an opportunity for input to ensure that rules are fair and can reasonably be followed. Work rules are directly related to work behavior and productivity. Employees who continually violate the rules are candidates for a disciplinary procedure.

When discipline is necessary within an organization, sometimes controversy and legal issues regarding employee discipline can be forthcoming. Fire department management can prevent these types of issues by utilizing these general guidelines:

- Employees are given advance notice of disciplinary action.
- Disciplinary rules are reasonable and directed at behavior, not individuals.
- Offenses are promptly and properly investigated.
- Accurate documentation of all actions is maintained.
- Rules are enforced equally and consistently.
- Penalties are related to the severity of offenses.

discipline
■ the state of employee self-control, orderly conduct, and the adherence to established standards within an organization. Also refers to the actions imposed by an organization on its employees for failure to follow the organization's rules, standards, or policies.

Organizational Control

If all employees always did what was best for the organization, control would not be necessary. As we all know, employees are sometimes unable or unwilling to act in the organization's best interests, and a set of controls must be implemented to guard against undesirable behavior and to encourage desirable actions. Even when employees are properly trained and equipped to perform a job well, some choose not to do so. If nothing is done to protect the organization from undesirable behavior, severe repercussions may result. At a minimum, inadequate control can result in lower organizational performance and the risk of alienating other employees and the public.

Organizational control can be achieved by avoiding some behavioral problems or by implementing one or more types of control to prevent problems or undesirable circumstances. The three main categories of control examined here are specific actions, results, and personnel.[1]

CONTROL OF SPECIFIC ACTIONS

Specific action control attempts to ensure that employees perform certain actions that are desirable and avoid or not perform actions that are not desirable. One method of this is the use of behavioral constraints that render the occurrence of a certain action improbable. These constraints may include physical devices such as locks and key personnel identification systems, and administrative constraints such as the segregation of duties, which makes it difficult for one person to carry out an improper act. As an example, many of the payroll and purchasing functions within a fire department require more than one signature for approval, or approvals at different levels or sections within the organization.

Another type of specific action control is action accountability, which is a type of feedback control system in which employees are held accountable for their actions. The organization must first clearly define the limits of acceptable behavior, normally through policy and procedure manuals or departmental rules and regulations. Second, the organization must be aware of the behavior of its employees and have a method to monitor and record it. It can use performance appraisal systems, training records, and personnel actions for this purpose. Finally, the organization must have a system for rewarding desirable behavior and correcting behavior that deviates from the established standards. These systems are effective only if the employees understand what is required of them and feel that their individual actions will be noticed and either rewarded or sanctioned in some significant way.

A third type of specific action control is pre-action review, which involves observing or approving the work of others before the activity is complete or initiated. For example, through direct supervision, review sessions, and approval of proposals, the correction of potentially harmful behavior can be prevented before the damaging effects are experienced. Such mitigation of harm can often be done in a collaborative way with the employee as a learning opportunity. Another advantage of pre-action reviews is that they can be used when the project or effort is difficult to define as to what is expected or in dynamic situations that have the potential to change rapidly.

CONTROL OF RESULTS

Control can be established by focusing on results. The use of the results accountability control method requires that the employees understand the desired level of performance against which they will be measured. The use of **control of results** attempts to motivate employees to behave appropriately and to achieve the identified results. A management by objectives (MBO) system is an example of a system to control results. For these types of systems to be effective, the achievement of results needs to be recognized and rewarded in some way to reinforce this behavior.

CONTROL OF PERSONNEL

A third type of control is **personnel control** because it emphasizes a reliance on the personnel involved to do what is best for the organization, and it provides them the assistance and resources to do so. Personnel within an organization can be controlled by implementing training programs or improving job assignments, as well as increasing performance requirements. The performance of employees can be motivated in many positive ways by performance awards, bonuses, and recognition programs. Personnel controls can also be exerted through the discipline of personnel for inappropriate behavior and acts that are detrimental to the organization and coworkers. The goal is for the employees to exhibit self-control in their actions and to avoid the disciplinary process.

The implementation of a control system in an organization depends in part on the feasibility of various types of controls. The amount of control provided by each of the control systems depends on its design and how well it fits the situation in which it is used. Personnel controls are the most adaptable to a broad range of situations. To an extent, all organizations rely on their employees to guide and motivate themselves, and this self-control can be increased by training, performance management, and a reward and disciplinary process. Managers in organizations do not rely exclusively on one type of control; they use a combination depending on the situation and the personnel involved.

Disciplinary Actions

Discipline is the state of employee self-control, orderly conduct, and the adherence to established standards within an organization. Discipline is necessary to maintain appropriate conduct in the workplace. A disciplinary action invokes a penalty against an employee who fails to meet the established standards. Disciplinary action is usually not management's initial response to a problem. Normally, more positive ways are effective in convincing employees to adhere to the policies necessary to accomplish the work of the organization.

It is important that fire officers understand discipline and know how to administer it properly and fairly. Discipline can be positively related to performance, and in fact, if a fire officer tolerates unacceptable behavior, employees tend to resent the unfairness of this. Training supervisors and managers on when and how discipline should be used is crucial. Research has found that training supervisors in procedural justice as a basis for discipline results in their employees viewing the disciplinary action as more fair than discipline imposed by untrained supervisors. Regardless of the disciplinary approach used, training on legal aspects, counseling, and communication skills provides fire officers with some of the tools necessary to deal with employee performance problems. This training should be required of all fire officers when they are promoted and on an ongoing basis as inservice training throughout their careers.

When considering a disciplinary action, the supervisor should focus on the improper action and circumstances and not on the individual employee. Discipline is behavior driven, not personality driven. The purpose is to inform the firefighter of the improper behavior and to prevent reoccurrences. Such action may involve a number of steps, including additional training. Constructive or positive discipline is an attempt to correct behavior and to eliminate further errors by the employee or others. Discipline that is motivated by a desire to punish and to belittle the employee has negative effects regarding the employee and others within the organization. A number of dynamics surround any disciplinary action within an organization. Discipline is a private personnel matter and should always be treated as such. However, it is difficult to keep secrets within the fire service environment, and in many cases the employees discuss the discipline imposed with others and seek advice or at least sympathy from others regarding the action taken. Therefore, it goes well beyond the firefighter involved, and others throughout the organization are witnessing and evaluating how the disciplinary process works. Is it fair and properly communicated? How would they feel if they were the one being charged? They are reaching conclusions about the basic fairness and management of the fire department during these events. The implications of the department's action regarding one's disciplinary case may have far-reaching consequences for the organization.

The first obligation in the disciplinary process always lies with the management of the fire department, beginning with clear rules and regulations that are understood by the employees. This foundation is essential to the disciplinary process and ensures fairness to employees. In order not to violate a rule or policy, one first must be aware of it and understand its application in the work environment. In many cases training is required to ensure a clear understanding of the behavior expected. In most fire departments, this training covers the policies and regulations of the general local government and the specific regulations within the fire department. Volunteer fire departments that are not part of a government structure have bylaws and rules that have been agreed on by the membership of the company.

If the rules and regulations are to be understood and properly enforced, they must also be realistic and practical. Many organizations have rules and regulations that no one pays any attention to because they are outdated and do not reflect the current work environment. In an environment where some rules are actually enforced and others are not, enforcement of any of the regulations becomes difficult. A process for the periodic review and discussion of the regulations that govern the department should be conducted on a frequent basis and should include employees of all ranks, including union representatives. In this manner, the regulations will stay current and will be the result of a group process as opposed to top-down enforcement.

In any organization, disciplinary actions of various types and levels of seriousness are necessary from time to time. The actions can range from reporting to work late to major criminal offenses. Most disciplinary action can be placed into four general categories: rules violation, unsatisfactory performance, personal issues, and illegal acts.[2]

Rules Violation

Rules violation is either a failure to follow or avoidance of established organizational rules and regulations, which cover a host of actions that are expected by employees to make the work environment productive and safe. Fire departments have specific regulations to protect both employees and the public. Regulations may be included in a fire department manual or general order book specific to that department. In addition, regulations may be imposed from outside the fire department, such as OSHA requirements or a medical protocol policy by a state agency for the administration of IV drugs. Although these regulations emanate from outside the fire department, fire department employees must adhere to them in the performance of their duties.

Unsatisfactory Performance

Unsatisfactory performance occurs when an employee fails or is incapable of performing the minimum requirements of the position held. Accumulated unsatisfactory performance may result in suspension or dismissal from the fire department. The disciplinary aspect would occur only after repeated attempts at training and motivation have proven unsuccessful. Please refer to Chapter 8 of this text for a review of the performance appraisal process.

Personal Issues

Personal issues from the off-duty life of an employee can often have a detrimental effect on his or her on-the-job performance. Employee assistance programs can help deal with these issues, and supervisors should refer troubled employees to these programs. At some point the personal problem may become an unsatisfactory performance issue that must be dealt with in that manner. These types of situations can be difficult for supervisors to deal with, because the main cause of the deficient performance is beyond their control.

Illegal Acts

Illegal acts are criminal offenses committed by an employee either on or off duty. A wide range of activities can fall into this category, from minor motor vehicle violations to felonies. Fire departments should require that any employee charged with a criminal violation immediately report that violation to the department. The adjudication of these acts may be difficult at times due do the interaction of the fire department discipline process and that of the criminal justice system.

Regardless of the type of violation or unsatisfactory performance that has occurred, a system must be in place to deal with these issues in a forthright and timely manner. Some general procedures to follow with regard to the administration of the disciplinary process produce the desired results with consistency and a clear understanding of the process by those involved. The noted management expert Douglas McGregor developed what he called the *red-hot stove rule* as a guide to an organization's actions in these circumstances.[3] According to this approach, disciplinary action should have the following consequences, which are analogous to touching a red-hot stove:

Immediate.   If disciplinary action is to be taken, it should be timely so that the individual clearly can understand the action and events that led to the discipline. The discipline relates to the behaviors involved and does not depend upon personality or organizational position. The stove burns whoever touches it, immediately.

With Warning.   Employees need to have a clear understanding of the regulations and the consequences for not adhering to them. As you approach a stove, you are warned by the heat to not touch it.

Consistent.   If discipline is to be perceived as fair, then similar circumstances must be dealt with in a consistent manner. This area is one of the biggest complaints in the disciplinary process—that it is not consistently applied. As with a hot stove, each person who touches it for the same period of time is burned to the same degree.

Impersonal. Disciplinary action is impersonal; it is based on behavior and not on the personal characteristics of who is involved. The stove burns whoever touches it, without favoritism.

The red-hot stove guidelines have merit and serve as general principles and concepts. However, the administration of discipline in an organization is just not that simple and clear-cut. Many variables are involved in the discipline cases that a fire service personnel manager deals with. Although the disciplinary charging document may be the same, the circumstances surrounding the disciplinary situations rarely are. In most cases, unique aspects characterize every incident. Do you treat a loyal 25-year firefighter the same way you would a recruit firefighter for the same offense? Most likely not, and therefore a supervisor quite often finds that he or she cannot be completely consistent and impersonal in the disciplinary process. However, one should strive for consistency and impartiality to ensure that essentially the same offense, given the same circumstances, is handled in a similar manner. In our example, loyal 25-year employees would be handled in a similar manner as would recruit firefighters.

It is never just the employee that is receiving the disciplinary action involved in the process. Many other employees are aware of the situation and waiting to see how the organization responds. How the organization acts in response to these situations will set precedents and establish the climate for future disciplinary actions. Discipline, if properly applied, has potentially positive aspects for the organization as well as for the individual employee. Consider the following:

- Discipline may alert marginal employees to their low performance and result in a positive change in behavior.
- Discipline may send a signal to other employees regarding expected levels of performance and standards of behavior.
- If the discipline is perceived as legitimate by other employees, it may increase motivation, morale, and performance.

Progressive Discipline

The principle of progressive discipline is widely accepted by organizations, arbitrators, and the legal system. Progressive discipline means that there is a hierarchy to the severity of the penalty for certain offenses that require disciplinary action. In essence, a supervisor responds to a first offense with some minimal action and to subsequent offenses with more severe actions. The overall goal of progressive discipline is to put the employee on notice and to correct problems at an early stage, before more severe discipline is necessary. This process documents the occurrence of a problem that requires disciplinary action and identifies the corrective action necessary to prevent future problems. A strong element of fairness here too permeates most cases in which the employee is appropriately warned before more formal action is required. With almost all employees, the first step in this process is sufficient to correct the situation.

Different organizations have various levels to the steps within their progressive discipline policy. In general, they follow the steps in a particular order, although certain offenses may result in termination or suspension without using the preceding steps. In general, the following steps to the progressive discipline policy are present:

Warning.   Notification is issued to the employee by the supervisor in private to inform him or her of inappropriate behavior that needs correction. Often, a warning can be followed by additional training to ensure a complete understanding of the actions involved.

Verbal Reprimand.   A face-to-face notification is used to inform the employee more strongly of the inappropriateness of his or her actions. The supervisor may make a note in the file that the verbal reprimand was issued and of the circumstances involved. This written report may be removed from the file after a certain time frame.

Written Reprimand.   A significant psychological effect can come from seeing the results of your behavior in writing. The written reprimand should identify the behaviors that led to this action and a plan of improvement for the employee to follow. The written reprimand can have two stages: One is to make it a part of the permanent personnel file, and the other is to remove it after a certain period of time.

Suspension.   The employee is removed from duty for a period of time. This suspension may be with or without pay, depending on the circumstances and the local personnel law and procedures. If the suspension is to have the proper disciplinary effect, then it should be without pay.

Termination.   The employee is separated from service with the fire department. This step is the most severe in the disciplinary process, but unfortunately necessary in some cases.

The progressive disciplinary process can be graphically illustrated as follows:

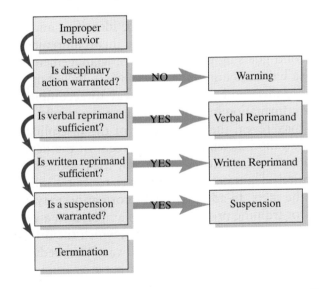

Even with the use of the progressive discipline system, it still is necessary to maintain proper levels of consistency in the actions of supervisors. The same actions that result in a verbal reprimand for a firefighter at Station 6 should not result in the suspension of a firefighter at Station 10. Often, consistency is hard to maintain with the same supervisor; it is even more difficult when magnified over a number of fire stations and work locations and then multiplied by three or four shifts. Many departments have provided general guidelines to inform supervisors as to when certain disciplinary actions are necessary and to recognize the proper level of discipline given general parameters. For example, a firefighter reporting to duty late on two occasions in a 6-month period must receive a verbal reprimand.

The supervisor of the officer who issues written reprimands and more serious personnel actions should review the circumstances and actions with the officer who took the action. This step ensures that departmental policy has been followed and that the firefighter who received the punishment got what he or she deserved—no more and no less. Most fire departments have an appeals process for written reprimands and more serious actions that may involve different levels of supervisors or disciplinary hearing panels to determine the facts and circumstances, particularly when a dispute is involved. Many actions of the disciplinary process are subject to the grievance procedure as pursuant to the collective bargaining agreement with the union. This issue is covered in Chapter 11, which deals with labor relations.

As an additional check to ensure that the disciplinary process is properly administered, certain actions should be reported to a central location, such as to the fire department personnel officer. Generally, a report would be mandatory, beginning at the written reprimand level, to ensure consistency and appropriateness of the actions. Fire departments and other government agencies are sometimes accused of disproportionately disciplining certain workers, such as women or racial minorities. In cases when the supervisors are largely white and male, the sensitivity to this issue may be increased. Quite often those employees without power in the organization may feel that they are singled out for unfair treatment. This unfair treatment may or may not have a factual basis; it might be only a perception. To ensure this is not the case and to be prepared to deal with it if it is, the management of the department must know what is occurring. By having these actions centrally reported at some point, trends can be identified and reviewed to determine the facts. Some supervisors may not like this requirement and resent what they feel is the second-guessing of their actions. Management must not only ensure that actions be balanced and understood by all, including why this review is necessary, but also understand that the only way to accomplish this is to gather and review all the information in order to observe the trends.

This analysis may reveal that 43 percent of the disciplinary actions for the year occurred in Battalion 2, which has 20 percent of the department's assigned personnel. The department would want to know the reasons behind the statistics, which may be entirely legitimate and appropriate, but confirmation can come from a review of the facts. In addition, this information may serve as a valuable needs assessment for training programs. For example, if in a year's time a disproportionate number of vehicle accidents occur in which the driver has violated the department's response policy, then one would want to schedule additional training in these areas. This type of information can serve as an incentive to the proactive management of the fire department.

Disciplinary Proceedings

All organizations have a series of rules and regulations designed to ensure reasonable predictability of employee behavior and to ensure their safety in the work environment. Established work rules help fulfill human needs for security, order, predictability, and avoidance of physical harm. Because most employees understand that these needs will not be satisfied in a chaotic environment, they are willing to accept these restrictions on their behavior. In addition, they expect that those who repeatedly violate these rules will be properly dealt with by supervisors to ensure that this environment is maintained. A process is needed to manage people who violate regulations and who must be formally disciplined by the organization. The process of progressive discipline addresses this need, and this section explores how this process is applied in the work environment.

Due process is an important concept regarding disciplinary actions. Basically, due process means that employees have a procedural route that provides the opportunity to explain their actions and confront disciplinary charges. Appropriate due process procedures tend to strengthen and create credibility as well as fairness within the system of discipline. In many cases the first argument that an employee makes is that he or she was denied due process within the disciplinary process. This circumstance is easily avoided by giving employees ample opportunity to speak and express themselves during each step of the process. Ensuring due process has many benefits for management, the union, and the employee. Its major elements are as follows:

due process
■ means that employees have a procedural route that provides the opportunity to explain their actions and confront disciplinary charges.

- *General notice.* The employee must know and understand the employer's expectations and workplace rules.
- *Specific notice.* The employee must be given notice of any failure to meet the employer's expectations. The employee must be advised of and understand the consequences of such actions.
- *Fair administration.* Rules and discipline must be administered fairly and consistently throughout the work environment.
- *Appeals.* There must be a viable appeals process.

After an event occurs that requires some level of discipline, the first step is a meeting between the supervisor and the employee to discuss the situation. How this meeting is conducted is important to the success of the disciplinary and motivational aspects of the employee and organization. This type of meeting can be confrontational in certain circumstances and always has heightened emotions because of the possible consequences. The supervisor must be thoroughly prepared to deal with the situation in a forthright manner and in full consideration of the rights of the employee and the organization. The following techniques have been proven to be of value in these circumstances:

1. Be fully prepared with the facts of the situation. Preparation may require some investigation on your part and the interviewing of coworkers or witnesses to the event. Review the personnel file for previous violations or disciplinary problems.
2. Conduct the meeting in a private office and minimize interruptions. Privacy is even more important for these types of meetings than it is in other circumstances. Often these meetings can become emotional, and privacy is crucial.
3. Communicate a clear understanding of the purpose of the meeting, and ensure that the employee understands the rule or regulation that was violated. Review the purpose of the regulation and what can happen if it is not followed.
4. Encourage the employee to explain the situation fully as you listen closely and encourage employee feedback. Do not prejudge who or what was right

and wrong; keep an open mind. Make notes to record the employee's version of what occurred.

5. State the reasons for the disciplinary action and why it is necessary for the organization. If the employee's version of the events indicates that formal action is not necessary, then state this variation and discuss what led you to this point. Do not press a case if the facts indicate otherwise; it is not a time for ego to be driving your decisions.

6. Do not berate or belittle the employee in any way, regardless of the anger the action caused. Do not conduct the interview until emotions are under control.

7. Discuss a plan of improvement if appropriate, or explain that formal charges will be placed due to the serious or repeated actions of the employee. Review this process with the employee and highlight employee rights in the disciplinary process.

8. Do not make any promises or speculate as to the final outcome of the disciplinary action if it will be referred for a formal hearing. Do not withhold any information from the employee; fully disclose everything regarding the case.

9. Attempt to conclude the meeting in a positive way. Try to express confidence in the employee regarding the future and the employee's ability to change the behavior to an acceptable level. If formal charges are to be presented, the supervisor will still have to work with the employee while this process takes place, which could be several weeks or longer. Always focus on behavior and not the person.

10. Document what occurred in the meeting and ensure that all of the facts and pertinent information are available in case further steps in the disciplinary process will be necessary.

11. Advise the supervisor of the actions that have been taken and any recommendations that will be made regarding formal charges.

The majority of career firefighters in the United States are represented by a local union or affiliate of the International Association of Fire Fighters (IAFF). One of the most basic jobs of unions is to represent members who have been charged with violations of departmental rules or regulations. In most cases they do this job very well, and it is a benefit to the employee to have this support. When a supervisor indicates to a firefighter the desire to have a meeting to discuss a violation, the employee may request representation by the union shop steward. The supervisor then has several responsibilities in this circumstance. Incidentally, under the "duty of fair representation," public sector unions may have to represent employees in disciplinary cases whether or not they are union members.

One of the first things to realize is that the employee must request the union representation, and the supervisor has no obligation to offer it. If the request is made, then the supervisor must allow it and give the union steward time to meet with the employee privately before the meeting takes place. In *NLRB v. J. Weingarten*, the U.S. Supreme Court ruled that a union employee has the right to demand that a union representative be present at the interview, if the employee reasonably believes that the meeting may result in disciplinary action **(Weingarten rights)**.[4] The supervisor has the option to reschedule the interview if the representation cannot be arranged within a reasonable time frame, but one cannot conduct the interview without union representation if it is requested. To

Weingarten rights
■ the rights that union employees have to demand that a union representative be present at the interview, if the employee reasonably believes that the meeting may result in disciplinary action.

do so would jeopardize the entire disciplinary action. In addition, the supervisor does not have to bargain with the union representative during the interview, and the union representative may not limit the questions asked by the supervisor. Anyone in this situation who has questions should consult with his or her supervisor or with the person who provides legal assistance to the department.[5]

Under the terms of a collective bargaining agreement, discipline must be processed under the terms of that agreement. Typically the "management rights" clause of the agreement retains for management the authority to impose reasonable rules for conduct and to discipline employees for just cause. Over the years, numerous arbitration cases have clearly identified that the concept of just cause requires an employer to produce persuasive evidence of the employee's culpability, conduct a fair hearing, and impose a penalty appropriate to the proven offense. In most cases unions do not object to employee discipline, provided that all employees are informed of the rules, the rules are considered to be reasonable, and the discipline is applied consistently throughout the organization. The difficulty between management and unions is that in many cases the reasonableness and the consistency of the rules are in dispute. This is a situation to be avoided, and good fire service personnel managers consult regularly with union leaders and shop stewards regarding the disciplinary process.

Disciplinary actions are not pleasant supervisory tasks. Many people find them to be difficult to implement and deal with the repercussions that may occur. It takes a strong and confident officer to handle these issues properly. For a number of reasons, supervisors may avoid taking disciplinary action. Some of these reasons are organizationally related and others are personally related. It is important to recognize these factors and to prepare for and deal with the issues before beginning the disciplinary process. Some of the reasons discipline is difficult for supervisors are as follows:

Lack of Training.   The supervisor may not have the knowledge and skill to handle the disciplinary process.

Intimidation.   The supervisor is fearful of how he or she will be perceived by both coworkers and union representatives.

Lack of Support.   The supervisor may be concerned that higher levels of management will not support the supervisor in the disciplinary process.

The Only One.   The supervisor may think, "No one else is disciplining employees, so why should I?"

Guilt.   The supervisor may think, "How can I discipline someone when I have done the same thing?"

Rationalization.   The supervisor knows that the employee knew that it was wrong and will not do it again, so why bother with the disciplinary process?

Workload.   The supervisor may be concerned about the amount of work and time it takes to pursue a disciplinary case.[6]

As one moves up the progressive disciplinary process ladder, the consequences are more severe and the documentation and legalities of the process increase. The employer and supervisors have the clear obligation to be factual in all cases and to be objective, consistent, and impartial. Employees who may be charged with an offense have an array of legal rights, and they can be expected to use them. These actions are not taken to offend the supervisor; they are just the nature of the proceedings at this level. It is unfortunate, but many defense attorneys seem to spend more time attacking the organization than they do addressing the merits of their case. One's credibility and ability as an officer may be called into question. The key is not to overreact, but to stick to the facts and the circumstances of the disciplinary action. If the officer leaves the hearing room with the knowledge that he or she did the job well and did what was best for the organization, it is enough.

Discipline in the Fire Service Environment

Many good things come from the close working relationship of the fire and rescue service. However, with regard to the discipline of employees, that relationship can make a difficult task even more arduous. The working environment of a fire station is unlike most any other occupation in civilian life. The crew works together for extended shifts, in some cases up to 24 hours at a time. The team spirit and reliance on each other during emergency incidents, as well as the on- and off-duty camaraderie, are evident in most fire departments. In some cases the fire station atmosphere is one of a big family, in which firefighters often tend to protect the wayward son or daughter. These attributes of fire station life all have positive points, but with respect to the discipline necessary to maintain order and effectiveness, they do present some complications. This section reviews these issues and makes suggestions about how best to deal with them regarding discipline.

Much of the difficulty for fire officers in the administration of discipline relates to their ability to deal with peer pressure. Fire officers work hard to develop a team atmosphere in which everyone is working together and pulling for each other. In addition, strong social and personal relationships develop within the fire service environment as people fraternize off duty. Then an infraction occurs and the officer must take disciplinary action regarding one of the members of the team, placing him or her in a new role and one that requires exerting power over subordinates. For this reason, the military to a large extent still segregates officers from the enlisted troops. A close working relationship and familiarity make it difficult to enforce discipline when necessary. The fire officer also may be concerned as to what the other members of the crew may think. These feelings may be more intense if the firefighter to be disciplined is popular or an informal leader at the station. Will the crew still think of him or her as one of them? Unfortunately, some officers mistake being liked with respect.

If a supervisor has a reasoned and fair approach to discipline, then others within the work environment will respect it, even if they do not particularly like it. Your stature as a leader will be increased because you did the right thing in maintaining the rules and regulations of the department. It may not be verbally expressed, but firefighters want the rules to be enforced fairly because they know consistency improves their work environment and keeps the department on the right path. In my opinion, if the people clearly understand that an officer will not tolerate certain actions, then these actions are unlikely to occur. An officer who appears to waiver or go with the popular sentiment when it is easier to do so will have many more challenges. The obligation is to uphold the regulations of the department in a fair and impartial manner; to do otherwise reduces the respect and effectiveness of an officer. Fire officers also find that a fire company that is doing things properly and within regulations is motivated and feels better about itself as a whole, as opposed to a company that is cutting corners and sliding by.

Some of the peer pressure within the fire service is also caused by the influence of the union into the officer ranks. Unions are based on both solidarity and union members sticking up and looking out for each other. In many jurisdictions it is necessary for members to bond together within the union to protect themselves and their jobs. This is not a bad thing; I clearly understand the positive value that unions have in the fire service. In fact, I have been a member of IAFF Local 1619 for over 30 years and was very active in the local earlier in my career. It is common in many fire departments for the officers, and in some cases up to and including deputy chief, to also be in the union. In certain departments, placing someone on disciplinary charges is viewed as causing problems for a union brother or sister. Union officials and shop stewards are also sometimes officers within the fire department, which magnifies the aspects of peer pressure. Additionally, the changing roles and loyalties that sometimes occur in these situations can be very confusing for employees. This unique aspect of working within the fire service environment can complicate the role of a supervisor.

The close working relationship in a fire station also allows for one's actions as an officer to be in full view of subordinates. There are not many secrets within a fire station. If the officer has not been following department regulations, those actions will be readily apparent to all within the station. The example set will be negative and may in fact encourage others to disregard regulations they do not personally agree with. This situation normally occurs for two reasons: (1) The officer just does not follow the rules, or (2) the officer selectively enforces regulations and will not enforce rules that he or she does not agree with. The message to the firefighters is the same: The officer is not following the rules, so why should they? This type of action puts an officer in an extremely difficult situation. If the supervisor must initiate disciplinary action, the employee's first response may be, "Well, how about the number of times you have reported late for duty?" Some fire officers may be fearful that their inadequacies will be exposed, so they do not engage the disciplinary process when it is warranted. This results in a fire company that is in chaos and lacking in the basic disciplines necessary to be effective.

In Chapter 8 we discussed the "halo and horn" effect with regard to evaluating performance. A similar occurrence can occur in the disciplinary process. For example, a firefighter who is excellent at emergency incidents and does many aspects of the job well may not be subject to the same level of discipline as are others. When an infraction or rule violation occurs, it may get overlooked. The good work that someone does should be taken into account to an extent, but it also needs limits. Conversely, a firefighter who may be struggling with certain aspects of the job may be dealt with in a harsher way in the discipline process, because he or she may be perceived as unworthy of the position. Managers must be very careful in these circumstances and balance decisions in an appropriate manner. In reality, some minor violations may be offset by good performance, but within reason and for the good of the organization.

The nature of the fire service environment as it relates to response to emergency incidents is one that includes occurrences of injury and death. Unfortunately, many of the causal factors in these firefighter deaths and injuries are due to the failure to follow established regulations and standard operating procedures. For example, every fire department that has personal alert safety systems (PASS) requires that they be used on the scene of structural fires. However, the number of firefighter deaths in which the device was worn by the firefighter but not activated in accordance with policy is disheartening. Eventually, the NFPA standard had to be changed to require automatically activated PASS devices. The fire department has to do better to protect its people properly. In the fire service environment, no officer should ever disregard a safety violation or allow anyone under his or her command to do so. It should be paramount to do what is necessary to enforce the regulations at all times. An individual can contribute no greater benefit to the organization and coworkers than to keep them safe. At times it may require a no-nonsense, very strict policy with regard to the enforcement of safety regulations. Do this as a fire officer and you will never regret it.

An age-old solution to discipline problems in some fire departments is to transfer the individual with disciplinary problems to another station. Although this option may be appropriate in certain circumstances, usually it is not. A transfer to another station does not address the performance issues causing the problem. It merely moves the problem to a new location where it can infect another crew. In the long term, transfer serves no one well: neither the department nor the employee. The situations requiring disciplinary actions, whether unsatisfactory performance or rules violations, should be resolved where they occur and not shuffled off for someone else to deal with. This would be unfair to the receiving station and officer who get these problem employees. If a firefighter warrants disciplinary action, then it should be completely resolved before the individual is eligible to transfer to another station.

Personnel in the fire department are expected to be of good character and inspire a high degree of trust from the public. This expectation means that off-duty activities can impact an individual's ability to be a member of the fire department. This situation occurs when a firefighter is arrested for an off-duty incident and then is subject to disciplinary action on the job, and it can be especially controversial when the firefighter claims innocence and can prove it at trial. As with disciplinary actions, criminal violations come in different sizes and shapes. The fire department needs to have a policy in place to deal with these issues. The first policy is to require that all arrests be reported to the department within 24 hours. This gives the department notification that the event has occurred. The department may then choose to collect some preliminary information as to what occurred and must make a determination as to whether the individual charged will be allowed to work until the criminal issue is decided. If the firefighter is to be suspended, will it be with or without pay? If the firefighter is to be placed in a position that has no public contact, how is the transfer done? Because the wheels of justice tend to move slowly, it may be months or years before the criminal charges are adjudicated, especially if appeals are involved. All of these aspects must be considered in the decision.

In many cases, it may be best to take whatever action is necessary within the fire department irrespective of the criminal proceedings. Managers have the right to investigate the incident to determine the facts and to conduct a disciplinary hearing if appropriate. The level of evidence and procedure needed for the fire department to conduct a disciplinary hearing is quite different from that needed at a criminal trial. It is possible that a firefighter can be found not guilty in criminal court and still lose his or her job. It is best to take this approach when the evidence is clear and convincing. When the evidence is disputed or not clear, but a serious criminal charge is involved, then it may be best to suspend the employee without pay until the issue is finally resolved. At some point, if the individual is cleared, he or she will be awarded back pay and reinstated. This policy may not be totally fair in all cases, but sometimes it is best for the department in light of limited information. Other small criminal offenses may not require action on behalf of the department, such as minor traffic violations and trespassing. However, if someone in the department is charged with a crime, the department must determine the facts as to what occurred and make decisions as to how best to handle the situation.

These issues can also be found in civil suits in which the department is involved. For example, a firefighter is involved in a vehicle accident while responding to an emergency and a civilian is unfortunately severely injured. The fire department policy is to investigate vehicle accidents, make recommendations for preventing future accidents, and determine whether violations of the response policy contributed to the situation. A civil suit is filed against the fire department and the local government. Most local offices of law prefer that the fire department not investigate and find fault in this accident. They know that if they do and if the fire department driver is found to be at fault, then it will make the civil case more difficult to defend. Fire chiefs know that if they do not address vehicle accidents properly, the ability to prevent them and also maintain adherence to the fire department's response policy will be compromised. My suggestion in this case is for the fire department to do what is right for the department and to follow through with the accident investigation and appropriate action. This procedure allows the best management of the fire department and correction of safety issues as they arise. In most cases, if the fire department driver is at fault, that fact will come out at the civil trial anyway.

Another interesting aspect of discipline within the fire service environment is in the case of volunteer firefighters or a combination department of career and volunteer firefighters. The concepts and reasons for having a sound policy on discipline are the same regardless of whether the department is career or volunteer. The major difference is in the employment relationship in that volunteers are just what the name implies, volunteers. They do not usually have enough of a property right and interest in a job to be considered employees, although some paid on-call volunteers may qualify. Many volunteer fire departments maintain no government relationship and no involvement with unions. In many cases, choices as to the discipline may be limited to extra work assignments at the station or suspension from the department. Verbal and written reprimands may also be issued to volunteer firefighters to advise them that their actions need improvement. Many volunteer officers may hesitate to take these actions because they are voted into office by the membership. Some volunteer fire departments are reluctant to suspend volunteers because they need all the personnel they can get to respond to emergencies. This need must be balanced against the overall good of the department in order to maintain a proper level of discipline. In many cases, the suspension or termination of a volunteer who will not comply with company policy is the best route for the long-term success of that organization.

Combination fire departments present the unique situation of volunteer and career firefighters working side by side. When a serious rule violation occurs, the career firefighter may be suspended without pay, whereas the volunteer is only suspended for the same infraction. This difference can cause friction between career and volunteer firefighters in a situation that frequently has too much friction to begin with. A sense of equity must be established so that the employees and volunteers are treated essentially the same, given the same circumstances. This need for equity applies to the formulation of rules also. This balance is often difficult to sustain, but to the extent it can be maintained, the department will operate much more smoothly.

The preceding paragraphs attempted to capture some of the unique aspects of discipline in the fire service environment. Fortunately, employment selection procedures lead to the hiring of firefighters of good character who are able to perform the job well. Discipline, although necessary, should be an occasional activity within the fire department. If a department must constantly deal with violations of regulations, then its officers must examine why and determine the root cause. The answer may be in employee selection, training, motivation, working conditions, regulations, or in all of the former categories. The discipline system of the department is necessary after the fact; the goal is to have adequate performance and behavior up front with minimal disciplinary action.

Exit Interviews

It is good policy to hold exit interviews with those who leave the fire department, either voluntarily or involuntarily. The exit interview is a discussion between a representative of the employer, usually the fire service personnel manager, and the person who is leaving the organization. Its purpose is to identify the characteristics of the department and general work environment that might hamper effective performance or cause high turnover within the organization. For obvious reasons, the exit interview is most useful and valid for employees who are leaving the organization on a voluntary basis. The candid observations these employees may offer in these circumstances can be valuable. They are more likely to express what they really feel without having to be concerned with department politics or consequences.

The interview should be conducted in private and as one of the last steps in the process of separation from the department. The interviewer must remember that the employee is doing the department a favor by agreeing to be interviewed, and should thank that person and explain that the purpose of the interview is to learn why he or she decided to leave the organization. In addition, the interviewer would want to obtain the employee's opinion of his or her treatment within the job and of the training and management practices of the department. It is important to encourage the employee to speak candidly, with the assurance that nothing said will be placed in the personnel file or affect references for future employment. If these interviews are properly conducted, they can be a valuable tool for assessing the effectiveness of the department's programs. Information received in these interviews is one part of this assessment from a particular viewpoint and should be confirmed by other observations and sources before a final opinion is reached.

In the exit interview, one should spend most of the time listening to what the employee is saying. A rule of thumb is that in these circumstances the interviewer should spend approximately 80 percent of the interview time listening. An interviewer need not defend certain programs or people or be offended by comments and appraisals that are revealed. Keeping the conversation as constructive as possible and encouraging the employee to make positive suggestions as to how the work environment can be improved are helpful strategies. Some people may have an "ax to grind" and will use this opportunity to do so. An interviewer should listen to what the employee has to say and realize the importance of collecting a number of these interviews and other collaborating information before taking action on any issue.

Summary

The purpose of discipline is to correct behavior that does not meet the standards established within the organization. It is not designed merely to punish people for their actions. Proper levels of discipline are important to the effectiveness and efficiency of a fire department, and everyone within the organization has a responsibility to maintain order in this process. Discipline is a complicated issue and requires a good understanding of the objectives to be achieved by the level of discipline and of the process necessary to implement it fairly. The unique work environment within the fire service poses special challenges in the administration of discipline. These challenges can be overcome by management that is attuned to these issues and by having a clear understanding within the department as to the required levels of performance and behavior.

Progressive and positive discipline is an opportunity for the department to correct behavior of employees before more drastic or permanent measures are necessary. A program of reasonable, fair, and impartial discipline benefits all employees and is an essential component of a successful fire department.

Review Questions

Use these questions to review the material in this chapter and for discussion purposes.

1. Discuss three types of organizational control that are available to fire department management. Which type do you feel is the most effective to use within the fire service?
2. Identify the four general categories of disciplinary actions and discuss how they have been implemented within your fire department. What would you improve on?
3. What are the steps involved in progressive discipline? Discuss examples of disciplinary situations that would occur at each specific level.
4. You have scheduled an interview with a firefighter to determine the facts regarding an incident that may result in disciplinary action. The firefighter requests that he or she be represented by a union official at this meeting. What do you do?
5. Name and discuss three situations in the fire service work environment that may complicate disciplinary actions or make them more difficult to implement.

Case Study

You are the deputy fire chief of a combination fire department that is approximately 60 percent career and 40 percent volunteer. The department is engaged in an effort to monitor departmental driving regulations more strictly regarding stopping at intersections and speeding. This has been a major issue in the past, with several vehicle accidents being the fault of the fire apparatus drivers.

Two drivers are placed on disciplinary charges for failing to adhere to the driving regulations after repeated warnings. One is a career firefighter and one is a volunteer firefighter. After the disciplinary hearing, both drivers are suspended for

3 days. This means that the career firefighter will lose 3 days' pay, whereas the volunteer will not be allowed in the station for 3 days. The firefighters' union challenges this on the basis of inconsistent application of the disciplinary process.

What are the relevant issues here, and how would you resolve this situation? Explain how you would prevent problems like this in the future.

Fire Service Professional Qualifications

This chapter contains information that will assist fire officers in meeting the job performance requirements (JPRs) of NFPA 1021, Standard for Fire Officer Professional Qualifications:

4.2 Human Resource Management—Fire Officer I

    4.2.5 Apply human resource policies and procedures given an administrative situation requiring action, so the policies and procedures are followed.

5.2 Human Resource Management—Fire Officer II

    5.2.1 Initiate actions to maximize member performance and/or to correct unacceptable performance, given human resource policies and procedures, so that member and/or unit performance improves or the issue is referred to the next level of supervision.

    5.2.2 Evaluate the job performance of assigned members, given personnel records and evaluation forms, so that each member's performance is evaluated accurately and reported according to human resource policies and procedures.

Endnotes

1. Philip B. Dubose, *Readings in management* (Upper Saddle River, NJ: Prentice Hall, 1988).
2. Nancy Grant and David Hoover, *Fire service administration* (Quincy, MA: National Fire Protection Association, 1994).
3. Wayne F. Cascio, *Managing human resources*, 7th ed. (New York: McGraw-Hill, 2006).
4. *NLRB v. Weingarten* (1975). 420 U.S. 251, 95 S. Ct. 959.
5. Cascio, *Managing human resources*.
6. R. Wayne Mondy and Robert M. Noe, *Human resource management*, 6th ed. (Upper Saddle River, NJ: Prentice Hall, 1996).

Suggested Readings

Cascio, Wayne F. 2006. *Managing human resources*, 7th ed. New York: McGraw-Hill.

Dubose, Philip B. 1988. *Readings in management*. Upper Saddle River, NJ: Prentice Hall.

French, Wendell. 1989. *Human resources management*, 2nd ed. Boston: Houghton Mifflin.

Grant, Nancy, and David Hoover. 1994. *Fire service administration*. Quincy, MA: National Fire Protection Association.

Mathis, Robert L., and John H. Jackson. 2003. *Human resource management*, 10th ed. Mason, OH: Thomson Learning.

Mondy, R. Wayne, and Robert M. Noe. 1996. *Human resource management*, 6th ed. Upper Saddle River, NJ: Prentice Hall.

NLRB v. Weingarten. 1975. 420 U.S. 251, 95 S. Ct. 959.

Sovereign, Kenneth L. 1989. *Personnel law*. Upper Saddle River, NJ: Prentice Hall.

10

Health and Safety

Source: MFR

KEY TERMS

general duty, *p. 252*

NFPA 1500, *p. 261*

NFPA 1582, *p. 260*

post-traumatic stress disorder, *p. 280*

risk control measures, *p. 271*

safety culture, *p. 269*

special duty, *p. 252*

workers' compensation, *p. 249*

OBJECTIVES

After reading this chapter, you should be able to:

- Describe approaches to safety.
- Describe the Occupational Safety and Health Administration.
- Understand fire service health and safety issues.
- Describe NFPA health and safety standards.
- Describe safety in the fire service environment.
- Understand how to access employee assistance programs.
- Describe violence in the workplace.

Ensuring the physical and mental well-being of employees in the workplace is the primary goal of the health and safety management process. In today's world, and particularly in the fire and rescue service, this complex activity requires the expertise of specialists from many disciplines, such as industrial hygiene, occupational medicine, and safety engineering. Concerns in health and safety management reach beyond physical conditions in the work environment to the emotional and mental well-being of firefighters and to the concept of employee wellness. Wellness programs in the fire service are designed to prepare firefighters so that their mental, physical, and emotional capabilities are strong enough to withstand the hazards and stresses found in their work environment and life in general.

Involvement in health and safety management represents an organization's commitment to a number of compelling influences, the most basic of which is a sense of social and humanitarian responsibility. Government intervention, pressure from labor unions, legal liability, and the public in general account for much of the progress made in health and safety issues. Increased experience and knowledge have led the management of organizations to appreciate how health and safety management are directly related to organizational effectiveness and profitability. The most valuable resource in any organization is people, and their well-being is vital to the success of organizations, regardless of size or work environment. The fire and rescue service has a long history of high rates of firefighter injury and death. Since the early 1980s, tremendous progress has been made in the protection of personnel and safe operating practices within the fire service. Although substantial reductions in the injury and death rates of firefighters have been realized, much work remains to be done toward further improvement.

Approaches to Safety

The management of health and safety of the workforce in today's society is a complex and highly regulated activity. To understand current health and safety standards and practices, it is useful to explore some of the history of this issue. Hazards to employees increased sharply with the machinery and factories of the Industrial Revolution. Long work hours, noisy and unhealthy conditions, unguarded machinery, and other problems of this era posed real risks to employees. As time passed, major efforts contributed to the health and safety of employees, which evolved into three general approaches: compensation, prevention, and uniform standards and enforcement.[1] Let us briefly examine each of these historical trends in safety and health.

COMPENSATION

The first approach was to compensate employees for job-related injuries or illnesses. The employee could bring suit against the employer for losses due to injuries sustained during employment. This approach seldom was a satisfactory remedy because the courts at that time generally favored the employer, and many employees could not afford the expense of a legal action. The inadequacy of legal remedies and the general attitude during the late nineteenth century, which stressed employer prerogatives more than worker rights, underscored the need for a better and more equitable system. Consequently, social activists and unions sought the greater protection of employees through state **workers' compensation** laws.

workers' compensation
■ a system that establishes an insurance program that compensates employees in the case of work-related injuries or their survivors in the case of their death. Benefits are provided regardless of fault or financial condition of the employer.

State workers' compensation laws essentially establish an insurance program that compensates employees in the case of work-related injuries or their survivors in the case of their death. These laws were founded on the belief that accidents and the disabilities of employees are a social matter, and the state has a duty to take care of the injured. In 1902, Maryland passed an act providing for a cooperative insurance fund. This legislation was the first to represent the compensation principle. Many of the earlier workers' compensation laws were declared unconstitutional as a denial of due process, but were reversed in later court decisions. By 1925, twenty-four states had passed workers' compensation laws, and all states now have such a system.[2]

Workers' compensation laws are designed to provide benefits regardless of fault or financial condition of the employer. This provision reduces delays and the complications of litigating employee claims and relieves the court system of such cases. Employers contribute to the fund that provides compensation to employees involved in work-related accidents and injuries. The benefits paid are related to the nature and severity of the injury. The employee is then barred in most cases from suing the employer because workers' compensation addresses the claims. Liability is then limited, and employers pay premiums to an insurance fund that covers the claims based on past experience of employee safety. Approximately 88 percent of the U.S. workforce is covered by workers' compensation, which provides three major benefits: (1) payments to replace lost wages, (2) payments to cover medical bills, and (3) financial support for retraining if necessary. For lost wages, the full salary is not provided, but a percentage up to a maximum amount.[3] The system works reasonably well in each state because it gives employees an expectation of benefits in the event of a work-related injury, and it provides incentives for employers to manage risks and improve their safety record. However, the obvious problem with workers' compensation is its reactive nature, in that someone has to be injured first before it is utilized.

PREVENTION

Employers quickly realized that much more was necessary than just taking care of injured employees. A proactive system would be the first step to prevent the needless injury and death that was occurring in the workplace, benefiting employers and employees alike. In the early 1900s, several groups began to consider these issues. A number of professional safety associations conducted research on the incidence and causes of industrial accidents, and established safety codes and promoted their use within industry. The American Society of Safety Engineers (ASSE), organized in 1915, focused on the technical aspects of accident prevention, such as machine guards and safety equipment. Another group that contributed to improved safety was physicians who specialized in industrial medicine. They concentrated on the control of the work environment to reduce illness and disease. A major driving force in the issue of workplace safety at the turn of the twentieth century and still a force today is the labor unions. They were active in pressuring employers to reduce the risk of injury to their members, whether they worked in factories, coal mines, meatpacking plants, or textile mills. Many of the major advances in workplace safety can be traced to union concern and action on behalf of its membership.

The focus today is still on the prevention of accidents and injuries to workers. Employers and employees clearly recognize the goal of preventing accidents and thereby avoiding pain and suffering, as well as the financial losses of claims, lost workdays, and insurance costs. Maintaining health and safety is a major objective of all organizations.

UNIFORM STANDARDS AND ENFORCEMENT

Even the best efforts in preventing accidents and injuries as well as compensation for those injured were not achieving the desired levels of health and safety. In many cases, workers' compensation systems provided no real incentives for employers to reduce hazards; no significant legal penalties were enforced against employers who were not serious about health and safety issues. The injury and death statistics for workers continued to rise in spite of the efforts regarding compensation and prevention. Injuries increased 29 percent from 1961 to 1970, with 14,000 deaths resulting from job-related accidents. This increase, coupled with several high-profile workplace accidents, aroused public and government concern.[4] Consequently, the federal government was compelled to act, and in 1970 Congress enacted the Occupational Safety and Health Act (OSHA), which required employers to improve workplace health and safety standards through inspection and sanctions for noncompliance. It also made the standards uniform across the United States with regard to health and safety for virtually all employers engaged in business that affects interstate commerce. Without question, OSHA has been responsible for marked improvement of the safety and health issues that affect U.S. workers.

Compliance with OSHA regulations has required employers to spend an increasing amount of resources on safety. Ensuring compliance with the OSHA regulations has benefited many organizations in a number of positive ways. Improved organizational safety and health are good for business, and the following trends have emerged:

Employee Relations. Businesses with good safety records have an effective method to retain and attract employees and to enjoy improved labor relations with employee unions.

Reduced Liability. Effective safety programs reduce insurance costs, litigation costs, and other costs associated with the results of accidents and injuries.

Profitability. Employees can only produce while they are on the job. Productivity of the workforce is enhanced by increased safety. Reductions in lost work hours and reduced medical costs can dramatically add to the bottom line.

Public Relations. A good safety record and concern for the environment can provide companies with a competitive advantage. They will be viewed as a good neighbor and an asset to the community.[5]

The 2007 worker fatality statistics from the Bureau of Labor Statistics indicate that the number of workers killed on the job in the United States has annually dropped to an historic low.[6] The number of worker deaths dropped to 5,488 in 2007, which is the lowest number since the Bureau began keeping track of this in 1992. The 2007 statistics are down 6 percent from 2006, in which there were 5,840 worker fatalities. However, this report did indicate significant increases in some types of fatal injuries: Workers who died in falls and the number of workplace homicides increased 13 percent. Fire departments need to be aware of the strategies for reducing worker deaths and injuries and pursue their implementation at every opportunity.

Occupational Safety and Health Administration

The cost of work-related injuries and their social consequences led the federal government to pass legislation regulating private and public employers. OSHA was passed in 1970 to ensure that working conditions for all Americans meet minimum health and safety standards. Under the provisions of this act, OSHA in the Department of Labor was charged with the responsibility of establishing health and safety standards, conducting inspections, and issuing citations and penalties for noncompliance.

OSHA was designed to reduce hazards and prevent injury and illness from work-related occurrences. The health and safety practices recommended by professional associations, medical groups, and unions were in many cases advisory in nature; however, OSHA standards are mandatory. OSHA established three agencies to administer and enforce its provisions. OSHA has the authority to establish health and safety standards, to conduct inspections of employer premises, and to issue citations and fines for OSHA violations. The Occupational Safety and Health Review Commission (OSHRC) consists of three members appointed by the president of the United States to hear employer appeals of OSHA citations. If the employer is not satisfied, he or she may appeal the OSHRC finding to the federal court system. The National Institute for Occupational Safety and Health (NIOSH) conducts research on the causes and prevention of work-related illnesses and injuries and provides information on which new standards may be based. NIOSH also trains OSHA inspectors and other enforcement personnel on how to enforce OSHA standards properly. In 1998, NIOSH was charged with the responsibility of fully investigating the cause and circumstances of every firefighter death in the United States. This development was the result of an effort led by the International Association of Fire Fighters (IAFF).

OSHA regulations apply to private industry and in some cases to government agencies. Under the law, the employer has a **general duty** to provide a place of employment "free from recognized hazards" and a **special duty** to comply with and enforce all OSHA standards and regulations. OSHA law applies to any business, regardless of size, that affects interstate commerce. However, some smaller businesses with fewer than ten employees are exempt from regular safety inspections. OSHA does not establish federal standards with which state and municipal agencies must comply. Currently, twenty-five states have chosen the option of designating a state agency to administer the health and safety plan. If this option is selected, then the state must develop standards that are equal to or greater than the OSHA standards, staff the agency with qualified personnel, and submit requested reports on compliance to the Department of Labor.

OSHA provides for on-the-spot inspections by OSHA representatives, called compliance officers or inspectors. During a visit, the OSHA inspector will most likely meet with organizational officials, review safety records and documents,

general duty
■ under OSHA regulations the duty an employer has to provide a place of employment that is free from recognized hazards.

special duty
■ under OSHA regulations the duty an employer has to comply with and enforce all OSHA standards and regulations.

and conduct an inspection using a variety of equipment to test for compliance with the standards. After the inspection, citations may be issued for any violation of the standards or provisions of the act. In most cases, whether a citation is issued depends on the severity and extent of the violations and whether the employer was aware of them. OSHA citations have been issued to fire departments, both for violations at fire stations and for violations on emergency incidents. In many cases, the emergency scene violations involve confined spaces, trench rescue operations, or respiratory protection standard requirements, such as the "two-in/two-out" regulations.

The five types of violations ranging from most severe to minimal, including a special category for repeated violations, are as follows:

Imminent Danger.   When there is reasonable certainty that conditions will cause death or serious physical harm if not corrected immediately. If the condition is serious enough and the employer does not cooperate, the OSHA inspector may obtain a federal injunction to close the business until the condition is corrected.

Serious.   When a condition could probably cause death or serious physical harm, and the employer should be familiar with the condition, OSHA issues a serious-violation citation.

Other than Serious.   These types of violations could have an impact on an employee's health or safety, but probably would not cause death or serious harm.

De Minimis.   A *de minimis* condition is one not directly and immediately related to an employee's safety or health. No citation is issued, but the condition is made known to the employer.

Willful and Repeated.   Citations for willful and repeated violations are issued to employers who have been previously cited for violations and have not corrected the conditions. The penalty for a willful and repeated violation can be severe, including large fines and imprisonment for company officials.

Both employers and employees have rights and responsibilities under OSHA requirements. To be effective, they must work together in the interest of improving workplace safety and health. The following chart is a summary of some of the more important responsibilities that employers and employees have under OSHA.[7]

Since 1970, OSHA has been working to save lives and prevent injuries and illness in the workplace. Since then the workplace death rate has been cut in half, but unfortunately about 13 workers still die every day on the job in the United States. The total number of workplace fatalities for 2006 was 5,840, or approximately 3.9 per 100,000 workers. In 2006, there were 4.1 million injuries/illnesses among private sector firms with 11 or more employees. The rate of injuries and illness for every 100 workers dropped from 8.9 in 1992 to 4.4 in 2006, the lowest on record. Back injuries are the most prevalent of all workplace injuries, accounting for approximately 1 million workers' compensation claims each year. The number of workdays lost to injuries has increased throughout the last 10 years, especially with repeated motion injury claims. Nearly one-third of all occupational fatalities are the result of vehicle accidents, and homicide has increasingly become a major cause of death in the workplace.[8] These rates and data apply to general occupations; the injury and death rates for dangerous occupations such as firefighting are higher.

OSHA RESPONSIBILITIES

EMPLOYER RESPONSIBILITIES

1. Provide a hazard-free workplace and comply with OSHA rules, regulations, and standards.
2. Inform all employees about OSHA and make copies of standards available for them to review on request.
3. Furnish employees with safe tools and equipment.
4. Establish, update, and communicate operating procedures to promote health and safety.
5. Report fatal accidents to the nearest OSHA office within 48 hours.
6. Keep OSHA required records of work-related injuries and illnesses, and post summary figures each February.
7. Cooperate with the OSHA compliance officer and accompany or consult with during inspections.
8. Do not discriminate against employees who properly exercise their rights under OSHA.
9. Post OSHA citations of apparent violations and abate cited violations within prescribed time period.

EMPLOYEE RESPONSIBILITIES

1. Read and be aware of OSHA regulations at the job site.
2. Comply with all applicable OSHA standards.
3. Follow all employer safety health rules and regulations, and wear or use prescribed protective equipment while working.
4. Report hazardous conditions to the supervisor.
5. Report any job-related injury or illness to the employer and seek treatment promptly.
6. Cooperate with the OSHA compliance officer conducting an inspection.
7. Exercise your rights under the act in a responsible manner.

In order to ensure that OSHA regulations are implemented, specific standards have been established regulating equipment and the work environment. National standards that have been developed by engineers and quality control groups, including worker input, are often used in this process. Four OSHA standards, in particular, that affect the fire service are the Hazard Communication Standard, the Bloodborne Pathogen Standard, the Respiratory Protection Standard, and the Personal Protective Equipment Standard, as described in the following section.

HAZARD COMMUNICATION

This standard requires manufacturers, distributors, and users of hazardous chemicals to evaluate, classify, and label these substances. Employers must make available to employees information about hazardous substances in the workplace. This information is contained in Material Safety Data Sheets (MSDS), which must be kept readily accessible to those who work with chemicals and other substances. The MSDS also indicate antidotes or actions to be taken in the event of an emergency. Fire departments rely on this information when responding to emergency incidents at facilities that require the MSDS information. The OSHA "lock out/tag out" and other safety programs are also included with this standard.

BLOODBORNE PATHOGENS

This standard was issued to eliminate or minimize occupational exposure to hepatitis B (HBV), human immunodeficiency virus (HIV), and other bloodborne pathogens. This regulation protects employees who are regularly exposed to blood and other such body fluids. It requires employers with the most pronounced risks to have written control and response plans and to train workers in following the proper procedures. Many fire service and EMS personnel have suffered debilitating diseases acquired from bloodborne pathogens. It is imperative that a safety program be established and that it be followed without exception.

RESPIRATORY PROTECTION

The most current Respiratory Protection Standard, issued in October 1998, included many changes for the fire service. This standard requires that all who use respirators, including self-contained breathing apparatus (SCBA), have medical evaluations, annual fitness tests, and training in the proper use of the equipment. In addition to the aforementioned requirements, interior structural firefighting requires the use of SCBAs and a protective practice known as "two-in/two-out"—at least two firefighters must enter and remain in visual or voice contact with one another at all times, and at least two firefighters must be located outside. There is a limited exception to this for emergency rescue procedures. Fire departments across the United States have changed their operating procedures to be in compliance with this standard.

PERSONAL PROTECTIVE EQUIPMENT

If the work environment presents hazards or if employees come into contact with hazardous substances, then employers are required to provide personal protective equipment (PPE) to employees. This standard requires that employers conduct analyses of job hazards, provide adequate PPE to employees in those jobs, and train employees in the use of PPE and associated equipment. PPE safety requirements are critical to fire service personnel. Many of the National Fire Protection Association (NFPA) standards identify the required protective clothing and equipment necessary for fire service and rescue operations. The safety equipment and protective clothing by NFPA requirements have to be independently certified as to meeting the standard for the protection of the workers. This is accomplished by organizations such as the Safety Equipment Institute (SEI) and Underwriters Laboratory (UL). Many other standards do not require the independent third-party certification of safety equipment, which means that the manufacturer certifies its own products. Independent certification is a much better system for the protection of workers exposed to hazardous environments.

Some employers would argue that OSHA regulations are excessively detailed and costly and, to an extent, do not take workplace realities into account. OSHA on its 2008 Web page states that it is committed to a commonsense strategy of forming partnerships with employers and employees; conducting fair but firm inspections; developing sensible, easy-to-understand regulations and eliminating unnecessary rules; and assisting employers in developing top-notch safety and health programs. It appears that OSHA is attempting to work more closely with business as opposed to just making rules and conducting inspections.

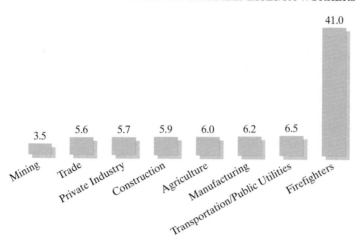

NUMBER OF JOB-RELATED INJURIES/ILLNESSES/100 WORKERS

Number of Job-Related Injuries/Illnesses/100 Workers. *Source: Survey of Occupational Injuries and Illnesses, 2006;* U.S. Bureau of Labor Statistics; and *IAFF 2006 Death and Injury Survey.*

Fire departments play a dual role with regard to health and safety in the workplace. One role is maintaining and improving the health and safety of its workforce, which is engaged in a dangerous occupation. Many of the workplace provisions of OSHA are designed to engineer-out the problem that is causing the injury or death. The control of the workplace and the equipment therein are important aspects of safety plans. This workplace control works for the fire department as long as firefighters are at the fire station, but the station is not where the danger lies. Firefighters respond anywhere at any time, so the ability to engineer-out problems is reduced because of the emergency response nature of public safety. Fire service agencies can engineer improvements in the safety equipment and protective clothing used, and many advances have been made in this area in the past two decades. Second, the fire department is an enforcement agency with respect to the fire code, hazardous materials, and other workplace dangers. By effective enforcement of its fire code responsibilities, the fire department can greatly improve the workplace safety of others. The programs of fire prevention, public education, and code enforcement are not OSHA regulations, but their effect in the workplace with respect to the saving of lives and reducing injuries is substantial.

JOB-RELATED FATALITIES/100,000 WORKERS

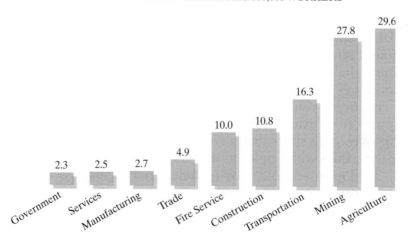

Job-Related Fatalities/ 100,000 Workers. *Source:* National Census of Fatal Occupational Injuries, 2006; U.S. Bureau of Labor Statistics; Public Safety Officers' Benefit Program; U.S. Fire Administration; and U.S. Bureau of Census.

Fire Service Health and Safety Issues

If a fire department is to implement an effective fire service health and safety plan, then it must first analyze the causes and circumstances of the injuries and deaths that firefighters suffer. Steady progress has been made since the early 1980s in reducing firefighter deaths and injuries. The rates are still too high, but the trend is encouraging and, I think, verification that the increased emphasis on safety in the fire service is working.

FIREFIGHTER DEATHS

Firefighter fatality statistics reveal a decrease of approximately 40 percent in the average number of annual firefighter deaths from 1977 to 2007. The number of deaths ranges from a high of 173 deaths in 1978 to the lowest total of 75 deaths in 1992.[9] This does not include the events of September 11, 2001, in which 343 firefighters lost their lives in a single day, the most firefighters ever killed in the line of duty in a single day or year.

The leading cause of fatal injury for on-duty firefighters has historically been stress-related heart attacks. Even though the total number of heart attacks has dropped since the early 1980s, it still accounts for over 40 percent of the deaths that occur each year. The NFPA reports that 80 percent of the firefighter heart attack victims for whom medical documentation was available had previous heart attacks, bypass surgery, or severe arteriosclerotic heart disease. These findings may indicate that better medical screening of firefighters as they age would be warranted. When stratified into age groups, the death rates for firefighters from 1992 to 2007 dramatically increase as the age of the firefighter increases.

As seen in the statistics, the majority of firefighter deaths occur when responding to or at the scene of emergencies. Historically, these two categories have accounted for approximately 65 percent of firefighter deaths in any single year. The number and proportion of firefighter deaths on the fireground have continued a downward trend since the late 1970s, when the NFPA started compiling these reports.

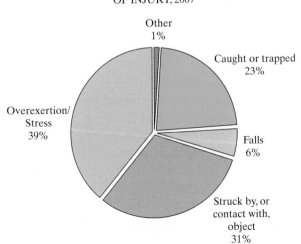

FIREFIGHTER DEATHS BY CAUSE OF INJURY, 2007

Other 1%
Caught or trapped 23%
Overexertion/ Stress 39%
Falls 6%
Struck by, or contact with, object 31%

Firefighter Deaths by Cause of Injury, 2007. *Source:* NFPA, 2008.

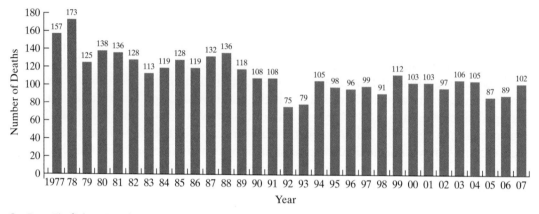

On-Duty Firefighter Deaths, 1997–2007. *Source:* NFPA, 2008.

The distribution of deaths of career and volunteer firefighters from 1977 until 2007 shows the general improvement in firefighter fatality rates but has stabilized in more recent years. Career firefighter deaths have declined during this entire period. Volunteer firefighter deaths have shown more volatility but have generally been in a downward trend. However, many experts point out that during this period the number of structural fires has declined and that firefighter deaths remain at an inappropriately high level.

FIREFIGHTER INJURIES

For the purposes of the NFPA reports, an injury is any physical damage a person suffers that requires treatment by a medical practitioner, such as a physician, nurse, paramedic, or EMT, within 1 year of the incident, regardless of whether treatment was actually received or physical damage occurred resulting in at least 1 day of restricted activity immediately following an incident. In the early 1990s, approximately 100,000 firefighters suffered on-duty injuries each year, but the trend is improving. In 2007, 80,100 firefighter injuries were recorded, which is one of the lowest totals since the NFPA began maintaining these statistics.[10] This news is good, but it is obvious that much more must be done to reduce the large number of firefighter injuries each year.

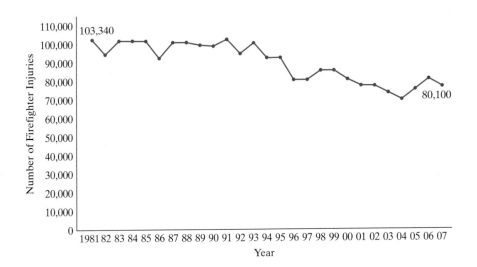

Total Firefighter Injuries by Year, 1981–2007. *Source:* NFPA Annual Survey of Fire Departments for U.S. Fire Experience, 1981–2007. *Note:* From 1994 on, number of exposures was collected separately.

In 2007, as in previous years, the majority of firefighter injuries occured while participating in emergency incidents. The category of "fireground" injuries accounts for approximately 50 percent of the total injuries in any given year. Firefighter injuries both on the fireground and at nonfire emergencies have decreased since 1987. However, fireground injuries in the most recent years have tended to stabilize at approximately 38,000 to 40,000 per year. In 2007, the leading causes of fireground injuries (27 percent) were "fell, slipped, or jumped." The category of "overexertion and strains" is responsible for an additional 25 percent of the firefighter injuries each year.[11]

As these firefighter death and injury statistics attest, the occupation of fighting fires and responding to emergencies is dangerous. The positive news is that the numbers of injuries and deaths are declining in almost all cases and categories. This decline has been the result of increased emphasis on and attention to firefighter health and safety issues, improved equipment, and a safer environment due to better building codes and automatic fire protection devices in facilities.

FIREFIGHTERS AND HEART DISEASE

Numerous fire service studies have identified that the most likely cause of firefighter fatalities is cardiac arrest. Approximately 45 percent of firefighter fatalities are caused by heart disease. Severe exertion stresses the heart and requires it to deliver more blood to working skeletal muscles. There is a nearly linear relationship between heart rate and work rate. Older individuals achieve lower maximal heart rates. As a result, older individuals cannot deliver as much oxygen to their working muscles as can younger people, so they do not have as great a work capacity. Physically active individuals of all ages generally have lower resting heart rates and greater physical work capacities than do sedentary individuals.

There are several contributing factors to heart attacks, the most important of which is the inability to deliver oxygen to the myocardium. Atherosclerosis, or a buildup of plaque along the walls of the blood vessels of the heart, impedes the flow of blood and slows the rate of oxygen delivery. If a piece of this plaque breaks off, or if a blood clot formed somewhere in the cardiovascular system flows to a coronary artery, blood can be completely choked off, and massive myocardial necrosis can occur. The same process, if it occurs in the brain, is called a stroke, and a clot in the pulmonary artery, called a pulmonary embolus, can be immediately fatal.

A heart attack is more likely to occur during episodes of high oxygen demand during strenuous exercise. It is also more probable in people who are not conditioned to physical exercise or who have other predisposing factors. Among these are high levels of blood cholesterol, smoking, being overweight, and heredity. The nature of the firefighter profession consists of prolonged periods of relative inactivity interrupted by brief periods of intense activity in dangerous conditions.

In June 2005, the National Fire Protection Association released its report titled *U.S. Firefighter Fatalities Due to Sudden Cardiac Death, 1995–2004.*[12] This report included data collected during a 10-year period of on-duty firefighter deaths related to "sudden cardiac death," an umbrella term that included fatal heart attacks as well as other heart-related deaths. The analysis did not include strokes or aneurysms, though they also involve the cardiovascular system. This study produced the following conclusions:

- During the study there were 1,006 on-duty firefighter fatalities of which 440 (43.7 percent) fell into the category of "sudden cardiac death."
- Autopsies or postmortem information was reported for 308 of the 440 victims of sudden cardiac death.
- Of the 308 firefighters, 134 (43.5 percent) had prior known heart-related conditions. These included previous heart attacks, bypass surgery, or angioplasty/stent placement.
- An additional 97 firefighters had atherosclerotic heart disease defined as arterial occlusion of at least 50 percent.

The NFPA report also identified the type of duty that the firefighters were performing when the heart attacks occurred. The results were as follows:

TYPE OF DUTY	PERCENT OF TOTAL
Fireground operations	35.2%
Responding to/returning from alarms	25.2%
Training	11.4%
Non-fire emergency response	10.7%
Administrative/station duty	10.5%
Other	7.0%

NFPA 1582
■ Standard on Comprehensive Occupational Medical Program for Fire Departments, the purpose of which is to identify the minimum medical requirements to be a firefighter at the NFPA 1001 level.

One of the most disturbing facts from this analysis is the number of firefighters allowed to continue in emergency response when they have known existing heart conditions. **NFPA 1582,** Standard on Comprehensive Occupational Medical Program for Fire Departments, identifies a category of existing medical conditions "that would preclude a person from performing as a member in a training or emergency operational environment by presenting a significant risk to the safety and health of the person or others." Among many others, these conditions include coronary artery disease, history of myocardial infarction, coronary artery bypass surgery, and coronary angioplasty.

The sudden incapacitation of a firefighter while engaged in emergency operations can result in risk of injury or death to other firefighters and civilians. If the provisions of NFPA 1582 were followed, one would expect to find few if any firefighters who had had previous heart attacks or undergone bypass surgery or angioplasty/stent placement on active duty. Prudent management would dictate that these individuals would have been reassigned away from emergency operational environments or retired. However, this is not the case and many fire departments are exposing firefighters to unnecessary risk. This has to stop, even when the firefighter wants to return to duty.

The NFPA study also showed that 93 percent of the firefighters aged 60 and over who were victims of sudden cardiac death were volunteer firefighters, and all but one of the firefighters aged 65 or over was a volunteer. This probably reflects the tendency of volunteer firefighters to remain active well beyond the retirement ages of career firefighters. Again, reasonable limits and medical screening must occur before firefighters are allowed to participate in emergency response.

In March 2007, *The New England Journal of Medicine* contained an article by Stefanos N. Kales et al. titled "Emergency Duties and Deaths from Heart Disease among Firefighters in the United States." One of this study's major conclusions was that fire suppression activity, which represents approximately 1 to 5 percent of the time a firefighter spends on duty, accounted for 32 percent of the deaths from coronary heart disease. The risk of death associated with firefighters from coronary heart disease was approximately 10 to 100 times as high as the risk associated with nonemergency duties. Clearly the cardiovascular demand of fire suppression activities is a major factor. The report also noted that many firefighters lack adequate physical fitness, have underlying cardiovascular risk factors, or have clinical coronary heart disease.

When comparing cardiac-related fatalities of firefighters to other occupations, an increased risk to firefighters is clearly demonstrated. The following chart shows the average percentage of cardiac fatalities from several different occupations.[13]

Cardiac Related Fatalities by Occupation

OCCUPATION	% OF ON-DUTY DEATHS
Firefighters	45%
Police	22%
Overall average all occupations	15%
Construction	11.5%
EMS	11%[8]

NFPA 1500
■ Standard on Fire Department Occupational Safety and Health Program, the purpose of which is to specify the minimum requirements for occupational safety and health as well as safety guidelines for those exposed.

In a 2006 needs assessment of the U.S. fire service, NFPA found an estimated 73 percent of firefighters worked in fire departments that did not have a program to maintain basic firefighter fitness and health, as required in **NFPA 1500**, Standard on Fire Department Occupational Safety and Health Program. In rural communities (under 2,500 population), where nearly all firefighters are volunteers, 88 percent of firefighters did not have access to such programs. If the fire service is going to deal effectively with the issue of heart-related firefighter fatalities, then an investment in the health and safety of those who perform these duties must be made. Physical fitness programs, regular medical physicals and examinations, reduction in body fat composition, and no smoking policies are going to have to be enforced. When this day comes, then major improvement in the health and safety of firefighters will be evident.

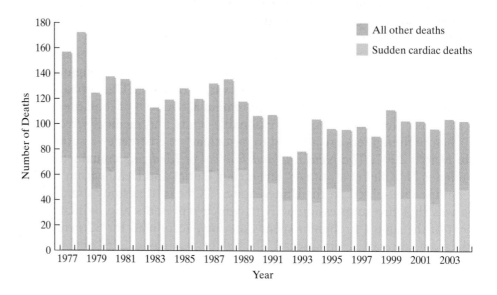

Type of Firefighter Fatalities by Year, 1977–2004. *Source:* NFPA, 2006.

FIREFIGHTERS AND CANCER

Many fire departments and some states have firefighter cancer presumptions within their workers' compensation systems. Recent studies have indicated that firefighters are more prone toward certain types of cancers. The *Journal of Occupational and Environmental Medicine* in November 2006 reported on a study conducted by the University of Cincinnati College of Medicine. In its meta-analysis of 32 previous studies, this study found that firefighters were at greater risk of prostate and testicular cancers, as well as the immune system cancers non-Hodgkin's lymphoma and multiple myeloma. Across the studies, firefighters had double the risk of testicular cancer as men in other occupations, a 50 percent higher risk of both multiple myeloma and non-Hodgkin's lymphoma, and a 28 percent higher risk of prostate cancer. In addition, five other cancers showed possible links to those involved in fire service careers.

The *Journal of Occupational and Environmental Medicine* also reported that firefighters face greater risk than other workers in developing these cancers:

Type and Percentage of Increased Risk

- Testicular cancer (102%)
- Multiple myeloma (53%)
- Non-Hodgkin's lymphoma (51%)
- Skin cancer (39%)
- Brain cancer (32%)

- Prostate cancer (28%)
- Stomach cancer (22%)
- Colon cancer (21%)[14]

There is no question that firefighters are exposed to many potentially cancer-causing chemicals released from burning materials. Even though firefighters wear a breathing apparatus and other protective equipment while fighting fires, they typically remove the gear when merely in the vicinity of the fire. At the scene of the fire, toxic substances such as benzene, lead, uranium, and asbestos can be inhaled or absorbed through the skin, and at the fire station idling apparatus can expose firefighters to diesel exhaust. Additionally, harmful substances can enter through the skin, and as firefighters sweat the pores in their skin open and they are more likely to absorb chemical residue.

Large-scale single events can also have cancer-related consequences in addition to long-term exposures. Since September 11, 2001, 283 World Trade Center (WTC) rescue and recovery workers have been diagnosed with cancer, with 33 of these resulting in fatalities. Many of these cases are blood-cell cancers such as leukemia, lymphoma, Hodgkin's lymphoma, and myeloma. Medical knowledge states that the cancers can strike 3 to 5 years after exposure to toxins such as benzene, a cancer-causing chemical that permeated the WTC site from burning jet fuel. Cancer clusters such as these have been a concern of many fire departments.

The finding of an association between firefighting and specific types of cancer raises significant issues that should encourage additional study as well as improvements in protective clothing. Protection against exposure to carcinogens for firefighters can be difficult. Often the result of the exposure does not manifest itself until years after the event or it is smaller exposures over an extended time frame. Wearing full protective clothing and showering upon return from the emergency are always good practices. Research is ongoing with respect to the causes of cancers in firefighters and in the development of protective equipment to protect from such exposures.

NFPA Health and Safety Standards

The first edition of NFPA 1500, Standard on Fire Department Occupational Safety and Health Program, was developed in 1987 utilizing the NFPA consensus code process via a technical committee chaired by Chief Alan Brunacini of the Phoenix Fire Department. Although some aspects of the standard are debatable, this standard unquestionably led a paradigm shift in the way that fire departments viewed the health and safety of their employees and volunteers. Prior to the development of this document, no standard existed for an occupational safety and health program for the fire service. The purpose of the NFPA 1500 standard is to specify the minimum requirements for an occupational safety and health program for a fire department and to identify safety guidelines for those members involved in rescue, fire suppression, emergency medical services, hazardous materials operations, special operations, and related activities.

FIREFIGHTER DEATHS BY TYPE
OF DUTY, 2007

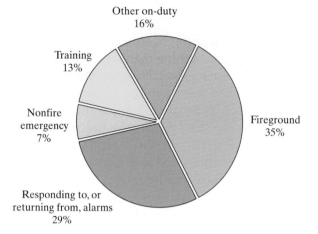

Firefighter Deaths—
Local Career vs. Local
Volunteer, 1977–2007.
Source: NFPA, 2008.

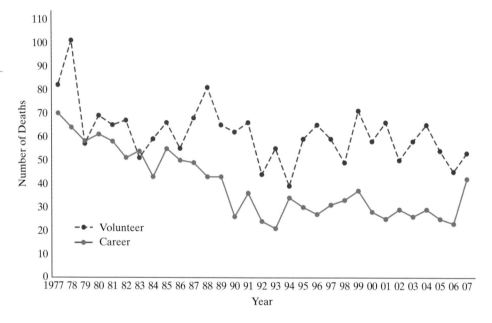

The NFPA 1500 standard is intended to be implemented in a logical sequence, based on a balanced evaluation of economic as well as public safety and personnel safety factors. When this standard is adopted by a jurisdiction, the authority having jurisdiction shall set a date or dates for achieving compliance with the standard and may use a phased-in approach. Most fire departments take a serious approach to the implementation of this standard, but in some cases the financial resources to implement the standard fully are just not available. By using a phased-in approach, the department can budget for safety improvements over a number of years to eventually be in full compliance. This implementation plan must be based on objectively assessed risks, established priorities, and reasonable time frames for compliance.

The implementation of the fire department safety and health plan is an integral part of the overall risk management plan of the community. Through it, the efforts of the fire department are well coordinated with those of other agencies of government, and the priorities within the fire department health and safety plan are evident to all concerned parties. Input to this plan can come from a number of sources, especially the firefighter's union. Unfortunately, in a few jurisdictions the implementation of NFPA 1500 has become a source of contention and dispute between the union and the fire department management. This should not be the case, because the union and fire department management should want the same thing: higher levels of health and safety. Normally, the funding and the timing of the safety improvements are the issues debated. By working together, both union and management have a much better opportunity to achieve the full compliance with the NFPA 1500 standard that everyone desires.

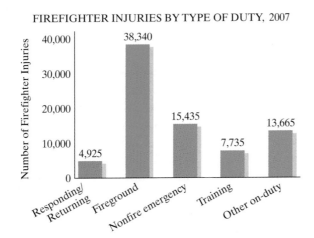

FIREFIGHTER INJURIES BY TYPE OF DUTY, 2007

Firefighter Injuries by Type of Duty, 2007.
Source: NFPA Firefighter Injuries for 2007, 2008.

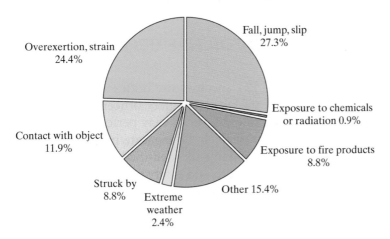

FIREGROUND INJURIES BY CAUSE, 2007

Fireground Injuries by Cause, 2007.
Source: NFPA Firefighter Injuries for 2007, 2008.

The NFPA 1500, Standard on Fire Department Occupational Safety and Health Program, includes the following information and sections:

Administration
Organization
Training and Education
Vehicles, Equipment, and Drivers
Protective Clothing and Protective Equipment
Emergency Operations
Facility Safety
Medical and Physical
Member Assistance and Wellness Program
Critical Incident Stress Program

The elements of the NFPA 1500 standard are extensive and by reference incorporate many other NFPA standards. The program represents a comprehensive approach to the health and safety of fire service personnel. The question is, "Have the safety improvements embodied in the NFPA 1500 standard resulted in fewer firefighter deaths and injuries?" The answer is "Yes," given that firefighter deaths and injuries have been substantially reduced. Since the early 1980s, much progress has been made in reducing the number of firefighter deaths—a decrease of approximately 40 percent in the average number of firefighters deaths that occur each year. In the late 1970s, an average of 151 firefighters died each year. In the 1980s, the number of firefighter deaths dropped to an average of 126, and in the 2000s the average dropped even further to 95 per year. Firefighter injuries have been reduced, but not as significantly as have firefighter deaths. In 1987, there were approximately 102,600 firefighter injuries; in 2007, there were 80,100 injuries.[15] This decrease may not all be directly attributed to the implementation of NFPA 1500, because other factors also have had an impact. However, NFPA 1500 has undoubtedly had a dramatic effect on the health and safety of the occupation and will continue to do so. The changes in protective clothing and equipment, operating procedures, and, most important, attitudes about safety have been remarkable. All firefighters owe a debt of gratitude to those who pioneered this effort and have led the way on this very important issue.

In addition to NFPA 1500, a number of other NFPA standards are directly applicable to improved firefighter health and safety. Several of the most relevant standards regarding safety are identified in the following paragraphs.

NFPA 1521, STANDARD FOR FIRE DEPARTMENT SAFETY OFFICER

The 2007 edition of NFPA 1521 specifies the duties and responsibilities of the health and safety officer and the incident safety officer as separate positions. In smaller fire departments the same individual may perform both functions, and in larger departments they will most likely be separate positions. This standard expands on the safety officer requirements identified in NFPA 1500 and distinguishes the responsibilities of a health and safety officer from those of an incident scene safety officer. In addition, it demonstrates how each of these positions is integrated into the department's risk management program.

The NFPA 1521, Standard for Fire Department Safety Officer, includes the following information and sections:

Administration
Organization
Functions of the Health and Safety Officer
Functions of the Incident Safety Officer

NFPA 1561, STANDARD ON EMERGENCY SERVICES INCIDENT MANAGEMENT SYSTEM

This standard contains the minimum requirements for an incident management system to be used by fire departments to manage all emergency incidents. The standard is intended to be integrated with emergency management systems that apply to multiple agencies and large-scale situations. The 2005 edition of the standard emphasizes that incident management systems are pertinent to much more than managing fires and includes information on firefighter accountability, rapid intervention teams, and interagency cooperation.

The NFPA 1561, Standard on Emergency Services Incident Management System, includes the following information and sections:

Administration
System Structure
System Components
Roles and Responsibilities

NFPA 1581, STANDARD ON FIRE DEPARTMENT INFECTION CONTROL PROGRAM

This standard addresses the infection control issues of fire departments and emergency medical services. These issues are paramount to the safety of firefighters today, given the increasing number of viruses and infections encountered on the emergency scene. The standard is complementary to federal guidelines and regulations applicable to infection control. The 2005 edition includes new information on decontamination and efforts toward a proactive program regarding infectious diseases.

The NFPA 1581, Standard on Fire Department Infection Control Program, includes the following information and sections:

Administration
Program Components
Fire Department Facilities
Emergency Medical Operations Protection
Cleaning, Disinfecting, and Disposal

NFPA 1582, STANDARD ON COMPREHENSIVE OCCUPATIONAL MEDICAL PROGRAM FOR FIRE DEPARTMENTS

NFPA 1582 profiles the medical condition necessary to perform all of the anticipated tasks assigned to a firefighter trained to the NFPA 1001 level safely and effectively. The purpose of the standard is to identify the minimum medical requirements to be a firefighter. The implementation of the medical requirements contained in this standard assist in ensuring that firefighters will be medically capable of performing their required duties, which reduces the risk of occupational injury and illness.

The NFPA 1582, Standard on Comprehensive Occupational Medical Program for Fire Departments, includes the following information and sections:

Administration
Medical Process
Category A and Category B Medical Conditions

For further information on these NFPA safety-related standards, as well as explanations and commentary, please refer to the *Fire Department Occupational Health and Safety Standards Handbook* by Stephen N. Foley. This excellent reference source is published by the National Fire Protection Association.

Safety in the Fire Service Environment

In the early 1970s, when I was a firefighter working in one of the busiest stations in our department, we paid little attention to safety. We liked to say at the time that our job was "fighting fire, saving lives, breathing toxic gases, and dodging falling timbers." Our view of the job was to "get in" on the fire as quickly as possible and perhaps even beat out one of the other companies. Although no one wanted to get burned, a little burn on the ears or the wrists showed that you were really in there. Generally, drivers drove as fast as they wanted, and we used breathing apparatus only when we had to because the last one to put it on was the toughest. One day an order came out saying that we had to use seat belts if the apparatus was equipped with them. We solved this by removing the seat belts from the engine. Our logic was that they would have slowed us down too much getting on and off the apparatus. In hindsight, that more accidents and injuries did not occur seems to have been a matter of luck in the face of such stupidity.

Safety is very much about having a safe attitude. Safety is also about many other things, but to me it primarily means an awareness and appreciation of conducting oneself in a safe manner. As described in the preceding paragraph, it is clear that many firefighters clearly did not have a safe attitude at that time, and no one was enforcing such an awareness. The conditions and attitudes about safety that existed during that time frame in the fire service have since dramatically changed for the better. However, there is still an undercurrent of the "macho mentality" that has existed in the fire service since its origins. Being a firefighter requires a certain toughness; it is not a job for either the physically or the mentally weak. The goals are to be smart and tough and to expose firefighters to dangers only in relation to the consequences and benefits of such actions. This balance is where the proper attitude about safety is so important and is of the most benefit to firefighters and their families.

Source: MFRI Archives.

Any fire department can have excellent equipment and technology, health and safety plans, and the latest protective clothing and apparatus, but until the proper attitude regarding safety is instilled in each and every firefighter, those aspects will not be effective in reducing injuries. They must all be used properly and at the right time to be of value. A safe attitude is about knowledge, control, decision making, organization, maturity, and many other things that cannot be purchased. Effective leaders constantly pay attention to their environment and are aware of circumstances that indicate a need for improved safety. They act to correct safety violations at every opportunity, not just some of the time depending on who is involved. This takes strong leadership, but to ignore safety violations will at some point result in tragedy.

As with most positive organizational attitudes, they need to start at the top. The regard for safety, the value of life and property, and the commitment to be safe should be of great importance to any fire chief. This attitude is then infused throughout the organization, is enforced, and eventually becomes embedded in the culture of that organization. A fire department that has achieved this organizational culture has reached the highest goal that any firefighter or fire officer could possibly attain.

In an article that appeared in the June 2007 edition of *Fire Engineering* magazine, Battalion Chief Mike Adler and Deputy Chief Mat Fratus discussed how to establish a **safety culture** within a fire department. They examined and discussed five elements that must be present in order for the progression to safe behaviors to exist within a fire department. The following five steps are important to establish a climate of consistently safe behaviors, which will result in a culture of safety:

safety culture
■ the elements necessary to establish a climate of consistently safe behaviors, which will result in a culture of safety by the human behavior and attitudes conforming to a respect for increased safety.

1. Duty to act—Identify the capability of your department and the expectations of the community. Determine what functions your department is properly trained, staffed, and equipped to perform.
2. Identify risks associated with your duty—Recognize hazards associated with response in terms of safety before the incident occurs. Begin to make decisions regarding the strategy and tactics deployed in the event of an emergency. Preplans and other similar types of information are important at this step.
3. Establish safety systems—Develop policies and procedures designed to provide safe levels of service to the public with reasonable and prudent risks to firefighters. Provide the necessary training, equipment, and other resources required to engage in emergency operations.
4. Establish a culture of safety—Human behavior and attitudes must conform to the goals of the department with respect to increased safety. Strong and proactive leadership will be necessary to "do the right thing" even if this clashes with the previously established culture. Having the proper procedures and equipment is not enough; the actions of the department must reflect the safety culture at all opportunities.
5. Consistent safe behaviors—The end goal is to have an established pattern of safe actions and culture at all levels of the department. Observation, review of accident and injury reports, and a mechanism to ensure adherence to policy are important steps at this level. A program to reward safe behavior is a very positive motivator. The quality of the leadership of the organization will determine the values that are necessary to maintain the "safety culture."[16]

In general, what causes firefighter deaths is no secret. The firefighter fatality reports since the early 1990s have clearly shown that stress- or heart-related causes and vehicle accidents account for approximately 70 to 75 percent of the firefighter fatalities in any single year. It seems logical and appropriate then that impacting this statistic would require a concentration of efforts toward reducing stress and improving firefighters' health.

The fire service sometimes seems to focus on the latest fad in safety equipment or with other issues that are portrayed as safety concerns, which may act as camouflage for another agenda. If firefighters are going to be safe, they need to concentrate on where the problem is the most pronounced: heart attacks and vehicle accidents. Certainly, improved medical screening and physicals on a regular basis would improve detection of heart disease in time for corrective action. Better health habits in general concerning body fat, diet, and proper exercise would help anyone, and especially firefighters (see following figure and chart). Any good firefighter safety program must include requirements for fitness and conditioning. The deaths from vehicle accidents statistically are more likely to involve volunteers. In fact, in approximately 42 percent of the vehicle accident deaths, the firefighter's own vehicle is involved, particularly where volunteers are allowed to respond in personal vehicles. Control, even in the form of a zero-tolerance policy, may be called for in these circumstances. The point is that about 70 percent of the firefighter deaths occur before the firefighter gets to the incident scene. With heart problems, the attack may occur on the emergency scene, but the reason for the attack occurred well before that moment. It is amazing that many fire departments almost ignore these two basic issues and concentrate their safety programs on occurrences that happen occasionally, even though they too are tragic. Dealing with issues of stress or heart problems and vehicle accidents first will do the most good for a department and its personnel. Then a follow-up with a comprehensive health and safety program works effectively to address many of the other issues and hazards that firefighters face.

Fire Department Vehicle Collisions and Resulting Firefighter Injuries While Responding to or Returning from Incidents, 2000–2007

YEAR	INVOLVING FIRE DEPARTMENT EMERGENCY VEHICLES		INVOLVING FIREFIGHTERS' PERSONAL VEHICLES	
	COLLISIONS	FIREFIGHTER INJURIES	COLLISIONS	FIREFIGHTER INJURIES
2000	15,300	990	1,160	170
2001	14,900	960	1,325	140
2002	15,550	1,040	1,030	210
2003	15,900	850	980	85
2004	15,420	980	1,150	220
2005	15,885	1,120	1,080	125
2006	16,020	1,250	1,070	210
2007	14,650	915	665	120

Source: NFPA Survey of Fire Departments for U.S. Fire Experience, 1990–2007.

Almost every occupation involves a degree of risk regarding the potential for injury, illness, or death. However, only a few professions are characterized by a willingness to accept a high degree of exposure to injury and death. The fire department has a responsibility to manage the level of risk to which its members are exposed in the performance of their duties. Many of these responsibilities are specified in OSHA safety standards, Environmental Protection Agency (EPA) regulations, and state and local requirements regarding safe work practices. The fire service has been relatively free from mandatory regulations as compared to the private sector, but this freedom will not be the case in the future. In fact, the OSHA Respiratory Protection Standard released in 1998 includes specific procedures for fire departments to adhere to at the scene of emergencies. Regardless of federal regulations, the legal system seems even more predisposed to challenge the decisions and actions that firefighters and fire officers make regarding the safety of their personnel.

The fire department and other public safety agencies are unique in that they cannot control the hazards they may face in their work environment. They must accept the situations as they find them when they respond to emergencies. However, fire departments can control their actions. Many of the risks that firefighters face at emergency incidents are predictable to an extent and can be overcome by adequate training, preplanning, experience, and by numerous other factors that go into making decisions under emergency conditions. The fire department does not always control the situation, but it always controls how its members react to it.

Source: MFRI Archives.

risk control measures
■ their proper application reduces risks to an acceptable level or fosters recognition that the department cannot handle the event safely and will revert to a defensive strategy. The major categories are administrative controls, engineering controls, and personal protection controls.

In the fire and rescue service, it is not possible to eliminate all of the risks. However, risks may be minimized by the application of **risk control measures.** Three general categories of risk control measures are administrative controls, engineering controls, and personal protection controls.[17] The proper application of these risk control strategies will reduce the risks to an acceptable level or foster recognition that the department is not staffed and equipped to handle the situation safely and revert to a defensive strategy.

ADMINISTRATIVE CONTROLS

Administrative controls are the ways in which one controls the behavior of people within the organization. This type of restraint is very important and strikes at the heart of the safety attitude discussed earlier in this chapter. Administrative controls can be training and certification requirements, emergency incident standard operating procedures (SOPs), general orders, driving regulations, prefire planning, and similar activities. They provide the general guidance and requirements for how firefighters act and react in certain situations. The control is in behavior and safe conduct. It reflects the professionalism and discipline within the department.

In order to be effective, the administrative controls must be consistently applied and enforced at all times. A policy or procedure that is not enforced is worthless and damaging to the desired safety attitude. Managers may find that their best opportunity to improve the health and safety within the organization resides within the effective use of administrative controls, because they are internal to the fire department and not something that has to depend on someone else.

ENGINEERING CONTROLS

Engineering controls are designed to remove or abate specific hazards in the work environment. For example, aerial platforms are engineered to provide better stability than ladders at various angles. Fire departments can govern the speed of fire apparatus by adjusting the engine, the transmission, or the differential. Computer-assisted personnel accountability systems currently help to track firefighters in emergencies. It is possible to provide a healthier environment in the fire station by having adequate ventilation and vehicle exhaust controls. These and any more examples of engineering controls demonstrate the importance of a safety program. Also, should the administrative controls fail, the engineering controls may still work. For example, no matter how fast the firefighter may want to drive to an emergency, the engine governor restricts the speed to a preset limit.

PERSONAL PROTECTION CONTROLS

Personal protection controls do not remove or abate the hazards that firefighters may face in the workplace. Instead, they provide a level of personal protection to a person so that he or she can resist the hazard and be protected up to certain limits. Many types of personal protection controls are used in the fire and rescue service, such as Level-A haz-mat suits, SCBA, PASS devices, and PPE. Firefighters must be well trained in the proper use and limitations of this equipment. When they are operating in the hazard zone, all that lies between them and the dangers of the environment may be that equipment. As such, the protective clothing and equipment must meet high standards and be regularly inspected to ensure that it can function properly when needed.

In order to have an effective health and safety program, all three types of controls must be present. Preferably, the administrative and engineering controls should be used first to limit and reduce the exposure before relying on the personal protection controls to protect personnel.

IAFF/IAFC JOINT LABOR MANAGEMENT WELLNESS/FITNESS INITIATIVE

In some jurisdictions the implementation of NFPA 1500 and other safety issues are the subject of collective bargaining agreements. At times, due to the nature of collective bargaining, these safety concerns may be used as bargaining issues within a process that is often adversarial. Sometimes the safety issues are forgotten in the power struggles between labor and management, which is unfortunate for a number of reasons. First, debate about reasonable safety issues should be minimal because both management and labor should want the same thing—improved safety. The difficulty usually surfaces regarding the expense of the safety-related changes and how existing personnel are treated in this process. The best opportunity for a fire department to have the resources and the ability to make safety improvements happens when the union and the management of the fire department work together. Therein lies the advantage of the *Fire Service Joint Labor Management Wellness/Fitness Initiative*.

In 1997, the International Association of Fire Fighters (IAFF) and the International Association of Fire Chiefs (IAFC) came together in a partnership to develop a wellness program that addressed the needs of the total individual in terms of the physical, mental, and emotional aspects of being a firefighter. Fire chiefs and the local union presidents from ten progressive fire departments, in conjunction with medical professionals and fitness experts, developed this program. The ultimate goal of this effort is to improve the quality of life of firefighters by investing in a wellness program designed specifically for their occupation.

The intent of the Fire Service Joint Labor Management Wellness/Fitness Initiative is to be a positive individualized program that is not punitive in nature. The initiative is designed to be a complete, long-term program that promotes good health in a positive way; it is not meant to be implemented in a piecemeal fashion. The program requires a sustained commitment from the union and the management of the department working together in the best interests of the firefighting personnel. In this way, the cooperation and resources needed for the program are more likely to be available, giving it the best chance for success. The May 2007 issue of *NIOSH Alert* noted that in the 131 investigations in fire departments with firefighter fatalities 39 percent had voluntary fitness programs and 8 percent had mandatory fitness programs. Unfortunately, this also means that 53 percent of the fire departments had no fitness programs. These numbers need to be improved to have the maximum beneficial effect on the health of firefighters, and mandatory programs will be necessary to accomplish this goal.

The Fire Service Joint Labor Management Wellness/Fitness Initiative consists of five major components, which are each described in detail in the program[18]:

Medical
Fitness

Medical/fitness/injury rehabilitation
Behavioral health
Data collection and reporting

The Fire Service Joint Labor Management Wellness/Fitness Initiative represents a comprehensive program specifically designed for fire and rescue personnel. It has a number of positive benefits for fire service personnel associated with it. First, it is well done and includes the work of a number of medical and fitness professionals. The program reflects the most modern thinking in this area. Second, it is a partnership of union and management designed to increase the buy-in and participation of the program. This total fire department support will be important in the attempt to acquire the resources necessary to implement the program. Third, its holistic focus considers the complete health and wellness needs of firefighters. This IAFF/IAFC program has an implementation guide including a video to introduce the program. This program represents real advancement in the area of health and safety for firefighters and should be highly considered if one is interested in progressing in this endeavor.

CANDIDATE PHYSICAL ABILITY TEST

One method to ensure that firefighter safety is improved is by ensuring that new firefighters who enter the department have the proper physical condition to withstand the rigors of firefighting activity. The Candidate Physical Ability Test (CPAT) is a validated physical ability test for preemployment testing of firefighter candidates resulting in a consistent test for hiring of firefighters. This program is a joint effort of the IAFF/IAFC Fire Service Joint Labor Management Wellness-Fitness Task Force. This task force worked on developing a holistic, positive, rehabilitating, and educational approach to wellness and fitness programs in the fire service. Since the 1999 release of the Candidate Physical Ability Test (CPAT) program, many jurisdictions have successfully implemented this valuable program.

The CPAT goal was to develop a fair and valid evaluation tool in the selection of firefighters to ensure that all firefighter candidates possess the physical ability to complete critical tasks effectively and safely. This CPAT program covers every aspect of administering the CPAT: from recruiting and mentoring programs, to providing recruits with fitness guidance to help prepare them for the CPAT, to setting up and administering the test. The entire validation process is discussed in detail, as well as the legal issues that departments might face when implementing the actual program.

Each of the original jurisdictions to implement CPAT has also served as a resource and has assisted fire departments in its region with the adoption of the program. The CPAT is a licensed program, and under the conditions of the license, jurisdictions must use the CPAT in whole and only for the purpose of candidate testing. This consists of, but is not necessarily limited to, recruiting, mentoring, preparatory programs, orientation programs, and proper program administration including using specified equipment, test parameters, qualified proctors, and test personnel. Any attempt by the licensee to use only a portion of the CPAT or to use the CPAT for testing incumbents is inconsistent with the scope of this license and is therefore prohibited by the policy of the IAFF and IAFC.

PEER FITNESS TRAINER CERTIFICATION

The IAFF/IAFC Task Force, which has worked to implement the Wellness/Fitness Initiative and the CPAT successfully, identified that there must be a firefighter in each department who can take the lead for physical fitness. This individual must have the ability to design and implement fitness programs, to improve the wellness and fitness of his or her department, and to assist with the physical training of recruits. This need for a department-level leader led to the development of the Fire Service Peer Fitness Trainer (PFT) Certification Program. Peer fitness trainers are vital in helping the IAFF/IAFC Task Force accomplish its fundamental mission of improving firefighter health, wellness, fitness, safety, and performance.

After careful consideration of the major personal trainer certifying agencies, the American Council on Exercise (ACE) was chosen to collaborate in the development and implementation of the PFT Certification Program for firefighters. ACE is a nonprofit organization dedicated to promoting the benefits of physical activity and protecting consumers against unsafe and ineffective fitness products and instruction. ACE is a standard-setting body for fitness professionals and the largest nonprofit fitness certification and education provider in the world.

The job of a fire service peer fitness trainer was designed so that peer fitness trainers would encourage safety and participation in fitness, through guidance and supervision of firefighter exercise programs. The Peer Fitness Trainer Certification Program supports the IAFF/IAFC Wellness/Fitness Initiative and CPAT programs through an integrated, multilevel approach by providing current wellness and fitness information that addresses the specific needs of firefighters through individual fitness assessment, individual program design, program implementation, and administration.

The purpose of the Peer Fitness Trainer Certification Program is to identify firefighters who have demonstrated the knowledge and skills required to

- Design and implement fitness programs
- Improve the wellness and fitness of their departments
- Assist in the physical training of recruits
- Assist the broader community in achieving wellness and fitness

The Peer Fitness Trainer program has proved to be very successful because using firefighters as the fitness trainers seems to have a beneficial and motivating effect on other firefighters. Any fire department that is serious about the safety and fitness of its firefighters should strongly consider implementing this program.

NATIONAL FIRE FIGHTER NEAR-MISS REPORTING SYSTEM

In the fire service, a *near miss* is defined as an unintentional, unsafe act that could have resulted in an injury, fatality, or property damage. Approximately 1 million of these near misses occur each year. By collecting and analyzing information on near-miss events, improvements can be made in firefighter education, operations, and training. Modeled after the Aviation Safety Reporting System, the near-miss reporting system is beneficial in identifying safety patterns of both systemic and individual origin. The goal of this voluntary, confidential, and nonpunitive reporting system is to improve firefighter safety by collecting, sharing, and analyzing near-miss experiences that can assist in developing strategies to reduce the number of firefighter injuries and fatalities.

The National Fire Fighter Near-Miss Reporting System, administered by the International Association of Fire Chiefs (IAFC), is a voluntary, confidential, non-punitive, and secure reporting system with the goal of improving firefighter safety. The project is funded by grants from the Department of Homeland Security's Assistance to Firefighters Grant Program and Fireman's Fund Insurance Company.

The mission of the National Fire Fighter Near-Miss Reporting System is to provide data and case studies for efforts related to preventing firefighter injuries and fatalities by collecting, sharing, and analyzing near-miss reports to support improving safety in the fire and emergency service. The reports and the data analysis of the system are improving the knowledge and experience in a positive way for firefighters and fire officers. The mistakes and near misses of others serve as powerful examples and provide lessons backed by real-world experience. These reports should be a strong component of fire service training programs.

The National Fire Fighter Near-Miss Reporting System's annual report for 2007 identified the most frequently occurring contributing factors in the reporting system as

- Situational awareness
- Human error
- Decision making

It is interesting to note that these factors are the strengths of the crew resource management system that many fire departments are now using to better manage emergency situations and to improve safety. Firefighters trained in the concepts of Crew Resource Management (CRM) learn how to deal with minimizing the effect of human error and maximizing human performance. CRM acknowledges that safety equipment and technology are very important, but in the end analysis it is the people who use this equipment that are the most significant factor. Firefighters trained in CRM develop skills that enhance communication, maintain situational awareness, improve decision making, and advance teamwork at all levels of the organization.

Through its accumulation of data, the Fire Fighter Near-Miss Reporting System provides great insight into methods and procedures to increase safety. The basic concept is "those who fail to understand the past are doomed to repeat the same mistakes in the future." The fire service has to understand the root causes of safety issues and then diligently work to make improvements. The system used to provide analysis was copied from the U.S. Navy Human Factors Analysis and Classification System, which provides a good framework in which to classify the reports. This system has the following four categories:

Level 1: Unsafe Acts. This level contains two categories: errors and violations. Errors are the result of a lack of skill, education, or training; poor decision making; and the like. The most recent report indicates that 80 percent of all near misses are the result of errors or mistakes that occur within the existing procedures. This would seem to indicate that training needs to be increased within fire departments. Violations are willful disregard for rules and regulations and occur less frequently.

Level 2: Preconditions to Unsafe Acts.   This level refers to substandard conditions or practices of the individual involved. It intends to determine how ready and prepared the firefighter was in relationship to the event. Adverse mental states due to, among other things, loss of situational awareness, channelized attention, distraction, and misplaced motivation are all contributing factors at this level.

Level 3: Unsafe Supervision.   This level of evaluation seeks to determine the role of the supervisor in all aspects of the operation. Inadequate supervision and planning, failure to recognize or correct safety issues, and supervisory violations are included in this category. The leading entries in this category are lack of guidance and failure to correct unsafe acts.

Level 4: Organizational Influences.   The management of the organization has a definite impact on the ability of firefighters to operate in a safe and prudent manner. This could be due to resource issues, organization culture and climate, as well as management ability. The major factors at this level are inadequate training, poor chain of command, recklessly aggressive culture, and ineffective risk management plans.[19]

In my opinion, the Fire Fighter Near-Miss Reporting System is an excellent tool that all fire departments should use. A host of information on its Web site (http://www.firefighternearmiss.com) includes analysis and statistics, drills and training material, best practices, and much more. A new resource recently added is *The 2007 Near-Miss Anthology,* which contains a collection of report synopses for easy review. I strongly encourage study of this material on the Web site.

Employee Assistance Programs

A member or employee assistance program (EAP) is a comprehensive approach that many organizations have taken to deal with the numerous employee difficulties and challenges that may arise on or off of the job. EAPs are designed to deal with issues such as stress, emotional or mental health matters, alcohol and substance abuse, marital and family difficulties, job performance concerns, financial troubles, and a host of other obstacles that may affect performance while on or off the job. The purpose of EAPs is to assist employees who are having difficulties to cope with the underlying issues of the problems and to help them remain productive members of the organization. These programs are especially important to fire departments due to the stress and the traumatic circumstances to which firefighters often find themselves exposed.

Alcoholism and drug abuse are two of the major issues that employee assistance programs deal with. The costs to an organization in terms of lower performance and productivity, accidents, theft, and other related concerns each year are staggering. According to the National Council on Alcoholism and Drug Dependence, the effects of alcoholism and drug abuse have the following impacts:

Alcoholism

Approximately 18 million Americans have a serious drinking problem.
Annual deaths due to alcohol number about 110,000.
Of all hospitalized patients, about 25 percent have alcohol-related problems.

Alcohol is involved in 47 percent of industrial accidents.

Approximately half of all vehicle fatalities involve alcohol.

Drug Abuse

Lateness occurs 3 times as often.

Early dismissal requests are 2.2 times as often.

Absences of 8 days or more occur 2.5 times as often.

Sick benefits are 3 times the normal levels.

Workers' compensation claims are 5 times as likely.

Accident involvement is 3.6 times as often.

Production is one-third less than that of fellow workers.

Overlay these findings with the critical nature of firefighting and the dependence among the entire crew on one another, and the problem is of major concern. If fire service personnel managers truly believe that people are their greatest organizational resource, then they must protect and assist this resource whenever they face unresolved challenges. Fire department EAPs need to be comprehensive programs designed to deal with a number of issues and include health care professionals who are familiar with emergency services work. Employee assistance programs have been proven valuable in having a place where employees can turn in crisis and receive assistance designed to maintain their value in the work environment. These programs have reduced costs and improved productivity in a number of areas, such as the following:

- Reducing absenteeism and tardiness
- Lowering employee turnover
- Reducing accidents and injuries
- Lowering health care claims
- Reducing employee grievances

EAPs are cost-effective efforts to retain experienced and valuable employees who may otherwise have to be discharged. These programs also demonstrate a clear commitment by the organization regarding the support it provides to employees. The purpose of the programs is to provide counseling and rehabilitative services to get the firefighter back to full duty as soon as possible. EAPs are somewhat limited in what they can do; they cannot assist everyone, especially those who may resist treatment. The work environment of a firefighter involves dangerous circumstances, and the reliance on each member of the team is important to citizens and to other firefighters. Firefighters influenced by drugs or other illegal substances cannot be permitted to work until these issues are resolved. Fire departments must have programs designed to detect employee use or abuse of substances, and these detection programs should be separate from the EAP program.

The goal of EAPs is to get employees who may need these services to use them voluntarily. The organization can do several things to ensure that employees recognize the benefits of the employee assistance system within the fire department. Confidentiality is one of the most important aspects in encouraging employees to avail themselves of the program. If the employees do not have a high degree of confidence in the confidentiality of the program, then they will not use it for fear of reprisals or damage to their careers. EAP services should be conducted by medical professionals who are trained and certified to provide this type of care. These medical professionals need to have a background in post-traumatic stress and similar issues common to emergency service personnel. Legal requirements affect the administration of EAP programs and the confidentiality of clients, which should be respected at all times. Supervisors need training in how to refer employees to the EAP, and employees need to know how they can self-refer to the EAP. Some initial financial support from the fire department to provide for counseling and other services is also helpful. At a certain point, the employee's health benefits must bear the costs of treatment. Clear guidance and limits are also required to indicate when EAP services will be allowed and in what circumstances the employee faces disciplinary action with no availability of EAP services.

Without question, EAPs have been valuable in quite literally saving the careers of many firefighters and officers. The services and humanitarian values of these programs have allowed personnel to continue to contribute in a positive way toward the organization even when faced with personal difficulties. I have witnessed the results of these programs and been amazed at the complete recovery and turnaround that one can make in one's personal life and career. In addition, the effect on the employee's family life has been remarkable. In fact, many EAP programs also allow immediate family members to use the program, because major unresolved family issues so often affect employee work performance. EAPs are a valuable asset to the personnel management of any fire department.

In addition to EAPs, many organizations have instituted "wellness programs" designed to maintain or improve employee health before the problem arises. These programs encourage self-directed lifestyle changes. Early wellness programs were aimed primarily at reducing the cost and risk of disease through early detection and screening programs. More current programs emphasize healthy lifestyles and environment, including reducing cholesterol, reducing blood pressure, and other such health-conscious issues. Organization-sponsored support groups have been established for individuals dealing with health issues such as weight loss, nutrition, and smoking cessation. These programs have been effective, and cost–benefit analyses by organizations tend to support the continuation of the wellness program efforts. In addition, the programs send a strong message that the organization really cares about the health and well-being of its employees.

JOB STRESS

Firefighters are routinely exposed to events that create stress, both on the job and in their private lives. Emergency response is not an easy endeavor, and firefighter-related stress can be elevated by repeated emergency response, limited control over the work environment, and a high level of uncertainty due to the very nature of emergency service. Workplace stress is the harmful physical and emotional responses that can happen when there is a conflict between job demands on the employee and the amount of control an employee has over meeting these demands. In general, the combination of high demands in a job and a low amount of control over the situation can lead to stress.

Stress in the work environment can have many origins or be caused by a single event. Many different signs and symptoms indicate when employees are having difficulty coping with the amount of stress they are experiencing:

Physical. Headaches, grinding teeth, clenched jaws, chest pain, shortness of breath, pounding heart, high blood pressure, muscle aches, indigestion, constipation or diarrhea, increased perspiration, fatigue, insomnia, frequent illness

Psychosocial. Anxiety, irritability, sadness, defensiveness, anger, mood swings, hypersensitivity, apathy, depression, slowed thinking or racing thoughts; feelings of helplessness, hopelessness, or of being trapped

Behavioral. Overeating or loss of appetite, impatience, quickness to argue, procrastination, increased use of alcohol or drugs, increased smoking, withdrawal or isolation from others, neglect of responsibility, poor job performance, poor personal hygiene, change in religious practices, change in close family relationships

Some stress is normal. In fact, it is often what provides us with the energy and motivation to meet our daily challenges both at home and at work. Stress in these situations helps one rise to a challenge and meet goals such as deadlines. However, as with most things, too much stress can have negative impacts. As a supervisor it is beneficial to be able to understand the symptoms of stress, so that one is in a better position to obtain help and assist one's employees. Some of the relevant behavior changes to look for follow:

- *Instincts*—Do not disregard your instincts as they are often accurate indicators, but be objective and do not react to one event.
- *Behavior*—Sudden changes in behavior that affect job performance in a negative manner can be a sign of emotional distress. This could be excessive arguing with coworkers, unwarranted absences, isolation from coworkers, and so on.
- *Life interference*—Behavior interferes with job performance and ability to handle basic human needs, such as eating, hygiene, and so forth.
- *Frequency*—Uncharacteristic repeated behavior by employee.
- *Emotionally difficult situation*—Events that may precipitate behavior change due to traumatic situations.
- *Department disturbance*—Major rule violation and conflict within the department driven by anger.[20]

By *being aware of these criteria*, you will be in a better position to assist personnel under your supervision by directing them to the proper EAP services or other programs to provide them the opportunity to deal effectively with the stress issues that are adversely affecting their behavior.

POST-TRAUMATIC STRESS DISORDER

post-traumatic stress disorder
■ a debilitating condition that affects people who have been exposed to a major traumatic event.

Post-traumatic stress disorder (PTSD) is a debilitating condition that affects people who have been exposed to a major traumatic event. PTSD is characterized by upsetting memories or thoughts of the ordeal, "blunting" of emotions, increased arousal, and sometimes severe personality changes. Specific causes for the onset of post-traumatic stress disorder are not clearly defined, although experts suspect it may be influenced by the severity of the event, by the person's personality and genetic makeup, and by whether or not the event was expected. Firefighters, emergency response personnel, and those directly involved in the event or families who have lost loved ones in the traumatic event are most like to experience PTSD.

Most of the normal and regular stress events that firefighters encounter diminish over time and do not require treatment. However, due to the culture of firefighters and the desire not to appear weak in any way, often when treatment is necessary it is avoided. Obviously, the issue that is causing the behavior is not addressed, and the firefighter continues to be adversely affected. Consultation with a mental health professional for diagnosis and a plan of treatment is always advised. Many of the responses to trauma, such as shock, terror, irritability, blame, guilt, grief, sadness, emotional numbing, and feelings of helplessness, are natural reactions. For most people, resilience is an overriding factor and trauma effects diminish within 6 to 16 months. It is when these responses continue or become debilitating that PTSD is often diagnosed. The fourth edition of the *Diagnostic and Statistical Manual of Mental Disorders* outlined three forms of the disorder:

- Acute: onset within 6 months of the event and lasting less than 6 months
- Chronic: symptoms lasting 6 months or more
- Delayed: onset at least 6 months after the event

With appropriate medication, emotional support, and counseling, most people show significant improvement. Behavior therapies can help reduce negative thought patterns and other issues common to PTSD. The firefighter in treatment typically moves back and forth through three recovery phases:

- *Phase One*—The elimination and/or management of dangerous behaviors and/or relationships. Becoming less fearful of thoughts, feelings, and dissociative episodes
- *Phase Two*—Resolution of traumatic memory processing. Developing a narrative account of the trauma without becoming re-traumatized
- *Phase Three*—Personality reintegration and rehabilitation[21]

Successful treatment depends in part on whether or not the trauma was unexpected, the severity of the trauma, if the trauma was chronic (such as for victims of sexual abuse), and the individual's inherent personality and genetic makeup. However, prolonged exposure to severe trauma such as experienced by victims of prolonged physical or sexual abuse may cause permanent psychological scars. Fire departments need to protect their employees from the damaging realities of PTSD and have a mechanism within the department to recognize and provide treatment for such individuals.

Violence in the Workplace

Violence in the workplace can have devastating effects on organizations and on the quality of life of employees. It is becoming increasingly important for employers to take into account the possibility of violent acts occurring and to take action to minimize their effect. Recent data from NIOSH indicate that people at work are increasingly exposed to lethal violence. Homicide accounts for 17 percent of all deaths in the workplace and is the leading cause of death for women in the workplace.[22] The fire service is well aware of this trend because it responds to these types of events almost every day. An increasing amount of violence is also directed toward firefighters and EMS personnel as they perform their jobs.

Additionally, we witnessed the horror of four fire officers murdered at a fire station in Jackson, Mississippi, by a disgruntled firefighter in 1996.

The majority of incidents of workplace violence are not fatal assaults but everyday occurrences of physical violence, verbal threats, and forms of harassment. The fire department must protect employees from violence as well as from the damage to morale, health, and productivity that the fear of violence produces. One of the first steps to reduce workplace violence is to recognize some of the risk factors involved. For violence and other traumatic situations to develop, a number of conditions generally have to exist in the workplace, which can be categorized as follows:

Individual Characteristics. Personality disorders and paranoid thinking; drug or alcohol problems; major life stress such as divorce, illness, or death; helplessness; and isolation

Precipitating Events or Conditions. Termination, job change, harassment by coworkers or supervisors, unresolved grievances, and so forth

System Characteristics. At-risk organizations such as those handling large sums of money or working in the public, or organizations that tend to punish and expel workers and poorly manage situations such as downsizing, terminations, poor communications, and so forth[23]

Individual warning signs that an employee may be prone to violent acts include behaviors, verbal threats, and physical actions.

BEHAVIORS

Be alert for changes in the employee with elements such as irritability and short temper. Notice verbal and nonverbal reactions by individuals that may indicate anger or hostility. Listen and pay attention to the employee's words, actions, and unspoken messages.

VERBAL THREATS

Take seriously an employee's remarks about what he or she may do. Experts indicate that people who make such statements usually have mentally committed to the act for a considerable period of time. It may take very little provocation to trigger the violence.

PHYSICAL ACTIONS

Employees who demonstrate violent physical actions at work are dangerous. The organization must investigate with trained experts and intervene. This may involve security personnel or local law enforcement. Failure to do so may be interpreted as permission to conduct additional or more serious physical actions. No action of this type should ever go unaddressed in the work environment.

Many organizations may assume they are free from the risks of violence, but these acts can occur anywhere and no one is totally safe. It is important that employers consider the possibility of these acts and take action to minimize the hazards involved. They can use a number of strategies to prevent or reduce the occurrence and severity of violence in the workplace.

1. Employers who conduct effective background checks can often identify people who have been prone to violence in the past.
2. Periodic review of security policies and procedures helps minimize the organization's vulnerability to violence and other forms of crime.
3. Training of employees and managers in how to handle confrontational personnel actions, such as termination and discipline, can be helpful.
4. A method whereby employees may alert others in the workplace of a potentially dangerous situation is in place.
5. The organization should have a clear, well-communicated policy on how to handle violence and threatening behavior in the workplace.
6. A zero-tolerance policy should be diligently enforced in which all threats are taken seriously, with reports of violence and threats immediately addressed and resolved.
7. Employee assistance programs must be available to help troubled employees and address concerns before violence or threats arise.
8. All employees should know how to report violent acts or threats of violence without fear of criticism or reprisal, with all incidents documented.

Although no definitive strategy will ever be appropriate for all workplaces, changing the way work is done in certain settings can minimize the risk of workplace violence. When a violent workplace incident occurs, a plan to manage the aftermath of the incident is essential. Employees need assistance with the psychological consequences of workplace violence. A debriefing after a serious incidence of violence and then any necessary follow-up care are necessary to minimize further damage from the incident. Members of EAPs and critical incident debriefing teams need to be trained in advance for these situations. To prevent further incidents and to show support of the victims, employers should encourage the prosecution of offenders, which may mean the accommodation of employees to make court appearances and other actions necessary to assist in the prosecution of these acts.

Workplace security is fostered by a cooperative working relationship between the organization and its employees. By taking responsibility for policies that abate workplace violence, the organization can demonstrate that it considers these offenses to be serious and that it has plans in place to deal with an event if it occurs. This advanced planning sends a strong message throughout the organization. It provides guidance in more effective monitoring and reducing dangerous or unacceptable behavior and is able to intervene at an early stage to prevent workplace violence. To be successful, the fire department must change the way it thinks about workplace violence by shifting the emphasis from reactionary approaches to prevention, and by embracing workplace violence as an occupational safety and health issue.

Summary

Ensuring the physical and mental well-being of employees in the workplace is the primary goal of the health and safety management process. In today's world, and particularly in the fire and rescue service, such complex activity requires the expertise of specialists from many disciplines. Safety as well as health and wellness programs in the fire service are designed to prepare firefighters so that their mental, physical, and emotional capabilities are strong enough to withstand the hazards and stresses found in their work environment and in life in general. The federal government was compelled to act to improve occupational health and safety, and in 1970 Congress enacted OSHA. OSHA regulations affect virtually all places of employment in one way or another and have contributed many safety improvements.

As firefighter death and injury statistics attest, the occupation of fighting fires and responding to emergencies is dangerous. The good news is that the numbers of firefighter deaths and injuries are declining as the result of increased emphasis on and attention to firefighter health and safety issues, improved equipment, and a safer environment due to better building codes and sprinklered facilities. Standards enacted within the fire service, such as NFPA 1500, have greatly reduced the number of firefighter injuries and deaths. Joint labor and management efforts, employee assistance programs, and the response to violence in the workplace have all contributed to a safer work environment for firefighters.

Review Questions

Use these questions to review the material in this chapter and for discussion purposes.

1. What is the concept behind workers' compensation laws, and how do they work in the fire service environment?
2. Explain the reasons for the Occupational Health and Safety Act of 1970. Describe the organizational structure and authority of OSHA, and discuss how they relate to fire service safety programs.
3. Review the causes of firefighter injuries and deaths since the early 1980s. What programs would you recommend to reduce these deaths and injuries? If your fire department has implemented some of these programs, how effective have they been?
4. What is NFPA 1500? Explain its purpose and impact on safety in the fire and rescue service.
5. Identify and discuss three types of control strategies that can be used to reduce firefighter deaths and injuries.

Case Study

You are the new fire chief of a department that consists of 10 fire stations and 170 personnel. You have been selected by the city manager and are from outside of the department. Prior to your arrival as chief, the department has been criticized in the community and news media for the high number of vehicle accidents that have occurred during emergency responses. You note within your first few weeks with this department that the organizational culture tends to ignore the safety regulations that are in place. The rules are there, but no one really abides by them, and few consequences exist for noncompliance. You are new to this position and want to be respected and liked by the members of this department.

How would you approach this situation as fire chief? What policies and structure would you put in place? How would you handle challenges from firefighters who may not like the new safety policies?

Fire Service Professional Qualifications

This chapter contains information that will assist fire officers in meeting the job performance requirements (JPRs) of NFPA 1021, Standard for Fire Officer Professional Qualifications:

4.2 Human Resource Management—Fire Officer I

4.2.4 Recommend action for member-related problems, given a member with a situation requiring assistance and the member assistance policies and procedures, so that the situation is identified and the actions taken are within the established policies and procedures.

4.2.5 Apply human resource policies and procedures given an administrative situation requiring action, so the policies and procedures are followed.

4.7.1 Apply safety regulations at the unit level, given safety policies and procedures, so that required reports are completed, in-service training is conducted, and member responsibilities are conveyed.

5.2 Human Resource Management—Fire Officer II

5.7.1 Analyze a member's accident, injury, or health exposure history, given a case study, so that a report including action taken and recommendations made is prepared for a supervisor.

6.2 Human Resource Management—Fire Officer III

6.7.1 Develop a measurable accident and injury prevention program, given specific data, so that the results are evaluated to determine effectiveness.

7.2 Human Resource Management—Fire Officer IV

7.2.4 Appraise a member-assistance program, given data, to determine if the program when used produces the desired results and benefits.

7.7.1 Maintain, develop, and provide leadership for a risk management program, given specific data, so that injuries and property damage accidents are reduced.

Endnotes

1. Douglas T. Hall and James G. Goodale, *Human resource management* (Glenview, IL: Scott Foresman, 1986).
2. Arthur Larson, *Workers compensation law* (New York: Matthew Bender & Company, 1982).
3. Wayne F. Cascio, *Managing human resources,* 7th ed. (New York: McGraw-Hill, 2006).
4. Hall and Goodale, *Human resource management.*
5. Robert Pater, Safety leadership cuts costs, *HR Magazine, 35,* 1990.
6. U.S. Department of Labor, Bureau of Labor Statistics, *national census of fatal occupational injuries in 2007.* (Washington, DC: U.S. Government Printing Office, November 2008).
7. U.S. Department of Labor, *All about OSHA* (Washington, DC: U.S. Government Printing Office, 2006).

8. U.S. Department of Labor, *OSHA facts*, retrieved October 23, 2007, from OSHA Web page, http://www.osha.gov

9. Rita F. Fahy, Paul R. LeBlanc, and Joseph L. Molis, Firefighter fatalities studies, 1997–2006, *NFPA Journal*, July/August 2007.

10. Michael J. Karter and Joseph L. Molis, Firefighter injuries for 2007, *NFPA Journal*, November/December 2008.

11. Ibid.

12. Ibid.

13. Stefanos N. Kales et al., Emergency duties and deaths from heart disease among firefighters in the United States, *The New England Journal of Medicine*, March 22, 2007.

14. Grace K. LeMasters et al., Cancer risk among firefighters: A review and meta-analysis of 32 studies, *Journal of Occupational and Environmental Medicine*, November 2006.

15. Arthur E. Washburn et al., *2006 fire fighter fatalities* (Quincy, MA: National Fire Protection Association, 2007).

16. Paul J. Antonellis, Jr., et al., Understanding post traumatic stress disorder, *Fire Engineering*, December 2006.

17. U.S. Fire Administration, *Risk management practices in the fire service* (Washington, DC: Federal Emergency Management Agency, 1996).

18. IAFF and IAFC, *The Fire Service Joint Labor Management Wellness/Fitness Initiative* (Washington, DC: International Association of Fire Fighters, 1997).

19. National Fire Fighter Near-Miss Reporting System—Annual report 2006, *FireRescue* Magazine, February 2007.

20. Paul J. Antonellis and Dabney Stack, Help me help you, *FireRescue* Magazine, November 2006.

21. American Psychiatric Association, *Diagnostic and statistical manual of mental disorders*, 4th ed. (Washington, DC: American Psychiatric Publishing, 2000).

22. National Institute for Occupational Safety and Health (NIOSH), *Workplace violence statistics* (Washington, DC: U.S. Government Printing Office, 2005).

23. Mark Braverman and Susan Braverman, Workplace violence: Nature of the problem, *USA Today Magazine*, January 1997.

Suggested Readings

Alder, Mike, and Fratus, Mat. 2007. The impact of department culture on fireground safety. *Fire Engineering*. June.

American Psychiatric Association. *Diagnostic and statistical manual of mental disorders*, 4th ed. 2000. Washington DC: American Psychiatric Publishing.

Antonellis, Paul J., and Dabney Stack. 2006. Help me help you. *FireRescue* Magazine, November.

Antonellis, Paul, Jr., et al. 2006. Understanding post traumatic stress disorder. *Fire Engineering*, December.

Braverman, Mark, and Susan Braverman. 1997. Workplace violence: Nature of the problem. *USA Today Magazine*, January.

Cascio, Wayne F. 2006. *Managing human resources*, 7th ed. New York: McGraw-Hill.

Carrell, Michael, and Christina Heavrin. 2001. *Labor relations and collective bargaining*, 6th ed. Upper Saddle River, NJ: Prentice Hall.

Foley, Stephen N., ed. 1998. *Fire department occupational health and safety standards handbook*. Quincy, MA: National Fire Protection Association.

Hall, Douglas T., and James G. Goodale. 1986. *Human resource management*. Glenview, IL: Scott Foresman.

IAFF and IAFC. 1997. *The fire service joint labor management wellness/fitness initiative*. Washington, DC: International Association of Fire Fighters.

Kales, Stefanos N., et al. 2007. Emergency duties and deaths from heart disease among firefighters in the United States. *The New England Journal of Medicine,* March 22.

Karter, Michael J., Jr., and Paul R. LeBlanc. 1998. 1997 U.S. firefighter injuries. *NFPA Journal,* November–December.

Larson, Arthur. 1982. *Workers compensation law.* New York: Matthew Bender & Company.

LeMasters, Grace K., et al. 2006. Cancer risk among firefighters: A review and meta-analysis of 32 studies. *Journal of Occupational and Environmental Medicine,* November.

National Fire Fighter Near-Miss Reporting System—Annual report 2006. 2007. *FireRescue* Magazine, February.

National Institute for Occupational Safety and Health (NIOSH). 1995. *Workplace violence statistics.* Washington, DC: U.S. Government Printing Office.

Pater, Robert. 1990. Safety leadership cuts costs. *HR Magazine, 35.*

U.S. Department of Labor. 2006. *All about OSHA.* Washington, DC: U.S. Government Printing Office.

U.S. Department of Labor. 1999. *OSHA facts.* Retrieved July 14, 2007, from OSHA Web page, http://www.osha.gov

U.S. Fire Administration. 1996. *Risk management practices in the fire service.* Washington, DC: Federal Emergency Management Agency.

Washburn, Arthur E., et al. 2007. *2006 fire fighter fatalities.* Quincy, MA: National Fire Protection Association.

Labor Relations and Collective Bargaining

Source: MFRI

Source: MFRI

KEY TERMS

arbitration, *p. 306*

collective bargaining, *p. 289*

duty of fair representation, *p. 299*

fact finding, *p. 306*

good faith bargaining, *p. 301*

impasse, *p. 303*

mediation, *p. 306*

union certification, *p. 299*

OBJECTIVES

After reading this chapter, you should be able to:

- Describe unionization.
- Describe labor legislation.
- Describe union and bargaining unit organization.
- Understand the process of contract negotiations.
- Know how to handle grievances.
- Employ enlightened labor relations.

The right of workers to organize and bargain collectively with employers has a long and at times turbulent history in the United States. The fire and rescue service is no exception; in fact, the career fire service has one of the highest percentages of union membership within the public sector. A successful fire service personnel manager must understand how unions are formed, how they operate, and the laws and regulations that govern them.

It is generally agreed that not only economic factors but also sociological and psychological factors are involved in belonging to a union. Companionship and the respect of peers in the workforce are powerful reasons to belong to a union. The ability to have an active voice in controlling the work environment and a better understanding of workplace dynamics is also a motivating factor toward union activity. Fair treatment and the desire to deal with dissatisfaction regarding management can contribute to union participation. These factors suggest that the desire to improve wages and benefits is only one of the reasons that workers join unions. The motivation to join a union is an attempt to fulfill a variety of human needs, including those for security, community, esteem, and a sense of involvement and fair treatment within the work environment.[1] This chapter explores the history, legal aspects, and recommended practices for improving the labor relations within the fire department from these various perspectives.

Unionization

Unionization is the effort by employees to act as a single unit when dealing with management over issues relating to their work. When properly constituted and recognized, a union has the legal authority to negotiate with the employer on the behalf of employees to improve wages, working hours, benefits, conditions of employment, and other issues, as well as to administer the labor agreement. When a union represents employees and bargains collectively, the terms and conditions agreed on become a contract between the union and the employer. Most labor agreements give bargaining unit members additional rights not granted by the law or to nonunion employees, such as seniority, layoffs, grievance, and arbitration procedures. **Collective bargaining** is the process by which union representatives for employees in a bargaining unit negotiate employment conditions for the entire bargaining unit. Labor relations is a continuous relationship between a defined group of employees represented by a union or association and an employer.

collective bargaining
■ the process by which union representatives for employees in a bargaining unit negotiate employment conditions for the entire bargaining unit.

Why do employees join a union? Although many people believe that employees are initially attracted to unions over wage and benefit issues, according to noted labor attorney Jonathan A. Segal, it is the "soft" issues that motivate many employees to unionize. Pay and benefits are important and will always remain so, but many of the other factors that move people toward union membership and keep them there follow.

Recognition.   Many employees feel overworked and underappreciated. If the company they work for will not recognize them, the union will.

Lack of Control.   Many managers do not empower employees to make decisions about their jobs, but collective bargaining can provide them with more control.

Job Insecurity.   Most nonunion employees work "at will," whereas union employees may be terminated only for just cause.

Representation.   If supervisors and human resource professionals do not represent employee needs and support them, then employees will seek assistance from a union.

Double Standards.   Many employees see the perks and working arrangements that management employees have and resent this, especially if they perceive the company as not paying them fairly.

Respect.   Unionized employees must be managed with a degree of respect and see themselves as more equal to management in union environments.

The process of collective bargaining in the private sector is different from that in the public sector. Although many of the practices involved with public and private sector labor relations are identical, their underlying legal frameworks are not. The nature and purpose of government as opposed to the management of a private company present major philosophical and legal differences. In the public sector, the employer is the public at large, which acts by establishing laws enacted by their elected representatives. In general, public employees have no right to withhold services from their fellow citizens, which forms the basis for the "no strike" clause in federal and most state collective bargaining laws, especially with public safety agencies such as fire and police departments.

The legislative history of federal sector labor relations essentially began with the passage of the Pendleton Act in 1883. Prior to then, government employees were appointed and subject to the political spoils system. The Pendleton Act created the merit system of competitive examination and qualifications for government positions. This act protected federal employees from being fired for failure to make political contributions and in some cases prohibited political contributions by certain employees. The Pendleton Act affected approximately only 10 percent of the federal employees at the time, but it allowed for the expansion of the merit system and was the foundation of the present federal civil service system. Many state and local governments followed the federal government's lead and instituted civil service merit systems for their employees. However, during the New Deal era, major legislation that defined the rights of the labor movement in the United States, such as the Taft–Hartley Act, the Wagner Act, and the creation of the National Labor Relations Board, still excluded federal employees and employers. Not until a series of executive orders in the 1960s were federal employees granted the right to join unions and to bargain collectively. In 1978, Congress passed the Civil Service Reform Act, which incorporated these executive orders into legislation and also created the Federal Labor Relations Authority (FLRA). This legislation clarified such issues as unit determination and scope of bargaining. Federal employees still may not bargain collectively over wages and benefits, both of which are established by Congress.

It is much more difficult to determine the status of state and local government collective bargaining. Each state bears the responsibility for developing its own laws that regulate collective bargaining for state agencies and for local jurisdictions within the state. In our system of government, both the federal and state governments have sovereign powers. Local governments are created and regulated by state governments, so they have no sovereignty. With respect to collective bargaining, this structure means that local government has to have legislation enacted at the state level allowing them to enter into collective bargaining agreements. In many states progressive labor legislation is in place at the local level, including binding arbitration for local government employees, such as firefighters and police officers. In general, a state with strong private sector unions tends to have stronger public sector unions. Conversely, "right-to-work" states have weaker unions in both the public and the private sectors.

Two of the fastest growing and largest unions with state and local employees are the American Federation of State, County and Municipal Employees (AFSCME) with 1.3 million members and the National Education Association (NEA) with 2.7 million members. The IAFF represents the vast majority of career firefighters in the United States at the federal, state, and local levels. The IAFF, with approximately 270,000 members, has its headquarters in Washington, DC. As a percentage of the existing workforce, the IAFF has the greatest percentage of union membership when compared to other unions.

Union membership as a percentage of the workforce has been decreasing for many years. The height of union membership peaked in 1945 at approximately 35.0 percent of the workforce; it was 22.0 percent by 1980 and is still declining.[2] In 2006, 12.0 percent of employed wage and salary workers in the United States were union members, according to reports from the U.S. Department of Labor's Bureau of Labor Statistics (BLS). The number of persons belonging to a union fell by 326,000 in 2006 to 15.4 million. The union membership rate for government workers (36.2 percent) was substantially higher than for private industry workers (7.4 percent). Within the public sector, local government workers had the highest union membership rate, 41.9 percent. This group includes several heavily unionized occupations, such as teachers, police officers, and firefighters. Despite the decline in organized labor's share of the workforce, the BLS study showed that median weekly earnings for union members who work full-time were approximately one-third higher than the median for nonunion workers.

Highlights from the 2006 BLS study on union member data follow:

- Workers in the public sector had a union membership rate nearly five times that of private sector employees.
- The unionization rate was higher for men than for women.
- Black workers were more likely to be union members than were white, Asian, or Hispanic workers.[3]

An interesting paradox for unions is their strong record of acquiring better pay and working conditions for their members while simultaneously losing members as a percentage of the workforce at an increasing rate. This decrease in membership is due to many factors involving the nature of business today and society in general. Because of their success, unions have reduced or eliminated many of the conditions that led to their foundation. Improvements in working conditions coupled with state and federal legislation, such as the Fair Labor Standards Act and the Occupational Safety and Health Act, have had the effects intended. Much legislation now establishes minimums for wages, benefits, and working conditions below which no employer is allowed to go. Enlightened management has seen the need for better labor relations and treatment of the workforce in general. The partnerships between employees and employers are at an all-time high; often it is hard to determine the difference. For example, the major stockholder of United Airlines is the pilots' union. The effects of the global economy have also been pronounced in that many entry-level and dangerous jobs have been moved overseas. In addition, the structure of corporations and the employment relationships are different as we have begun the twenty-first century. The major manufacturing and industrial base of the United States is decreasing and being replaced by the white-collar jobs of the information society. These younger workers are more mobile and less tied to certain companies or unions. All of these factors combine to diminish the role of unions in society as compared to the middle of the twentieth century.

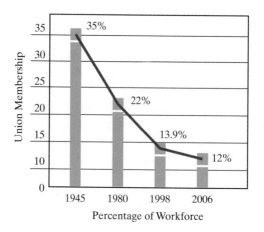

Although unions may have outlived aspects of their role in terms of their original objectives, they still play a substantial role in the workforce. This presence will continue in the public sector, where unions have experienced their largest growth in the last 30 years. Although less conflict and a reduction in the hostilities today mark labor relations, the need for a deterrent effect to check management on occasion is still valuable. In my opinion there will continue to be more cooperation and partnerships as both management and labor work for each other's success, knowing that will assist them in achieving their objectives.

Labor Legislation

The history of unionism in the United States has been closely involved with labor legislation. Much of the progress of the labor movement and the improvement in the working conditions for employees can be tied to legislative victories, which has not always been the case. The legal principle that dominated early labor relations was that of individual rights. Every individual was free to negotiate employment terms and to change employers at will. Unions were viewed as a criminal conspiracy that abridged the rights of individuals. In 1805, the Philadelphia Cordwainers (shoemakers) tried to establish a standard rate for each type of shoe and asked all cordwainers to refuse to work except for this wage rate. The courts at that time ruled "the collective action of a few persons in pursuit of their selfish interests contravened the interests of citizens in general, and was therefore a criminal conspiracy."[4] The courts continued to find collective activity of this type illegal well into the nineteenth century. Even though the 1842 case of *Commonwealth v. Hunt* allowed the association of workers, unions were legal only if their self-interest did not interfere with free-market competition.

Severe economic disruption and massive unemployment in the 1920s and 1930s as well as major scandals in working conditions and the use of children for employment aroused public sympathy for the plight of workers. During the next decade, several pieces of major labor legislation that were passed at the federal level established the relationship between employees and employers. These major legislative acts still set the stage for how unions are recognized and the relationship for the collective bargaining process. It is important to understand this history and the requirements of these acts regarding labor relations.

CLAYTON ANTITRUST ACT

Public criticism of the use of the injunction against labor unions caused Congress to pass the Clayton Antitrust Act in 1914. This act sought to limit the court's injunctive powers against labor organizations. It stated that neither the labor organizations nor their members were considered illegal combinations or conspiracies in restraint of trade. Unfortunately for the unions, the courts continued to apply the Sherman Antitrust Act after passage of the Clayton Antitrust Act by narrowly interpreting its provisions.

RAILWAY LABOR ACT

Passed in 1926, this act was the first national labor law to be found constitutional, giving employees the right to choose whether to be represented by a union and to engage in union activity. It also encourages the use of arbitration and mediation to resolve labor disputes. Congress passed this act to prevent serious economic consequences due to labor unrest in the railway industry. The act originally applied to the railway industry, but was later extended to other industries.

DAVIS–BACON ACT

Passed in 1931, the Davis–Bacon Act attempted to support the fledging union movement. This act put into place a requirement that companies using federal funds for construction projects use the "prevailing wage rate" of the area as a minimum wage rate on their construction projects. The Davis–Bacon Act meant that companies who used union labor would not lose their competitive advantage when bidding on federal jobs by paying their workers union wage.

NORRIS–LAGUARDIA ACT

The Norris–LaGuardia Act, passed in 1932, declared union membership to be a legal right of employees. This act prevented arbitrary court injunctions against lawful union activities unless they presented a clear and present danger to life or property. Unions must be given the opportunity to respond to the charges before an injunction is issued. In addition, this act outlawed "yellow-dog" contracts, which employers used to get employees to agree not to join a union or participate in union activities.

WAGNER ACT, OR NATIONAL LABOR RELATIONS ACT

The Wagner Act, passed in 1935, was the first piece of national legislation that placed unions on a more equal footing with management. The act firmly established employee rights to organize and to bargain free from an employer's interference. It required employers to bargain with a union over wages, hours, and conditions of work, if a majority of employees desired such union representation. By legislating the recognition of employee representatives and protecting the right to strike, Congress forced the employer to share the decision-making power with employees. The Wagner Act regarded the following as unfair labor practices:

- Interference with the efforts of employees to organize
- Domination of the labor organization by the employer
- Discrimination in the hiring or tenure of employees to discourage unions
- Discrimination against or discharging an employee for exercising the rights under this act
- Refusal to bargain collectively with a representative of the employees

This act also established the National Labor Relations Board (NLRB), which was charged with two major functions: (1) to establish procedures for holding bargaining unit elections and to monitor the election procedures and (2) to investigate complaints and prevent unlawful acts involving unfair labor practices.

FAIR LABOR STANDARDS ACT

Passed in 1938, the Fair Labor Standards Act established minimum wage rates and a maximum workweek beyond which overtime must be paid. This act also outlawed the employment of children under the age of 16 in industries involved in interstate commerce.

TAFT–HARTLEY ACT, OR THE LABOR–MANAGEMENT RELATIONS ACT

Passed in 1947, the Taft–Hartley Act relates to the Wagner Act in that it clarified and corrected some of the problems that still plagued labor relations in the United States. This act outlawed "sweetheart contracts" that employers used to establish their own union without consulting with the majority of employees. It specified exclusive representation, in that when a majority of employees desire a specific union, that union represents all employees in the bargaining unit, whether or not they are members. The Taft–Hartley Act also stated that a "closed shop," which requires that employees be union members at the time of hire, is illegal. However, a "union shop," in which employees are required to be a part of the bargaining unit as a condition of continued employment, is legal.

The Taft–Hartley Act created the Federal Mediation and Conciliation Service (FMCS), which offers assistance in contract settlement and maintains a listing of arbitrators. Under the Taft–Hartley Act, bargaining unit certification elections were forbidden from being held more than once per year and employees could initiate decertification elections. Additionally, this act empowered the president of the United States to order an 80-day cooling off period if a strike threatened the public welfare.

LANDRUM–GRIFFIN ACT

In 1959, Congress passed the Landrum–Griffin Act in response to labor racketeering and abuses by unions, many of which were directed at their own members. By providing standards for union conduct, it eliminated much of the union's flagrant abuse of power and protected the democratic rights of employees to a degree. The Landrum–Griffin Act required that union members have the right to vote for union officers and dues increases, rights of free speech in union matters, and the reporting of financial transactions by unions. The Landrum–Griffin Act also clarified provisions for closed shops in the construction industry. It supported state right-to-work laws that outlawed union shop clauses requiring union membership after hire and other issues. Twenty-one states have adopted such laws, which continue to be a source of irritation between labor and management.

EMPLOYEE RETIREMENT INCOME SECURITY ACT (ERISA)

In 1974 Congress passed the Employee Retirement Income Security Act, also known as the Pension Reform Act. The act requires annual reporting to the Department of Labor on pension plans, vesting benefits for plan members, portability between pension plans, and service credit requirements. This act was passed to rectify abuses found in pension plan management.

CIVIL SERVICE REFORM ACT

Federal employee labor relations are governed by the provisions of Title VII of the Civil Service Reform Act of 1978. This act created the Federal Labor Relations Authority (FLRA), which oversees labor–management relations within the federal government. The FLRA monitors the creation of bargaining units, conducts elections, decides representation cases, determines unfair labor practices, and can seek enforcement of its decisions in the federal courts.

NATIONAL PARTNERSHIP COUNCIL

In 1993, President Bill Clinton signed Executive Order 12871, which provided a significant change in how the federal government approached labor relations. The purpose of this order was to alter the relationship by which managers and the unions worked together to reach decisions. This order created the National Partnership Council (NPC) to advise the president on labor–management issues. The NPC was composed of union leaders, representatives from the Federal Labor Relations Board and the Federal Mediation and Conciliation Service, and executive branch directors and managers. The concept of the order was to establish and promote labor–management partnerships at various levels of the federal government to encourage a more cohesive and productive workforce.

STATE AND LOCAL GOVERNMENT LAWS

In the traditional sense, many people did not view government as an employer but as a representative of the people, supplying certain necessary services. Therefore, people employed by the government were not considered employees but public servants, and they were protected from the arbitrary actions of private employers by already existing systems. This traditional view has passed, and now more than two-thirds of the states have enacted legislation granting collective bargaining rights to the public sector. State legislation usually includes bargaining over wages, hours, terms of employment, and working conditions. Unfair labor practices and limits on or prohibitions of the right to strike also are legislated and usually enforced by an administrative agency. Local governments have also widely adopted collective bargaining laws or, by practice, recognize and bargain with employee organizations.

Union and Bargaining Unit Organization

The American Federation of Labor and Congress of Industrial Organizations (AFL–CIO) is the central trade union federation in the United States. The American Federation of Labor was founded in 1886 by twenty-five labor groups representing skilled trades. Samuel Gompers of the Cigarmakers Union was the first president of the AFL, a position he held until his death in 1924. He is recognized as one of the single most important individuals in trade union history. In 1938, a number of AFL-affiliated unions broke off from the AFL and created the Congress of Industrial Organizations (CIO) to promote the organization of workers involved in mass production and other unorganized industries. John L. Lewis, president of the United Mine Workers, was elected the first president of the CIO. The rivalry generated by these two federations, which was fierce at times, was conducted during a period of major growth and influence of labor unions. After tiring of engaging in bitter and costly disputes, both the AFL and the CIO recognized the need for cooperation and unification. In 1955, a merger agreement was ratified, and the AFL–CIO became a reality under the leadership of its first president, George Meany.[5]

The AFL–CIO represents the interests of labor and its member national unions. It does not engage in collective bargaining, which is the responsibility of the national and local unions. The AFL–CIO promotes the means by which member unions can cooperate to pursue common objectives and labor-friendly legislation. The federation is financed by its member national unions (approximately ninety-five) and is governed by a national convention, which meets every 2 years. The IAFF is one of the national unions that belong to the AFL–CIO.

A national union is composed of the local unions that it charters. The local union, not the individual worker, holds the membership in the national union. In most cases, the national union is supported financially by each local union, whose contribution is based on its membership size. In some industries, the bargaining contract is negotiated at the national or regional level. The national union is active in organizing workers within its jurisdiction, assisting local unions with contract and legal issues, providing educational and research services, as well as being a focal point for national political action to benefit their members. As an example, the Prince George's County Professional Fire Fighters Association is Local 1619 of the IAFF. A portion of an individual's union dues paid at the local level goes to the IAFF to support its operations.

The most basic element in the structure of labor organizations is the local union. To the individual union members, this level is the most important, because it is where they turn when help is needed. The members of the local elect their leadership and vote on contracts and other issues of importance to the local union. The handling of grievances, resolution of health and safety issues, and many other activities are best conducted at the local level. In addition, union locals hold meetings, publish newsletters, and otherwise keep their members informed. Many IAFF locals have considerable political clout at the local level, which is used to further their causes. The IAFF locals conduct their own contract negotiations that do not require approval at the national level. Some state-level firefighter associations assist in the passage of legislation at that level. These organizations normally include the leadership of the various firefighter locals of a particular state or region.

BARGAINING UNITS

The purpose of the National Labor Relations Act (NLRA) was to minimize industrial strife interfering with the national flow of commerce. Using the authority of the Commerce Clause of the U.S. Constitution, Congress legislated a federally protected interest in the legal process by which labor and management would bargain. This action created the National Labor Relations Board (NLRB), whose members are appointed for 5-year terms by the president of the United States. The NLRB, through a wide range of functions, is empowered "to protect the rights of the public in connection with labor disputes that affect commerce." The following basic principles guide the administration of the NLRB:

1. Encouragement of labor organizations and collective bargaining
2. Recognition of majority representation
3. Establishment of proper administrative action for enforcement and the imposition of sanctions and punishments, if warranted

One of the NLRB's major functions is to conduct the process in which a union is chosen to represent employees. This procedure includes a certification election to determine whether the majority of employees want a union and which union they want to represent them. In the public sector, a state collective bargaining agency, instead of the NLRB, may discharge some of these duties.

Before collective bargaining can occur, it is necessary to identify appropriate criteria for the formation of bargaining units to be represented by the union. The NLRB may use several criteria to determine an appropriate bargaining unit. Generally, these are a demonstrated mutual interest in wages, working conditions, and benefits. The basic concept is that the greater the similarity of the working conditions, the greater the opportunity for the unit's members to agree on priorities, and thus the collective bargaining process works better. The following criteria are frequently used in deciding what constitutes a bargaining unit:

- Community of interest
- History of bargaining
- Desire of employees
- Prior union organization
- Relationship of the unit to organization
- Statutory considerations
- Public interest

Public and private sector organizations differ when the subject of bargaining units is considered. The two most commonly utilized criteria in the public sector are to divide employees either by agency or by occupation. Agency bargaining units are established within each state or local government agency as a separate bargaining unit. This format has the advantage of working within the existing agency management structure, but can create a large number of bargaining units. If the criteria are occupational, the basis is usually the jurisdiction's job classification system, such as firefighter or police officer. This type of system creates fewer bargaining units, which are preferred by both labor and management.

The structure of the bargaining unit is important to both the union and the management. Who is included is as critical to the process as the number of different bargaining units that one union may represent. For example, fire departments generally have one bargaining unit for firefighters and a separate bargaining unit for officers. Both bargaining units can be, and usually are, represented by the same union. In other departments, the firefighters and the fire officers are organized into separate bargaining units and are represented by different locals of the IAFF, such as in New York City. Fire departments are also unusual when compared to other industries in the number and rank of supervisors allowed in the bargaining unit. Many fire departments have everyone up to deputy fire chief in a bargaining unit and represented by the union for pay and benefits; this has the potential to divide loyalties and cause confusion regarding management roles within the fire department.

Unions tend to want as many members as possible because numbers contribute to their strength. In many fire departments, the union attempts to represent as many bargaining units as it can. Paramedics are included in the bargaining unit with firefighters in certain jurisdictions, and in others they are not. Some IAFF locals represent the administrative or clerical bargaining units in the fire department as well as the firefighters and officers. The more units and the more people the union represents, the more powerful it becomes.

RECOGNITION AND CERTIFICATION

Once appropriate bargaining units have been established by the federal or state labor relations agency, unions are free to organize employees for the purpose of collective bargaining. To a large degree, recognition and certification procedures are generally the same in all states. An employer may voluntarily recognize a union as the exclusive bargaining agent for employees in that bargaining unit without a recognition election if the union can demonstrate—by acquiring signed authorization cards from the employees—that a majority of the employees in the bargaining unit want to be represented by that union.

If voluntary recognition does not occur, the union can win recognition through an election process, which has several general steps. The first step is for the union to persuade employees to sign cards that authorize union representation in collective bargaining. If the bargaining unit is under the authority of the NLRB, then 30 percent of the employees in a unit must sign the authorization cards. In many occupations, competing unions may vie for the authorization cards of employees. In most all fire departments, however, the IAFF has no real union competition. An election is then requested because the authorization cards have demonstrated substantial interest from the employees to be represented by that union.

The election campaign can be a controversial period, especially when the management of the company wants to keep the union out. This turmoil is less pronounced in the public sector because the unions are able to exert considerable political pressure to prevent a real campaign against them. However, management must be careful not to conduct activities that could be construed as unfair labor practices. Accordingly, employers should not

- Misrepresent the facts concerning the union and its officers
- Threaten employees with loss of jobs, transfers, reduction of benefits, or other intimidating activities
- Promise benefits or rewards to reject the union
- Conduct surveillance activities of union actions
- Interrogate employees to determine who has signed authorization cards
- Prohibit solicitation by unions for members[6]

The company can make its case to employees and should do so in a factual and dignified manner. Management can advise employees of the cost to join unions, discuss the history of strikes and work stoppages, state what the company is currently providing regarding pay and benefits, and offer other information that it may deem appropriate to provide employees with a clear understanding of the collective bargaining process. In the current economy, most unions do not go forward with an election unless they are satisfied they will be successful. The organization's past relationship with its employees, not the activities during the election process, will most likely be critical as to whether the union's certification campaign will be successful.

Union certification means that the NLRB or appropriate state agency conducts by secret ballot of the eligible employees an election of the union that is then certified to represent the employees of that bargaining unit. If a majority of the employees within the bargaining unit have chosen the union, then it will be certified as the official bargaining representative of the employees in that unit. The union is certified as such for at least 1 year. After this period, decertification is possible if a union has not served its members or if employee membership has changed and no longer supports the union. The same group of employees who vote in a union can vote one out, by means of the same general procedure used in the certification election. However, this would be an unusual step, especially after the union was successful in negotiating the first collective bargaining agreement.

Once the union has been certified as the exclusive bargaining agent, then employees represented by the union cannot individually agree with the employer regarding the wages and working conditions, but must do so through the union. The employer is obligated under law to make a good faith effort to reach written agreement as to the terms and conditions of employment. In addition, under the **duty of fair representation,** the union must represent the interests of the members and in disciplinary cases may be obligated to provide representation. The duty of fair representation means that the union has a duty to represent the interests of the union membership in a fair and impartial manner in consideration of the interests of all of the members. This can be difficult at times for the union, because as in any organization, competing interests exist among the members. In the grievance process, the union member does not have an absolute right to have the grievance pursued by the union. The duty of fair representation is breached when the union's conduct toward a member is arbitrary, discriminatory, or in bad faith. In many cases this issue occurs when a union member wants to pursue a grievance, but the union determines that the case lacks merit or is too expensive to pursue. The union must be careful in its decision, as evidenced by the

union certification
- an election of the union the NLRB or appropriate state agency conducts by secret ballot of the eligible employees that is then certified to represent the employees of that bargaining unit.

duty of fair representation
- union's duty to represent the interests of the union membership in a fair and impartial manner in consideration of the interests of all of the members.

Supreme Court case of *Bowen v. U.S. Postal Service*. In this case, the Court apportioned the damages due to wrongful discharge between the union and the employer. The Court reasoned that if the employee had been properly represented, the employer's liability would have ended at the arbitration decision, so all back-pay benefits from that point forward were caused and should be paid by the union.

TYPES OF UNION MEMBERSHIP

The two basic kinds of unions are craft and industrial. A craft union, such as the United Electrical Workers, is typically composed of members with a particular job skill. These members usually obtain their skills or trade through an approved apprenticeship program that is often managed by the union. An industrial union generally consists of the workers in a particular plant or place of business. The type of work they do and the level of skill they possess are not conditions that must be met in order to join the union. An example of an industrial union is the United Auto Workers.

After a union becomes the collective bargaining representative, one of its first objectives is a security clause. This clause is necessary to support the union in its role regarding contract negotiations and the representation of employees. It is a form of compulsory unionism that the employer must agree to before it is enforced via the bargaining agreement. The six basic types of union security clauses follow:

1. *Closed shop*—It requires union membership as a precondition of employment and continued membership as a condition of maintaining employment. The NLRA as amended no longer permits closed shops.
2. *Union shop*—It requires all new employees to become members of the union within a certain period of time and remain members until the expiration of the contract, with some specific exceptions.
3. *Maintenance of membership*—All members of the bargaining unit who are members at a specific time must remain members for the duration of the contract. Those employees who are union members at the time of the contract signing have a specified escape period during which they can resign from the union.
4. *Agency shop*—All employees in the bargaining unit pay regular union dues and initiation fees regardless of actual membership. This type of arrangement is popular with public sector unions.
5. *Exclusive bargaining shop*—Some of the "right-to-work" states allow only exclusive bargaining shop provisions. The company is legally bound to deal with the union, but the employees are not obligated to join or maintain membership in the union or to contribute to it financially.
6. *Hiring hall*—Section 8(f) of the NLRA permits the use of a hiring hall for employment referral in the building and construction trades only.

Once the type of the security clause is established between the employer and the union, the next goal for the union is a dues checkoff provision. This provision in the bargaining agreement allows the company to withhold union dues from members' paychecks and to send them directly to the union. It relieves the union from the process of collecting dues individually from each member.

Contract Negotiations

Collective bargaining is a complex process in which union and management negotiators maneuver to obtain the most advantageous agreement relative to their issues and objectives. Collective bargaining is the method by which a formal agreement is established between management and labor regarding wages, hours, working conditions, and similar matters and is a key aspect of labor relations. Labor and management conduct their negotiations within a web of government regulations and laws that limits their strengths and protects their rights. At its base, the relationship is about power and how best to use it to the advantage of the group that is represented for the long-term good. At its worst, the negotiating relationship is about personality and winning at all costs. Unfortunately, the history of labor relations contains countless examples of unnecessary grief and disruptions that served no one well. The focus of this chapter is on the process of how to negotiate a fair collective bargaining agreement successfully and not on the details or specific sections within that agreement.

Central to the concept of positive labor relations and especially with contract negotiations is the concept of **good faith bargaining.** In order for the process to work as envisioned, both the employer and the union are obligated to bargain in good faith. In fact, under the provisions of the National Labor Relations Board (NLRB), failure to bargain in good faith can result in an unfair labor practice. Therefore, it places a duty on both the employer and the union to bargain in good faith at all times. Good faith bargaining generally means that an open, honest, and reasonable attempt is put forth by both parties in the negotiation process toward reaching an agreement. Merely going through the motions without actually seeking reasonable agreement does not meet this requirement. This concept is not to be confused with either party to the negotiations having to make concessions. Hard bargaining on a major issue by itself does not necessarily exhibit bad faith because a party is not required to yield on a position fairly maintained.

Listed here are some of the circumstances and situations that can lead to an unfair labor practice because of the lack of good faith bargaining.

- *Surface bargaining*—This is going through the motions without any serious commitment to reaching an agreement.
- *Inadequate concessions*—There is unwillingness to compromise reasonably.
- *Inadequate proposals and demands*—Proposals should be met with counterproposals.
- *Dilatory tactics*—There is failure of one party or the other to meet at reasonable times and intervals.
- *Imposing conditions*—These conditions are so unreasonable they demonstrate "bad faith" by themselves.
- *Making unilateral changes in condition*—This is a rather obvious sign that the party doing it is not bargaining in good faith.
- *Bypassing the representative*—There is failure to negotiate with the designated party or parties.
- *Committing unfair labor practices during negotiations*—It strongly suggests that one side is not acting in good faith.
- *Withholding information*—Management, upon request by the union, is obligated to provide the union with information necessary for union representatives to discuss issues rationally and intelligently.

good faith bargaining
■ an open, honest, and reasonable attempt put forth by both parties in the negotiation process toward reaching an agreement. Failure to bargain in good faith can result in an unfair labor practice.

- *Ignoring bargaining items*—There is a failure to address mandatory bargaining items or demanding negotiations on a permissive item.[7]

One of the first things to realize is the perspective that labor relations is a year-long, continuous effort that must be made. The goal is to establish a positive long-term relationship between union and management within the fire department and within the local government. The actual contract negotiations are a subset of this relationship, albeit an important one. All is not won or lost during the negotiation of a bargaining agreement, but seeds of discontent planted here may be harvested at some future date. The narrow view that labor relations consists of only the contract negotiations misses many opportunities to build a relationship of trust and mutual respect. This relationship, if well established prior to the contract negotiations, is very beneficial during that process. Labor relations is a continuous effort within a fire department.

NEGOTIATION SESSIONS

A successful negotiation session often depends on the skill and knowledge of the negotiators. They must be knowledgeable about their positions as well as about the other side's positions and logic during the negotiation process. Successful negotiators propose workable, attainable, and realistic issues within the framework of the negotiations and stay on point throughout the sessions. Listening skills and the ability to communicate clearly are two very positive attributes. A thick skin is helpful at times, because the other side may resort to personal attacks either as a planned maneuver to satisfy a constituency or out of frustration. A good negotiator never responds in kind during these situations. Personal integrity and trust are important aspects of any negotiation process and once they are lost, they are extremely difficult to regain. Successful negotiators practice these points and are well prepared at all times.

The preparation by both sides for the contract negotiation will be important to the end success of the collective bargaining process. It is crucial to be well prepared to address the many potential issues that may be raised during the contract negotiations. Both union and management need to have a plan as to where they would like to be at the conclusion of the bargaining sessions. A primary step should be for both sides to review the current agreement and determine what is working and what may need improvement. An analysis of grievances filed between the contracts may be helpful to point out areas of the contract that require clarification and modification. In addition, one of the first things necessary is accurate statistical data on the comparative wages, benefits, and working conditions in the immediate area and region. These comparisons must be realistic and not slanted in one way or the other to assist the presentation. In many cases, it is good to reach an agreement between the parties prior to the analysis as to what cities or jurisdictions will be used in making comparisons. A number of state and national organizations regularly compile statistical data of this type.

Prior to the negotiations, both sides should attempt to anticipate the proposals that the other will likely put forth. This anticipation allows for the development of potential counterproposals and the collection of specific data that may be necessary. Each side should also have a plan as to the specific objectives it would hope to achieve during the negotiations. All of these documents are for planning purposes and must remain flexible. It is important to listen carefully to the proposals and reasons behind them during the bargaining sessions. Do not prejudge or become too attached to a certain position at this point; use the information only as general planning guidelines.

Within each negotiation team, clear decisions must be made about the amount of authority it will have. What concessions or agreements can be made during the negotiation sessions, and which ones require approval from a higher authority? Often when contract negotiations get to a certain point, things begin to happen rapidly, and the negotiation team must have some level of authority to respond in this atmosphere. If not, delays can become frustrating for the other side and inhibit the process. It is important to be clear as to what a team can agree to and to state it emphatically. If a negotiation team agrees during the process on a certain point and then has to retract it later, the bargaining process may be damaged. Effective planning and communication improve the collective bargaining process.

Labor laws impose a wide range of legal constraints on both labor and management during the negotiation process. Bargaining in good faith is a concept that improves the negotiation process. Good faith bargaining consists of the general guidelines under which certain activities or actions are indicative of either an actual intent to bargain in good faith or a refusal to do so. A true intent to reach an agreement, active participation, and the making of realistic counter-proposals are all indicative of good faith bargaining. Stalling tactics, sudden shifts in position when agreement seems close, rejection of items that are standard in most labor agreements, and the refusal to furnish data necessary for bargaining are all clear indications of bargaining in bad faith. However, bargaining in good faith does not mean that either party is compelled to agree to a proposal or make concessions. The obligation to bargain in good faith is something that each side has a right to expect of the other. When it does not occur, the bargaining process is disrupted, and even if the contract agreement is eventually reached, hard feelings may exist for quite some time afterward, thereby complicating good labor relations.

The NLRB has established three categories of bargaining issues: illegal, mandatory, and voluntary. Illegal issues are those that would conflict with the law and cannot be a part of the contract, even if both sides agree to do so. Mandatory bargaining issues consist of wages, hours, and conditions of work, and voluntary bargaining issues include all those that are lawful but not mandatory. It can be difficult at times to determine precisely the difference between mandatory and voluntary issues. This distinction can be further confused because of the disparity between the public and private sectors wherein the labor laws are applied in a different manner. The category of the bargaining issues is most critical when an **impasse** is reached or if one side is making the charge of an unfair labor practice.[8]

In many fire department contract negotiation scenarios, the parties engage in what is known as distributive bargaining. This occurs when the negotiations are seen as "win–lose" and the goals of each side of the table are in conflict. Each side wants to prevail to the maximum extent possible and "win" the contract negotiations. I do not recommend this style of negotiations, but I know that it is very prevalent. The key to successful bargaining lies in flexibility, understanding, and controlling the process. Anyone can refuse to bargain or reject proposals; the key to success is knowing when to say "yes." Just where this point is represents the essence of contract negotiations.

impasse
- occurs when the two sides negotiating an agreement are unable to reach an agreement and become deadlocked. Alternative strategies are then employed to move the negotiations toward a successful conclusion.

The key elements of the distributive bargaining process are described here:

- *Target point*—The optimal goal or objective for the negotiations. This is the point at which you would most want to conclude.
- *Resistance point*—The maximum or minimum beyond which there is no agreement possible. It represents the bottom line for either side.
- *Settlement range*—The difference between the resistance points of labor and management is the range where the actual bargaining occurs, because anything outside of this will be rejected.
- *Settlement point*—Agreement on a point within the settlement range is the settlement point. Each side wants this to be as close as possible to its target point.[9]

The ultimate test of strength in the negotiation process is either the point at which the union will strike or management's ability to take a strike and deal with the consequences. Obviously, with fire departments the ability to strike is severely limited or illegal, as binding arbitration is the most common step used for the final decision. It is important to make valid assumptions in this process because a miscalculation of the target and resistance points can lead to costly actions and frustration on behalf of both parties to the negotiation.

IMPASSE RESOLUTION PROCEDURES

In most cases, the bargaining process works, resulting in a contract that both sides can support and live with. This preferred method is the most productive and least disruptive to both parties affected by the bargaining agreement. However, contract negotiations can reach an impasse, and when they do, a number of options are available to bring forth a resolution of the issues in dispute. Before moving forward to discuss these issues, let us establish several definitions:

Strike	Refusal to work by employees.
Lockout	Refusal by management to allow employees to work.
Grievance	Formal dispute between an employee and management on the conditions of employment.
Mediation	Process whereby a neutral third party helps the union and management negotiators reach a voluntary agreement.
Fact finding	Commonly used in the public sector where an arbitrator hears arguments from both sides of the dispute and then makes a nonbinding recommendation.
Arbitration	An arbitrator conducts a formal hearing to hear the case and then makes a decision that is binding on both parties.
Final offer arbitration	The arbitrator chooses between the final offer from the union and the final offer from management.

The process of contract negotiations entails much more than the fire department; it also involves other government agencies and services. Typically, on the management negotiation team for the local government are fire department representatives, labor relations experts, and financial and personnel specialists. The union most likely has its principal officers, labor attorneys, and perhaps assistance from the state or national union office. These negotiations include a vast array of personalities and agendas, all of which are not predisposed toward a settlement at times. Someone on the management team is given the authority from the elected official, be that a mayor or county executive, to make a deal when the time is right. Normally, the team looks to the fire service personnel manager for direction on management issues and proposals from the union that may deal with the operation of the fire department. The fire department representative of the negotiation team will not have much to say regarding the pay and benefits that are offered, which is most likely the responsibility of the financial experts in concert with the representative of the elected official. When an agreement is reached, in most cases, it must still be approved by the legislative body of the jurisdiction.

The union negotiation team most likely presents a long list of negotiable items it would like to see in the contract. The union officers receive a great deal of pressure from the union membership during contract negotiations and are subject to reacting to this pressure at any point. Be aware that there is more going on than just what occurs during the negotiation sessions. Frustrations and side issues may arise, so it is important to stay on point as much as possible during the meetings between management and labor. At the conclusion of the negotiations the union membership gets the opportunity to vote on whether to ratify the contract as negotiated by the union negotiation team. In good faith bargaining the union negotiation team is obligated to support the ratification of the contract at this meeting.

One issue that frequently arises is whether the fire chief should be on the management negotiation team. In my opinion, it is generally better to exclude the fire chief from the team. The fire chief needs some space from the atmosphere that may be present during the early stages of the negotiations and must be able to intervene, if necessary, at the conclusion. An involvement too early in the process negates this ability. In addition, much venting and frustration are exhibited early during many negotiation sessions. Many fire chiefs have trouble sitting through this and not reacting or overreacting. However, it is critical that the fire department be represented at every negotiation session. The fire service personnel manager is well suited for this role to maintain communication and understand what is occurring during the negotiations. Quite often, the other members of the management negotiation team do not understand the nature of the fire service, with its shift work, operational needs, and other issues, causing them to agree to an item that they should not agree to. The fire department representative is there to help in these types of situations. To be quite honest, the fire service representative often must watch the other members of the management team, because it may be easy for them to agree to an item that they do not have to deal with. For example, they may be willing make a concession on a fire department operating policy in return for a union concession on a financial issue. Beware of the management negotiation team members who let the fire department know if it is needed at the contract negotiation session, because they may have their own agenda.

Often, contract negotiation difficulties emanate from issues outside the bargaining sessions. Management may be willing to agree to certain items with the fire union, but is concerned as to the impact it will create with other unions in the government. In a sense, management is always dealing with all of the unions at the same time. Another difficult circumstance is when, for political reasons, it is hard to obtain a contract agreement. This impasse can occur for a variety of reasons, and it is important to recognize when it does. It may require a political solution that is outside the realm of contract negotiations. An example would be an elected official who is antiunion and does not agree to a contract settlement, even though it is reasonable.

Many times, the negotiation dynamic simply comes down to the union wanting too much or management giving too little. Either way, the situation often requires outside assistance to reach an acceptable agreement. Fortunately, a number of techniques can be employed to move beyond a stalemate. In the public sector, especially public safety, strikes and lockouts are generally prohibited for the protection of the public. Because of this prohibition, public sector labor relations has some avenues available that are not found in the private sector.

If an impasse is reached in an attempt to obtain a settlement at contract negotiation, the first step is to enter into mediation. **Mediation** is when an experienced third party assists labor and management to reach an agreement. A mediator can be invited in from, for example, the Federal Mediation and Conciliation Service, an independent agency. The mediator takes an active role in getting each side to understand the other's proposals, develop alternatives, and remove the obstacles to an agreement. The mediator cannot force a settlement because he or she does not have the authority to decide the case, only to facilitate a successful resolution. The strength of mediators is in the trust and respect that they command so that they can work with both sides of the dispute. They need to be good communicators and to listen well and then restate the positions in a more objective way. Often an outside, experienced viewpoint is a good way to balance the issues and work more productively toward settlement.

In **fact finding,** a neutral party, either an individual or a panel, is appointed to determine the facts of a dispute. Fact finding is used much more in the public sector than in the private sector. The fact finder, usually an arbitrator, conducts a hearing and reports on the reasons for the dispute, the views and arguments on both sides, and a recommended settlement. Both union and management are free to decline the recommendations of the fact finder, but this often gives rise to public pressure to reach a settlement. Even if the fact finder's recommendations are not accepted, he or she may identify and frame issues in such a way that resolution is easier to achieve by the parties in the negotiation process. The cost of fact finding is similar to that of arbitration and it is often time consuming.

The third method to resolve an impasse in reaching a bargaining agreement is arbitration. **Arbitration** is the process in which union and management involved in a dispute agree in advance of a hearing to abide by the decision of an independent arbitrator. Arbitration is typically used in the private sector to settle grievance issues arising from contract administration or interpretation. Contract settlement arbitration is generally used only in the public sector, where strikes are illegal. In certain instances, the arbitrators can fashion their own solution to the dispute, which is called *conventional arbitration.* In other cases, the arbitrator must choose either the management or the union's last and best offer. This option, known as *final-offer arbitration,* can involve either the entire contract proposal or selected portions.

Voluntary arbitration is when the parties agree to submit some of the unresolved contract issues to arbitration. The parties in this circumstance can agree that the arbitrator's decision will be binding or it may be advisory. *Compulsory*

mediation
- when an experienced third party assists labor and management to reach an agreement. The mediator takes an active role in bringing the sides together, but does not have the authority to make a final decision or settlement.

fact finding
- a neutral party, either an individual or panel, appointed to determine the facts of the dispute. A recommended settlement is proposed but is not mandated on either the union or management.

arbitration
- the process in which union and management involved in a dispute agree in advance of a hearing to abide by the decision of an independent arbitrator.

binding arbitration is when the unresolved issues are submitted to an arbitrator or a panel of arbitrators. In this case the ruling of the arbitrator is binding on both parties; this type of arbitration is most commonly found in fire and police departments in which the employees are determined to be essential and employee strikes are prohibited legally. No matter which arbitration method is employed, arbitration can be costly to both sides of the dispute and can conclude with some very surprising decisions.

Although arbitrators are adept at the economic issues that may be in dispute, they may have difficulty with fire department operational issues that could be a part of the contract negotiations. Therefore, they often take a conservative approach to these issues. The best solution is for the parties to the agreement to resolve their issues because they have the most knowledge and the most to gain or lose. The union and management officials in the negotiation process are closest to the issues and should make every possible effort to resolve them without third-party intervention. Involving a third party means giving up some of the power to determine your own future, so this step should not be taken lightly. Mediation at times can be helpful, but arbitration should be seen as a last resort and only after considerable effort by both sides to reach an agreement.

Grievances

Once the collective bargaining agreement is approved, then it is the controlling document for the daily operation and activities within the organization subject to its conditions. Contract administration is a continuous process that involves active participation of the union and management. Often in the stress of reaching an agreement, vague or incomplete contract language can make the administration of the contract difficult. Because of the difficulty of writing an unambiguous agreement anticipating all of the applications to the workplace, disputes will occur over the interpretation and application of some sections of the contract. The most common method of resolving these disputes in an organized manner is the grievance procedure.

Whatever the subject matter of a particular grievance may be, it is usually more revealing to determine why the grievance was filed. One of the following circumstances is usually behind the filing of a grievance.

Clarifying Contract Provisions.   After the contract is agreed on, unforeseen circumstances change some operating conditions and a disagreement erupts over the interpretation of the contract language. Some bargaining agreements last 2 or 3 years, which is a substantial time frame.

Support for Future Negotiations.   Unions often encourage their members to file grievances in certain areas to provide support for certain issues during future negotiations. The union then points to this during contract negotiations as evidence of a problem that needs resolution.

Show of Power.   Sometimes employees and union officials file grievances to demonstrate their authority and influence. Union officials may feel a need to remind members that they are working for them and their interests. In some cases, even if the union officials know that the grievance is without substantial merit, they will proceed anyway if it is a popular concern of the membership.

Pay and Benefits. Employees may feel that they are due a special benefit or action and file a grievance to receive this benefit. This is usually beyond the base pay and may deal with shift differentials or special duty pay issues.[10]

Grievances can be filed over any workplace issue that is subject to the collective bargaining agreement, or they can be filed over the interpretation and implementation of the agreement. The most common type of grievance that eventually reaches the arbitration stage involves discipline and discharge of employees.[11] Many union members view the grievance procedure as the key to fair treatment in the workplace, and its effectiveness depends on the degree to which employees feel that they can use the process without recrimination or retaliation. The union has an obligation to represent its members in a fair and impartial manner. Under the duty of fair representation, the union has to represent all bargaining unit members, whether or not they are union members. All bargaining unit members must have equal access to and representation by the union in the grievance procedure process.

The grievance process begins with a charge that the contract between union and management has been violated. Either individual employees or the union itself can initiate this procedure. The employee may consult with the union shop steward before filing a grievance. Normally, the grievance must be filed within a specific time frame after the act that led to the grievance. Time limits also restrict the processing of each step of the grievance procedure so that the issues are resolved in a relatively quick time period. The grievance time periods for each step can be extended by mutual agreement. The grievance, when initiated, is presented to the first-line supervisor for resolution. If the first-line supervisor has the authority to resolve the grievance and does so, the process stops at this point.

If the first-line supervisor does not resolve the grievance, it goes to the next level, which is usually the manager of the first-line supervisor. In most fire departments, this manager would be at the battalion or division level. The union will have a more formal role and involvement at this step. After a meeting to hear the facts, the manager at this level resolves or denies the grievance. If the union and employee are not satisfied, they may proceed to the next level. The actual number of levels varies with different contracts and different-size fire departments.

At the next level, the chief of the fire department is normally responsible for making the decision on the grievance as the last step within the fire department. At this step, the fire chief would likely consult with the personnel director and the fire department's attorney. They should make a concerted effort to resolve the grievance within the department; however, this is not always possible. The resolution of grievances can set a precedent and allow the contract to be interpreted in that manner in the future. One must balance the resolution of the grievance against a number of issues and circumstances within the department. The goal should clearly be to resolve the issue, but only if it will not damage the department in the future and if the union is being reasonable in its position. Because the cost of arbitration is shared by the union and management, some incentive exists to settle, if possible. If the grievance is not settled at this level, it will proceed to arbitration.

The arbitrator is a neutral, mutually acceptable individual who may be appointed from an agency that recommends arbitrators, such as the American Arbitration Association (AAA), or from another source. The arbitrator conducts a hearing at which evidence is presented, and then he or she makes a ruling on the grievance that is usually binding on both union and management. The time necessary for this process to take place can be quite extensive, so if an issue goes to arbitration, it most likely will not be quickly resolved. In this case, the decision on an important issue involving the fire department is made by someone who may have little experience in this field. Arbitrators are quite adept at handling disciplinary and economic issues, but less so with fire department issues, especially those that involve the operation of the department. In these types of cases, the arbitrators are usually conservative in their opinions.

The following list is reflective of the criteria arbitrators use to make a decision in cases involving discipline or discharge.[12] These due process questions are important.

1. Did the employee know what the rule or expectation was and the consequences of not adhering to it?
2. Was the rule applied in a consistent and predictable way?
3. Were the facts collected and reported in a fair and systematic manner?
4. Did the employee have the opportunity to question the facts and present a defense?
5. Did the employee have the opportunity to appeal the decision?
6. Was there progressive discipline, except for severe cases?
7. Were there mitigating circumstances, and were they considered?

The levels and types of grievances filed within the department can reveal much about the current state of labor relations. Too many grievances may indicate a problem with the supervisory practices within the department or unclear language within the contract. Few grievances may indicate a fear of filing a grievance, a belief that the system is not effective, or that the representation is not adequate. Few grievances may also indicate a department with a high degree of effective management, communication, and trust. In the role as a fire service personnel manager, one should be aware of the number and types of grievances filed and analyze this information to see whether any trends are evident that may require corrective action. In this way, one can be proactive and resolve the source of the grievances for the betterment of the department and its personnel.

Enlightened Labor Relations

The history of labor relations shows an evolving relationship between management and labor and the balance of power between the two. Times have changed and will continue to change, especially as the United States continues to compete in the more global economy. Neither management nor labor should attempt to hold on to the practices of the past, because today's world does not reward or tolerate such entrenched behavior. The circumstances, people, and work environment of the future will continue to create new opportunities between labor and management that must be acted on in order to remain competitive in the world's marketplace. The President's Commission on Industrial Competitiveness summarized the necessary characteristics of organizations to foster a new era of labor relations.[13]

- Labor–management cooperation is essential. Cooperative relationships maximize productivity by involving labor and management in the decision-making process.
- Employee incentives must be strengthened to reward the efforts of individual employees and to emphasize the linkage between compensation and performance.
- Employers must be encouraged to strengthen their commitment to employee job security.
- Improved work skills must be developed by training and educating employees to compete effectively in the job market.

The goals of union and management are the same and generally cooperative with respect to maximizing total revenue or service, making the company or agency successful. The more success and revenue that the parent organization enjoys, the better the position it is in to agree to higher wages and benefits for the union membership. Unfortunately, union and management often take an adversarial posture in deciding how to divide the revenue regarding wages and benefits. This adversarial posture has historically established the tone of labor relations.

LABOR–MANAGEMENT RELATIONSHIP

The International Association of Fire Chiefs, in its report on *Labor–Management Conflict in the Fire Service,* identified several common characteristics that can cause conflict between fire chiefs and unions. Here is a synopsis of these conflicts.

Resistance to Change

Many conflicts are the result of change that is not communicated properly and/or change that threatens the status quo of benefits or working conditions. Inducing change in a fire department with a well-defined power structure can create conflict, often beyond what is reasonable due to the emotions involved. Resistance to change is a common issue in many fire departments.

Power Struggles

Fire chiefs and union presidents can be rivals for power and influence. A new fire chief may be viewed as a threat to the established power base, and actions may ensue to determine who has the most power and influence in the new relationship. During this period conflict will be high.

Confrontational Issues

Some situations involve unavoidable confrontation over certain issues or circumstances. For example, declining budgets and revenues might cause situations to arise that would never be necessary during normal times. The fire chief has to implement the policy of the jurisdiction whether he or she personally agrees or not. These confrontations may involve one major issues or an accumulation of issues over time.

Outside and Recently Appointed Fire Chiefs

Many confrontations are initiated during the first 2 years of the tenure of fire chiefs. Often, those chiefs appointed from outside of the fire department and not connected to the current power structure are frequently challenged with conflict. A chief appointed from outside of the department may have been selected to induce change, and the manner in which this is conducted is very important to the success of this effort. It is also not unusual that when the fire chief comes from outside there is a new alignment of power and position within the fire department, which can lead to conflict.

Personality Conflict

Some situations simply involve basic personality conflict between union officials and fire chiefs. These feelings may be very strong and difficult to resolve through mediation or other such steps. In many cases, these types of conflict can be protracted and have the potential to do considerable damage to the fire department until one party leaves.

In particular, since the mid-1990s and continuing to the present, a much more concerted effort is evident toward enlightened labor relations in which the needs of management and labor are more fully met in a cooperative relationship. This enlightened relationship has many positive benefits to management and labor.

It is my belief that much of the success of enlightened labor relations revolves around communication and trust. As Chief Randy R. Bruegman noted in an article titled "A Matter of Trust" (*Fire Chief* magazine, August 2002), trust in any relationship is very important to that relationship's long-term success. When dealing with labor negotiations, there is no question that trust between the parties is crucial and that a lack of trust has derailed many contract settlements. Chief Bruegman has listed actions that build or reduce trust as noted here.

Actions and Behaviors that Build Trust

- Communicating openly and honestly without distorting the information
- Showing confidence in the abilities of your personnel and treating them as skilled, competent associates
- Being inclusive; listening to and valuing what others have to say, even when you do not agree
- Keeping your promises and commitments
- Cooperating, collaborating, and looking for ways to help one another

Actions and Behaviors that Reduce Trust

- Acting more concerned about your own welfare than anything else
- Sending mixed messages so your people never know where you are
- Shirking responsibility for your own actions
- Jumping to conclusions and reacting without first checking the facts
- Making excuses or blaming others when things do not work out

The fire chief and the management of the fire department must develop open communications and trust within the union leadership. This openness is essential if they are to work together in a mutual way to address departmental challenges. The union leadership must also extend itself to be a part of this relationship and realize that the success of the fire department is essential to the success of the union. If this mind-set can be established, then the possibilities for cooperation and the improvement of the work environment are outstanding. If the union leadership feels the need to view the fire chief as the "enemy" or if the fire chief considers the union leadership to be "firefighters with a grudge," the relationship is doomed. With respect to this relationship, the management of the fire department and the union leadership must view one another as equals with different responsibilities in the process. This attempt at objectivity is difficult for many people and their egos, but it is an essential viewpoint that must be established if a truly cooperative atmosphere is going to prevail.

Retired Fire Chief Dennis Compton of the Mesa, Arizona, Fire Department, in a recently published text entitled *When in Doubt, Lead!*, makes several astute observations regarding a successful relationship between labor and management:

- Management shares authority and labor shares responsibility, or the process will simply not work.
- The most important strength labor and management have for the future is the relationship between their leaders.
- Labor and management leaders must continually work on the issue of trust: trust in the process and trust among the individual participants.
- The fire chief and the labor leader must require compliance with agreements made through the process and value participation in the process.
- If either party (labor or management) dominates the relationship, the process will deteriorate. The purpose is not just to get along—it is to make the organization stronger by working together to accomplish common goals.[14]

Of the many ways in which the relationship of communication and trust can be developed within the fire department, the most obvious is actually to talk to each other on a regular basis. Neither the fire chief nor the union president appreciates being surprised with an issue. It is important to keep each other informed of what the issues and concerns are, so that they can be resolved at an early stage, which takes regular communication that is honest and forthright. A highly developed system of communication between the fire chief and the union president is beneficial for both parties. For example, when I was a fire chief, I had a standing meeting with the union president once every 2 weeks, or more often if necessary. The meeting was just between the two of us, allowing us to have a confidential time to discuss issues and strategies in detail without having to worry about organizational roles.

Involvement is a key component of building trust and extending the communication process. It requires that union and management work in a partnership to develop policies and regulations that affect the work environment. This partnership can take the form of formal committees established pursuant to the bargaining agreement, such as a "health and safety committee," or it can be informal "quality teams" appointed to address specific issues. Areas of agreement and disagreement can be discussed and analyzed in this process, and even if full agreement cannot be reached, the issues will be framed for future work. The value of these groups is in the inherent success of the process that is taking place and not whether total agreement is reached. In fact, a degree of disagreement can be valuable as well. At times the group must decide to agree to disagree, as long as this disagreement sticks to the issues and not to the people or process. What one wants to avoid is the situation in which the department management independently develops a policy and then sends it to the union for review. Rather than cooperation and involvement, this action establishes an adversarial tone from the beginning.

The essential trust in enlightened labor relations comes from the manner and circumstances in which issues are dealt with over time. If someone agrees and gives his or her word on an issue, then that person must keep to this agreement and follow through on it. If for some reason either the union or management cannot maintain its original agreement, then this change must be made clear to the other party, along with a logical reason as to why it is necessary. Then a new agreement must be reached. The key point is the clear communication of the change in the agreed-on course of action and to have clear and logical reasons. This type of change is accepted if it is handled properly and occurs infrequently. If changes become the normal course of business, then the essential trust in the relationship is diminished. It must be understood that everyone has a boss. Occasionally, the union president may agree to an item and the union executive board or membership may overrule him or her. Likewise, the fire chief may encounter difficulty when an agreement is sent for approval to the mayor or county executive. Effective leaders are aware of their strengths and weaknesses and lay the groundwork ahead of time to ensure that their decisions are upheld at higher levels.

Successful fire departments have a good relationship with their union, which makes them both stronger. When a dysfunctional relationship characterizes union and management, both are weaker. Some politicians may want the union and management to be at odds, because they know that they can then play one against the other. These politicians have a fear of their working too closely together, because they know it makes both much stronger in the community and at the time for budget requests. Significant power and benefits lie in positive labor relations, unlike petty agendas or personalities that deride progress. It is also important to understand where and when criticism is due. Unfortunately, some unions tend to "beat" on the fire chief regardless of the issue. Some still hold the belief that a good union does not agree with management. This attitude is counterproductive and serves no one well. If the fire chief is responsible for the decision and deserves criticism, so be it. However, often he or she is following the policy of the government, and if the union wants to vent frustrations, it should be aimed at the right level.

The desired positive relationship of trust, communication, and respect takes time. It needs to be built day by day on small issues and circumstances that establish its full potential. It takes the involvement of everyone in the department because each person will lose or benefit from it. If the fire chief and the union president set the right tone for a professional relationship, then others will recognize and follow it. A substantial amount of effort must be directed at this relationship on a constant basis to ensure a long-lasting process that can withstand an occasional crisis but still function.

IAFC/IAFF LABOR–MANAGEMENT INITIATIVE GUIDING PRINCIPLES

In the interest of cooperating as national organizations and setting the example and tone for future labor relations, the leadership of the International Association of Fire Chiefs (IAFC) and the International Association of Fire Fighters (IAFF) have worked together to cosign the joint *IAFC/IAFF Labor–Management Initiative Guiding Principles*. These guidelines represent a core component of the organizations' ongoing initiative to strengthen relationships between firefighters and chief officers in the best interests of all parties.

The document articulates how the two international organizations will move forward with improving labor–management relations and offers a model and a tool for helping individual fire departments do the same. In 2008, the IAFF and IAFC presidents ceremoniously signed the *Guiding Principles* at both the IAFF Annual Convention and the IAFC's Fire-Rescue International conference so that each organization's membership could witness the commitment being made on this critical issue.

The International Association of Fire Chiefs (IAFC) and International Association of Fire Fighters (IAFF) have developed and agreed to the following guiding principles in the true spirit of cooperation for the enhancement of the fire/EMS service and the communities they serve:

- To recognize that labor and management have a mutual goal of ensuring the well-being and safety of fire/EMS personnel and providing high quality service to the public.
- To work together to improve communications, enhance training, increase participative decision making, and promote a labor–management relationship based upon mutual trust, respect, and understanding.
- To create labor–management partnerships by forming labor–management committees at appropriate levels, or adapting, as necessary, existing councils or committees if such groups exist.
- To provide systemic training to labor and management leaders on collaborative methods of dispute resolution, recognizing that this process allows management and union leaders to identify problems and craft solutions to better serve their members and the public.
- To promote these principles to our respective members at all levels of both organizations.

The benefits of this program are both numerous and very worthwhile. These guidelines provide leadership and guidance for those who employ these principles and hopefully will move labor–management issues and disputes toward a more successful and improved conclusion. Some of the benefits include the following:

1. Improved labor–management relations
2. Increased trust between the fire chief and union president
3. Improved interpersonal communications between labor and management
4. Increased involvement and cooperation from all members of the organization
5. Increased focus on goals benefiting both the organization and its members
6. Reduction in adverse actions that affect labor–management relationships

The Labor–Management Initiative is typically delivered in one of two primary ways: to the leadership of the fire department and the union within a single fire department, or regionally, in cooperation with state fire chief and firefighter associations. The program via the second method is delivered to representatives of approximately twenty fire departments at a single conference. For a fire department to participate, the fire chief and union president must attend the workshops. The IAFC and IAFF can also provide facilitators for local or regional training events, such as conferences, symposiums, and workshops.

As one can glean from the preceding information, this effort is in line with the concepts of enlightened labor relations. These principles can help make both the union and the fire department stronger and better managed. The end result will be improved working conditions for firefighters without the strife and disruptions that are abundant in the history of fire departments. This effort represents a significant step forward in an intelligent and proactive manner. All fire chiefs and union leaders need to understand and utilize these principles.

GETTING TO YES

The crux of contract negotiations or building a positive labor–management relationship is obtaining agreement, or "getting to yes." The process of obtaining agreement has undergone a great deal of study and thought. The concept of *principled negotiation* was developed at the Harvard University Negotiation Project by Roger Fisher and William Ury. The basis of principled negotiation is that issues are decided on their merits and that mutual gains are sought wherever possible. Where the interests of the parties conflict, the goal is a result based on a fair standard independent of the will of either side. The methods of principled negotiation are hard on the merits, but soft on the people. As such, this concept has a great deal of value in establishing and extending the desired positive labor relations.

The work of the Harvard University Negotiation Project focused on developing an alternative to positional bargaining. Its principled negotiation method can be summarized in four basic points.[15]

NEGOTIATION FOCUS	PRINCIPLED NEGOTIATION STRATEGY

People	Separate the people from the problem
	   –Understand attitudes and perceptions
	   –Communications is key
	   –Build relationships

Interests	Focus on interests, not on positions
	   –Understand real desires
	   –Identify multiple interests

Options	Generate a variety of possibilities before deciding what to do
	   –Use creative thinking
	   –Seek numerous options
	   –Explore shared interests

Criteria	Insist that the result be based on some objective standard
	   –Validity must be demonstrated
	   –Use expert advice

The first point recognizes that people have strong emotions, and these emotions become entangled with the merits of the problem. Once a person's ego becomes closely identified with a position, the situation intensifies. It is important not to express a personal position too strongly before the negotiations. For example, if a union leader at a union meeting tells everyone that he or she will never agree to this issue, the statement becomes difficult to back away from, even if it makes sense to do so. By keeping the people separate from the problem, the two sides can focus more on issues and circumstances, not on each other.

Focusing on the interests and not on positions drives attention to the heart of the matter. Often in negotiations, especially in the initial stages, various positions are bantered about. Both management and labor need to dig past the extraneous opinions to reach the underlying interests of the parties involved and attempt to address them. In the fire service, for example, a number of positions have been expressed about "two-in/two-out" procedures on the scene of emergencies. Although the stated position is clear, perhaps the real interest is adequate staffing for emergencies. If that matter could be addressed, the actual interests of the parties could be resolved.

In any problem-solving technique, developing a number of options to consider is recommended. Due to pressure and perhaps due to an adversarial relationship, this ability is sometimes inhibited in negotiations. When the consequences are high, the ability to develop options is hampered. It is good to set aside time specifically for this purpose and encourage creativity in the development of options. This list can then be narrowed down to options that advance shared interests and include possibilities for mutual gain.

Often the negotiations come down to who has the most power in the relationship. The person or group with the power can at times force the agreement by concessions of the other party. This type of negotiation erodes the positive relationship necessary for long-term success between union and management. When the power balance changes, and it eventually will, the tendency for payback is likely to emerge. By using objective criteria and standards, a fair solution is attainable. In this method, the focus is on the criteria and not on the relative power of the parties involved in the negotiation. The criteria form the focus rather than what either party is willing or not willing to do. An objective solution emerges that satisfies both parties.

The positive and negative aspects of positional bargaining as opposed to principled negotiation can be summarized in the following chart.

Principled negotiation methods focus on basic interests, mutually satisfying options, and fair standards that typically result in a wise and acceptable agreement. This method permits reaching consensus on a decision in an efficient manner, because it avoids wasting time establishing hard positions only to have to look for a way out later in the process. Principled negotiations have worked in many different settings and environments. Realistically, parties in a negotiation do not often have equal bargaining power, so the temptation to engage in pressure bargaining may be strong. If this occurs, one can attempt to keep the discussions in a principled negotiations model by refusing to participate in this type of tactic. It takes discipline, but it is vital to stay on the issues and not on the positions.

Fire service personnel managers should become well versed in these techniques and use them to obtain the type of agreement on issues that bodes well as a solution and for the long-term relationship.

POSITIONAL BARGAINING		PRINCIPLED NEGOTIATION
SOFT	**HARD**	**PRINCIPLED NEGOTIATION**
Participants are friends	Participants are adversaries	Participants are problem solvers
Goal is agreement	Goal is victory	Goal is a wise and mutual outcome
Make concessions	Demand concessions	Separate people from the problem
Trust others	Distrust others	Proceed independent of trust
Change position easily	Hold to your position	Focus on interests, not positions
Make offers	Make threats	Explore interests
Single answer they will accept	Single answer you will accept	Develop multiple options
Insist on agreement	Insist on your position	Insist on objective criteria
Yield to pressure	Apply pressure	Yield to principle, not pressure

Source: Roger Fisher and William Ury, *Getting to yes* (New York: Viking Penguin, 1981).

Summary

The motivation to join a union is an attempt to fulfill a variety of human needs, including those for security, community, esteem, and a sense of involvement and fair treatment within the work environment. Historically, union and management have operated as adversaries because many of their goals are in conflict. However, this conflict is detrimental to both management and labor, and modern labor relations have reduced this conflict to the benefit of both parties.

Collective bargaining is a complex process in which union and management negotiators maneuver to obtain the most advantageous agreement relative to their issues and objectives. The process of collective bargaining in the private and the public sectors is different. Although many of the practices involved with public and private sector labor relations are identical, their underlying legal frameworks are completely different. The very nature and purpose of government as opposed to the management of a private company present major philosophical and legal differences. The history of unionism in the United States is intertwined with labor legislation. Much of the progress of the labor movement and the improvement in the working conditions for employees can be tied to legislative victories.

The effectiveness of collective bargaining and contract administration is usually assessed by a measure of how well the process is working. In most cases the bargaining process works, and a contract results that both sides can support and live with. This preferred method is the most productive and least disruptive to both parties affected by the bargaining agreement. However, contract negotiations can reach an impasse, and when they do, a number of options are available to bring forth a resolution of the issues in dispute. Because of the difficulty of writing an unambiguous agreement anticipating all of the applications to the workplace, disputes may occur over the interpretation and application of some sections of the contract. The grievance procedure is the most common method of resolving these disputes in an organized manner.

The circumstances, people, and work environment of the future will continue to create new opportunities between labor and management that must be acted on in order for an organization to remain competitive in the world marketplace. As economic conditions in the world change substantially, so do union–management relations. Unions and management see cooperative relationships as instrumental in protecting jobs and organizations. The desired positive relationship of trust, communication, and respect necessary for effective labor relations takes time. It needs to be built day by day on small issues and circumstances that establish this relationship's full potential. The goal is to achieve enlightened labor relations as a framework for cooperation and the resolutions of differences.

Review Questions

Use these questions to review the material in this chapter and for discussion purposes.

1. How do you assess the strength of unions since World War II? What factors have been significant in the membership of unions during this period?

2. What is the significance of the Wagner Act of 1935, and what did it mean to unions at that time?

3. Identify and explain the steps necessary for a union to be properly recognized and certified

as the exclusive bargaining agent for a group of employees.

4. What are the three major types of contract and grievance impasse resolution procedures? Identify the strengths and weaknesses of each procedure.

5. Identify and explain three methods to improve the labor and management relationship. How you would apply them to your fire department?

Case Study

You are the fire chief of a large combination fire department in a suburban county within a large metropolitan area. Your department has approximately 800 career firefighters and officers and over 1,000 volunteer firefighters and officers. Some of the volunteer fire officers are career firefighters in the neighboring jurisdictions. The local union of the IAFF has filed charges against the volunteers who are career firefighters and union members in the neighboring jurisdictions. They have been accused of being members of a hostile organization because of their volunteer activities, which means there is

less need for career firefighters. This is a very emotional issue within the organization and has received extensive media coverage. The volunteer organizations oppose this action because it strips them of their membership and raises issues of individual rights to do what one wants on one's own time.

What are the union issues here, and what logic is behind this move to curtail volunteer membership? What are the issues on the volunteer side, and how will these firefighters approach this situation? Most importantly, how do you as the fire chief deal with this very volatile situation?

Fire Service Professional Qualifications

This chapter contains information that will assist fire officers in meeting the job performance requirements (JPRs) of NFPA 1021, Standard for Fire Officer Professional Qualifications:

4.2 Human Resource Management—Fire Officer I

4.2.5 Apply human resource policies and procedures given an administrative situation requiring action, so the policies and procedures are followed.

7.2 Human Resource Management— Fire Officer IV

7.2.2 Evaluate current employee/management relations and initiate the development

of a process that supports a positive and participative employee/management program.

7.3.2 Monitor local, state, and federal legislative activities, given fire service issues, in order to enhance the effectiveness of the fire department.

Endnotes

1. Wendell French, *Human resources management,* 2nd ed. (Boston: Houghton Mifflin, 1990).

2. Wayne F. Cascio, *Managing human resources,* 7th ed. (New York: McGraw-Hill, 2006).

3. Bureau of Labor Statistics, Union members in 2006, *Union Members Summary Report,* January 2007.
4. John Fossum, *Labor relations,* 3rd ed. (Plano, TX: Business Publications, 1985).
5. Wayne R. Mondy and Robert M. Noe, *Human resource management,* 7th ed. (Upper Saddle River, NJ: Pearson Education, 2005).
6. Randall S. Schuler and Susan E. Jackson, *Human resource management,* 6th ed. (New York: West Publishing, 1996).
7. Gary Dessler, *Management: Leading people and organizations in the 21ˢᵗ century,* 2nd ed. (Upper Saddle River, NJ: Prentice Hall, 2001).
8. French, *Human resources management.*
9. Michael R. Carrell and Christina Heavrin, *Labor relations and collective bargaining,*
10. 6th ed. (Upper Saddle River, NJ: Prentice Hall, 2001).
11. Ibid.
12. Raymond A. Noe et al., *Human resource management,* 2nd ed. (Boston: McGraw-Hill, 1997).
13. J. R. Redecker, *Employee discipline: Policies and practices* (Washington, DC: Bureau of National Affairs, 1989).
14. Mondy and Noe, *Human resource management.*
15. Dennis Compton, *When in doubt, lead!* (Stillwater, OK: Fire Protection Publications, 1999).
16. Roger Fisher and William Ury, *Getting to yes: negotiating agreement without giving in* (New York: Viking Penguin, 1991).

Suggested Readings

Carrell, Michael, and Christina Heavrin. 2001. *Labor relations and collective bargaining,* 6th ed. Upper Saddle River, NJ: Prentice Hall.

Cascio, Wayne F. 2006. *Managing human resources,* 7th ed. New York: McGraw-Hill.

Compton, Dennis. 1999. *When in doubt, lead!* Stillwater, OK: Fire Protection Publications.

Fisher, Roger, and William Ury. 1991. *Getting to yes: negotiating agreement without giving in.* New York: Viking Penguin.

Fossum, John. 1985. *Labor relations,* 3rd ed. Plano, TX: Business Publications.

French, Wendell. 1990. *Human resources management,* 2nd ed. Boston: Houghton Mifflin.

Mondy, Wayne R., and Robert M. Noe. 2005. *human resource management,* 7th ed. Upper Saddle River, NJ: Pearson Education.

Noe, Raymond A., et al. 1997. *Human resource management,* 2nd ed. Boston: McGraw-Hill.

Redecker, J. R. 1989. *Employee discipline: Policies and practices.* Washington, DC: Bureau of National Affairs.

Schuler, Randall S., and Susan E. Jackson. 1996. *Human resource management,* 6th ed. New York: West Publishing.

Swoboda, Frank. 1999. Membership trends are mixed for unions. *Washington Post,* January 26.

Varone, J. Curtis. 2008. *Fire officer's legal handbook.* Clifton Park, NY: Thomson Delmar Learning.

PGFD Archives.

KEY TERMS

effectiveness, *p. 325*

efficiency, *p. 324*

extrinsic rewards, *p. 337*

hygiene factors, *p. 337*

incentive systems, *p. 341*

intrinsic rewards, *p. 337*

motivators, *p. 337*

performance management, *p. 331*

productivity, *p. 322*

OBJECTIVES

After reading this chapter, you should be able to:

- Describe productivity.
- Describe increasing productivity.
- Understand performance management.
- Describe Kaizen principles.
- Describe motivation.
- Understand performance incentive programs.

Many in government service have often heard the phrases "more bang for the buck" and "doing more with less." These slogans are but two of many that embody the concepts of productivity and performance. The desire to have high-performing and efficient government services affects the fire and rescue service in many ways. The fire service is being increasingly challenged to prove that it is productive and that it is using limited tax dollars toward the best value for the community it serves. Many fire departments have experienced difficulties because of a lack of data and the inability to articulate how well they are achieving their mission. Some fire departments are having problems identifying both what their mission is and what level of service they expect to provide.

Measuring and managing performance are challenging enterprises and two keys to productivity and organizational effectiveness. Organizations can gain a competitive advantage by their wise and innovative use of human resources. This chapter focuses on productivity and performance improvement of the personnel within an organization as opposed to the use of technology and other methods to enhance performance.

Productivity

productivity
■ the ratio between inputs and outputs that indicates the value added by an organization. Productivity has to do with goods produced or services rendered in relation to cost in terms of time, finances, and other resources.

Jack Welch, past CEO of General Electric, understands that the best companies know, without a doubt, where **productivity,** real and limitless productivity, comes from. It originates from challenged, empowered, excited, and rewarded teams of people. It comes from engaging every single mind in the organization; making everyone part of the action; and allowing everyone to have a voice, a role, in the success of the enterprise. Doing so raises productivity increasingly.[1]

Recent economic and political conditions have forced public employers to pay more attention to the development of their employees and to increase productivity at every opportunity. The constant search for productivity improvements has meant that fire departments must have a better and more sophisticated understanding of employee motivation and satisfaction, job design, performance, and other related issues. As a nation, the ability to earn more money and to enjoy a higher standard of living depends in many instances on the improvement of productivity. The United States experienced productivity increases of less than 1 percent during the 1980s and into the early 1990s, but by the mid-1990s, U.S. companies were achieving rather strong yearly productivity increases of more than 3 percent. In many industries, the United States is now the global leader in productivity. Much of this success has been linked to improvements in human resource and personnel management techniques and practices. The strong relationship between productivity and systematically integrating effective personnel management with a flexible production system has been well established.[2]

Lack of productivity costs money and when costs are under threat, then so are jobs as well as benefits to employees. Regardless of one's position or the type of organization, the search for and the need to improve productivity is the ultimate goal. This is as true today as it was yesterday and as it will be tomorrow. Organizations waste financial resources as well as opportunities for improvement each and every day because of their inability or unwillingness to improve productivity constantly. This situation can be improved by a concerted effort at all levels including management and labor. Successful organizations have learned this lesson well and constantly work toward improving productivity in all levels of the organization.

By using the benefits of training, education, planning, and coordinating the activities of supervision and employees, solid improvements in productivity can be achieved. These activities form the basis of an effective and efficient organizational team. Managers need to display the best aspects of productivity improvement from their own actions by setting the proper example.

FUNDAMENTALS OF PRODUCTIVITY IMPROVEMENT

In any team approach to improving productivity, a number of fundamental aspects should be borne in mind:

- No organization is too small or too large, whether it is in the public or private sector, to rule out the possibilities of improving productivity.
- Productivity is a state of mind and should be promoted as such.
- Involve as many employees as possible in the search to improve productivity. Psychologically, it is sound to spread the degree of participation so that as many as possible feel that, to some extent, the success of the program rests with them.
- It should be a continuing program, for it is never finished. Improving productivity should therefore be a permanent, ongoing part of the organization at all levels.
- Concentration should be on controlling the controllable. Unless care is exercised, much time and money can be wasted in endeavoring to control the uncontrollable. Much of the success in improving productivity comes from judgment exercised in relation to categories in which various expenses lie and the respective opportunities offered.
- Improving productivity is not achieved by raising voices against people. Often there is too much talk about productivity and too little planning and concentrated effort.
- Improving productivity is the real key to national prosperity under a rising standard of living.
- Improving productivity requires resourcefulness, imagination, and enthusiasm.
- Productivity leadership flows from the top down. It rarely seeps up from the bottom.[3]

Productivity enables organizations to achieve their goals, save time and resources, and maintain satisfied employees and customers. The benefit to organizations, government, and to our general way of life is enhanced by increasing productivity. The more productive an organization is, the better its competitive advantage, because the costs to produce its goods and services are lower. Better productivity does not necessarily mean that more is produced; perhaps fewer people, less money, and less time were used to produce the same amount. It is useful to view productivity as a ratio between inputs and outputs that indicates the value added by an organization. Productivity has to do with goods produced or services rendered in relation to cost in terms of time, finances, and other resources. It is the quality or state of furnishing results, benefits, or profits. The "costs" of fire protection are the total system costs, which include many factors well beyond the fire department itself, including all funds spent on fire protection, such as fire insurance costs, water main costs to maintain fire flows, and costs of built-in fire protection, such as sprinkler systems and standpipes. Many fire service personnel managers tend to think of costs in terms of what the fire department directly controls, such as compensation, fire apparatus and equipment, and similar items. However, fire departments must consider the full effect of the larger costs of providing fire and rescue protection not only to the fire department but also to the community.

This systems approach should be applied when one is making decisions about the relative worth and value of productivity recommendations. Often what is put forth as a true gain in productivity is in reality just shifting the work somewhere else in the system, and the true productivity gain is suspect. However, the productivity and effectiveness of some decisions can have far-reaching consequences. An example is a community that adopts a residential sprinkler ordinance for all buildings, including one- and two-family dwellings. The inputs or resources needed by the fire department will be less over time as the systems are installed and perform, making the department more productive. Fire stations can be farther apart and fewer in number. Water main sizes may be reduced. Fire insurance costs will go down, as will medical expenses for those who may be injured. Increased flexibility can be accomplished in the building code, saving homeowners money. Many more examples can be listed, but the productivity gains from a single decision can have far-reaching impacts within the entire system or community and not just within the fire department.

Productivity at the organizational level ultimately affects profitability and competitiveness in a for-profit organization and total costs in a not-for-profit or governmental organization. The department of human resources is most often analyzed and examined concerning productivity in an organization. This is true for fire service organizations, because 80 to 90 percent of career fire department budgets are personnel-cost related. One method to measure organizational human resource productivity is by unit labor costs. Unit labor costs are computed by dividing the average costs of workers by their average levels of output. By using unit labor costs, one can see that an organization that pays relatively high wages and benefits can still be economically competitive if it can also achieve a high level of productivity. As an example, in fire departments in which firefighters are also cross trained as paramedics, the productivity is increased in most cases because the human resources respond to a larger number and to more of a variety of emergency incidents. This is true even when the firefighter/paramedics are compensated at a higher rate than firefighters.

PRODUCTIVITY AND PERFORMANCE

Productivity concerns two specific assessments of performance: efficiency and effectiveness. **Efficiency** is measured as a ratio of outputs produced to inputs consumed. In a service organization such as a fire department, efficiency implies an ability to perform well while making good use of available resources. The measurement of efficiency requires the identification of a performance outcome, such as the number of fire responses and the amount of resources used to produce the outcome, such as funds allocated for fire suppression. The efficiency ratio for our example is then

$$\frac{\text{Number of Fire Responses}}{\text{Funds Allocated for Fire Suppression}}$$

The ratio measures number of fire responses per allocated funds for fire suppression. Efficiency in this example may be increased in two basic ways: by increasing the number of fire responses with the same allocation of funds, or by having the same number of fire responses and using less of the allocated funds. The number of fire responses can be reduced in a variety of ways, such as changing dispatch

efficiency
■ is measured as a ratio of outputs produced to inputs consumed. This implies an ability to perform well while making good use of available resources.

policy, improving fire prevention, better educating the public, and reducing flammable materials in buildings. However, the efficiency definition lacks a measurement of quality.

Effectiveness is the degree to which a goal has been achieved. It is the quality of the output measurement as compared to a standard. Effectiveness is an important concept, but the assessment of effectiveness has proven to be one of the more intractable problems in organizational theory. Organizations are large, diverse, and fragmented. They perform many activities simultaneously. They pursue multiple goals and generate many outcomes, some intended and some unintended. Sometimes efficiency leads to effectiveness, but in other organizations, efficiency and effectiveness are not related. An organization may be highly efficient but fail to achieve its goals because it makes a product for which there is no demand. Likewise, an organization may achieve its goals, but be inefficient in the process of doing so and also fail as a business.

The overarching goals in the fire and rescue service are to save lives and prevent damage or harm to property and the environment. In order to measure the accomplishment of these goals fully, the fire department would have to have some method to calculate lives saved, fires prevented, property damage avoided, and so forth. However, what most fire departments measure are fire deaths and injuries, the number of fire responses, accumulated property damage, and so on. It is very difficult to calculate many of the things that the fire department really needs to measure to determine its effectiveness. Departments can easily measure fire and EMS responses and the number of fire prevention demonstrations, fire inspections, building code reviews of plans, and other measurements in order to show productivity. What cannot be realistically measured are the results of these activities, which are the most important data. The issue of not having fully measurable results is a problem shared with most emergency service–oriented agencies. Some suggested measures for fire department effectiveness follow.

Effectiveness Measures for Fire Prevention and Suppression

- The number of fires per 1,000 population protected
- The dollars of property loss per capita
- The dollars of property loss per $1,000 of market value of the property
- The dollars of property loss per fire
- The number of civilian injuries and deaths per 100,000 population protected
- The number of civilian injuries and deaths per 100 fires
- The number of firefighter injuries and deaths per 100 fires[4]

Unlike business organizations, which exist to maximize profits, the fire service exists to maximize service at a fixed cost. This fixed cost is usually the fire department budget within its government structure. Fire departments occasionally experience difficulty in justifying expenses, because their goals are difficult to measure and quantify. As stated in the preceding paragraph, the fire service normally uses indirect measurements to explain and clarify just what is being accomplished and with what degree of efficiency and effectiveness. In many cases, the effectiveness and value of factors such as better management, the use of improved technology, and other attempts at increased productivity are difficult to assess. The fire service in the United States knows that it is doing a better job, and most of its indicators, such as fire deaths, fire injuries, number of structural fires, and others, are all down from previous years. What the fire service does not know is the exact cause of this reduction and a way to prove it. The best the fire service can do is infer the causation and at least be pleased that it is doing something right.

The National Fire Incident Reporting System (NFIRS) developed by the U.S. Fire Administration (USFA) attempts to collect data on a national basis regarding fire department activity. Unfortunately, not all states currently participate in this system, and many participating states do not fully report all jurisdictions within their respective states. It is difficult to justify expenditures and fire safety programs when as a profession the fire service cannot even quantify its basic business. Before a fire department can perform analysis, determine causation, and establish cost–benefit relationships, the essential basic data must be present. On a national basis, this area still offers room for a great deal of improvement.

Regardless of the success of the NFRIS system, it is still necessary to collect data at the local level. It is important that the data associated with the productivity factors be collected and reported on a regular basis. Monthly, quarterly, and annual reports provide an excellent medium to gauge trends and develop action plans to deal with identified problems. Reports also give outside agencies such as governmental leaders and the media an opportunity to realize the efforts and trends associated with fire department response. To be effective the fire department must do more than just count the statistics; it needs to use this information in the development of future prevention and fire suppression activities.

To measure productivity, fire departments first need to define what will be measured, determine the "benchmark" from which progress or gains will be measured, and decide how the data will be collected and measured. This need not be an elaborate process, but it must be accurate and consistently collected over time. Adequate training of those responsible for the data collection and analysis will greatly improve the accuracy of the results. Here are some examples of the types of data that a fire department would want to collect:

Fire Suppression

- Fire extension after arrival
- People rescued from fire incidents
- Fire-loss statistics
- Amount of mutual aid requested
- Response times
- Activation of fire suppression systems

Fire Prevention

- Number of actual preventable structure fires
- Dollar loss associated with various fire causes
- Number of inspections
- Number of fire code violations
- Number of building plans reviewed

Emergency Medical Services

- Number of responses for BLS and ALS
- Number of transports
- Number of heart attack victims saved

- Number of blood pressure screenings
- Extrication time and transport time

Public Education

- Number of smoke detectors inspected/installed
- Number of groups receiving instruction
- Number of public service announcements
- Participation in home escape plans and drills
- Presentations to special risk groups

The preceding lists are just illustrative; your local circumstances and the local demographics will determine the need to collect certain types of data. In the data collection process, it is wise to be aware of the state and national databases and ensure that what your fire department collects at the local level can be uploaded to other systems in a compatible format.[5]

Increasing Productivity

Increasing productivity is invariably a good thing for an organization to do. Some fire departments have been justifiably criticized for not pursuing increases in productivity and for being very traditional and resistant to doing things in a new way. Increased productivity must be real, and it must represent an increase in the value and service that the fire department contributes to the community. It cannot be just a facade or give the appearance of increased productivity. Several approaches to improved productivity are as follows:

Redesign the Work.   Some work can be redesigned to make it faster, easier, better, or more rewarding to employees.

Outsourcing.   Contract with someone else to perform activities previously done by the employees of the organization.

Employee Enhancements.   Improve training by replacing outdated and inefficient processes, methods, rules, and procedures.

Technology.   Use technological improvements to improve the productivity of workers, such as bar code scanners and other such devices.

Replacing Workers with Equipment.   Certain jobs are not done as well by humans. The jobs may be repetitive, physically difficult, or require great precision.

Capital Equipment Improvements.   Typically the more spent on equipment per worker, the greater the output per worker.[6]

At the local jurisdictional level, a fire department can implement a number of methods to measure the services that it provides. These results allow for comparisons and provide a source of information with respect to the achievement of the department's goals.

COMPARING DATA FROM PREVIOUS YEARS

The most common method is to compare current data to data from previous years within the same department. The annual reports of many fire departments include activity data of various sources, such as number of responses, inspections, and injuries. The strength of this method is that the comparisons are normally valid in that the same jurisdiction uses the same definitions and categories for the data. These comparisons can be made over a substantial period of time. However, most fire departments do not correct for population changes and other demographics that change from year to year within a given community. The fallacy of this method is that the department is comparing only its current performance to its own past performance and experience, rather than to an established benchmark or independent standard.

COMPARING DATA FROM SIMILAR DEPARTMENTS

In this method, the fire department is compared to other fire departments in terms of some identified criteria, such as population, region, personnel size, or other factors. Depending on the accuracy and completeness of the data, valid comparisons can be made. Although these results may not be definitive, they can highlight certain factors that may warrant further examination and study. The key is determining how similar the departments actually are. Comparing apparently similar departments can be misleading and result in erroneous conclusions due to underlying differences not noted in the comparison.

As an example, the Phoenix Fire Department conducted a survey to determine the workload experienced by fire departments in delivering fire suppression and emergency medical service responses. The survey included 159 fire departments that serve populations of 100,000 or greater. The Phoenix Fire Department could then compare its experience to that of a number of similar fire departments.

Number of Fire Calls per Thousand Population

POPULATION	MEAN	MEDIAN	RANGE
100,000–249,000	21.5	14.4	0.7–115
250,000–499,999	18.4	12.4	4.2–99
500,000+	26.7	17.6	3.9–108
All	21.6	12.4	0.7–115

Source: Phoenix Fire Department.

Number of EMS Calls per Thousand Population

POPULATION	MEAN	MEDIAN	RANGE
100,000–249,000	48.2	41.9	0.5–129
250,000–499,999	54.6	55.9	17.5–117
500,000+	57.2	66.4	3.7–97
All	51.1	45.1	0.5–129

Source: Phoenix Fire Department.

INTERAGENCY COMPARISONS

This method attempts to compare the services within a given community. The data derived from these comparisons are usually qualitative and opinion driven. However, this information can be valuable in assessing how well the public views the work of various government agencies or how the agencies view themselves in comparison to one another. In most public opinion surveys, the fire service is regarded in a notably positive manner.

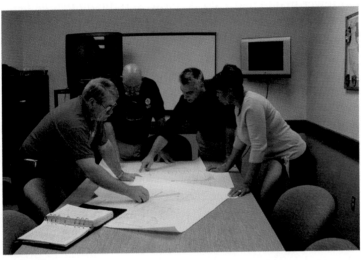

Source: MFRI Archives

USING MODELS

This method requires the use of models or computer simulations in determining optimal operational performance and then running a comparison with existing operating procedures. Information, alternatives, and recommendations can then be suggested for adoption by the local jurisdiction. The difficulty with the use of models is that they can handle minimal variables at one time, although this limitation is minimized with advances in computer technology.

Determining the causation and the interdependence of the many variables involved regarding issues of productivity and effectiveness of fire service organizations is a significant challenge for managers. The use of operations research techniques that link a technical approach to decision making with the power of computer simulations is promising. Such methods can address the specific area in which performance may be improved and also simulate what would happen if proposed changes were made to the system. In some of these advanced simulations, the differences in productivity can be assessed without actually making the changes in the real world. The limitation of this type of analysis is that it is normally restricted to a single operation or variable at a time. Fire departments must be extremely careful not to improve productivity and performance in one area while reducing it in another. As an example, a fire department could reduce staffing on a per-station basis and open more stations to reduce the response time. However, property damage may increase because of the inability of smaller-size companies to control fires. It is important that analysis results be combined with the experienced judgment of fire service managers to ensure that the conclusions are accurate and properly evaluated.

COST–BENEFIT ANALYSIS

A cost–benefit analysis helps determine whether the benefits of an alternative exceed its cost and choose the alternative that provides the maximum net benefit. *Tangible benefits* are those that can usually be related directly to their costs and ranked or evaluated accordingly—for instance, lower compensation costs, reduced inventory, and fewer maintenance costs. However, the *intangible benefits* of alternatives are also important and must be given full consideration, including customer satisfaction, organizational flexibility, and improved working relationships.

The costs associated with providing fire and rescue protection are increasing, and fire service leaders are under continuous pressure to reduce public expenditures wherever possible. In this analysis, the first look is at those costs that require excessive funding for the results achieved. By reducing the nonproductive costs, the total system can become more efficient. As an example, most fire departments in the United States have removed the street box system for the notification of emergencies. They were expensive to maintain and had little value when the methods of communication technology, including cellular phones, rapidly expanded. The street box system was productive at one time, but false alarms, increased maintenance, and advanced technology eliminated the need for this system. A separate system for the notification of fire departments was no longer needed because existing communication systems could be used, thereby reducing the costs to the total system.

ACCREDITATION

One effective way in which to judge and compare the performance of a fire department is by meeting an established standard for accreditation by an independent organization. Many previous methods of achieving accreditation involved evaluations by insurance organizations or associations. These evaluations served a good purpose, but their basic premise was to establish fire insurance premiums and rates for a jurisdiction, which serves the purposes of fire insurance underwriters, but evaluates the fire department from only a single perspective.

In 1998, a joint effort of the International Association of Fire Chiefs (IAFC) and the International City/County Managers Association (ICMA) resulted in the Commission on Fire Accreditation International. This commission provides a comprehensive system of fire and emergency service evaluation that can assist local governments in determining their risks and fire safety needs, evaluate the performance of the organizations involved, and provide a method of continuous improvement. Many local government officials find it difficult to justify an increase in expenditures unless they can be attributed directly to improved or expanded service delivery in the community. This effort is often hampered by the lack of a nationally accepted set of criteria by which a community can judge the level and quality of fire, EMS, and other services provided.

The Commission on Fire Accreditation International provides an excellent self-assessment process for fire and emergency service agencies to evaluate themselves. This self-assessment is then followed up by a peer assessment team that provides an extended on-site visit to the jurisdiction and then submits a report to the commission regarding its recommendation for accreditation status. The process seeks to answer three basic questions:

1. Is the organization effective?
2. Are the goals, objectives, and mission of the organization being achieved?
3. What are the reasons for the success of the organization?

As fire departments have moved into the twenty-first century, the ability to measure performance is critical for their success. In meeting the requirements of the Commission on Fire Accreditation International, fire departments can elevate the professionalism and level of service delivery within their jurisdictions.

Performance Management

Performance management includes the activities of an organization that seek a competitive advantage through the efforts of its employees and the functions of the organization. Performance management ensures that the work of individuals is in concert with the goals of the organization and that total organizational performance is maintained at a high level. Performance management is much more than appraising the performance of employees through an evaluation process. Individual employee appraisal systems are important, but, in this process, the motivation, incentives, and general success of the organization are as important. Performance management requires a commitment and desire to improve individual and organizational performance at every opportunity on a continuous basis throughout the year.

At a general level, in order to be proficient in the performance management process, a manager must be able to do the following three things well[7]:

1. Define performance.
2. Facilitate performance.
3. Encourage performance.

In order to monitor and improve performance, one must first be able to define what is expected of employees and the organization. This task involves the setting of goals at all levels within the organization. Research has shown that organizations that establish clear, specific, and challenging goals are the ones with higher levels of achievement and performance. On average, studies have demonstrated that productivity can be improved by up to 10 percent by using effective goal-setting strategies.[8] The establishment of the goals is not sufficient. They must be developed through a process of involvement within the organization and be realistic with regard to the level of resources and opportunity for success. In addition, the goals that define the performance need to be evaluated on a periodic basis, and this assessment must be shared with those employees responsible for the attainment of the goals. Therefore, the goals must be measurable and an agreed-on plan should be in place demonstrating the manner in which this measurement will be performed. Defining performance demands that management establish goals, formulate a plan as to how to measure accomplishments, and then communicate the goals and measurement plans throughout the organization.

performance management
- the activities of an organization that seek a competitive advantage through the efforts of its employees and the functions of the organization. The goal is total organizational performance to be maintained at a high level.

Performance is generally considered to be a function of both ability and motivation and can be expressed as such:

$$\text{Job Performance} = f\,(\text{ability})\,(\text{motivation})$$

Ability depends upon many factors such as education and experience, and it can be a very long-term process to make improvements in these areas. When compared to ability, motivation can be improved in a much quicker time frame. In general, a manager has to apply many more options when attempting to improve motivation. As a general guideline, there are seven basic strategies regarding improving motivation toward better performance:

- Positive reinforcement and high expectations
- Effective discipline
- Treating people fairly
- Satisfying employee needs
- Setting specific work-related goals
- Restructuring jobs
- Base rewards on performance

These basic strategies need to be adapted to specific work environments in order to be effective. Essentially, there is a gap between what the individual and organization are producing and the desired state of performance. The goal then is to move both individuals and the organization in a positive direction toward better performance at all times. This takes a concerted effort and planning and may be frustrating at times, but organizations that perform these seven guidelines well will prosper in the long term.[9]

Fire departments that strive to provide high levels of performance recognize that they must facilitate performance at every level of the organization. Nothing is more frustrating than wading through a bureaucratic structure that hinders rather than facilitates performance. Effective leaders are adept at removing the obstacles to high performance within their organization. Obstacles may include overly detailed operating procedures, poor use of technology, outdated equipment, or many other issues. In this context, leaders work to remove obstacles through organizational processes rather than through circumventing or ignoring policies and procedures.

Adequate resources must be available to accomplish the tasks necessary to attain these goals. Employees must be given a fair chance to achieve the goals; otherwise, they become frustrated and disenchanted with the process. Managers must always strive to provide adequate resources, but when that is not possible, goals can be adjusted accordingly. In order to facilitate performance, the organization must have employees who are motivated and capable of performing high levels of work. Fire departments have highly developed and competitive selection procedures for the purpose of ensuring the appointment of highly qualified firefighters and officers.

Encouraging maximum performance is different from facilitating performance. The encouragement of performance is at a much higher level and requires an active and persistent focus on the performance of the organization and the ability to improve it at every opportunity. Organizations that focus on performance also reward successful performance in meaningful and timely ways to encourage and reinforce this behavior. Fire departments that define performance and then proceed to facilitate and encourage high levels of performance are the leaders in their profession. They are recognized as providers of excellent public service in a productive and cost-effective manner. Because these organizations are a pleasure to work for, when positions become open, they are widely sought after. A high degree of satisfaction and pride is apparent within these organizations. This level of performance should set the standard.

Kaizen Principles

Kaizen refers to change and good. This Japanese management structure is based on the total quality management (TQM) principles, but goes further. In a management sense, Kaizen means continual and gradual improvements through evolution rather than through revolution. In a traditional organization, reexamination is a project. Within an organization that has adopted Kaizen, it is a way of life. Fire departments, which are particularly traditional and difficult to change, can be well served by the principles of Kaizen management. Continuous improvement and change are sought at every opportunity through a total team environment.

Ten Main Principles of Kaizen

1. Focus on customers, both external and internal.
2. Seek continuous improvement in products, services, and processes.
3. Acknowledge and bring problems to the surface.
4. Foster an environment of openness, honesty, and frankness.
5. Use teams as the primary unit of organization.
6. Manage projects cross-functionally.
7. Nurture supportive relationships at all levels of the organization.
8. Develop self-discipline, personal responsibility, and accountability.
9. Foster organizational communications in all directions.
10. Set up employees for success; enable their achievement.

It is fair to state that the preceding principles can serve as a way to achieve the desired level of performance in any organization that is seeking improvement, and all organizations should be. Many of the Kaizen principles involve team activities, for which fire departments are already structured. Fire departments and most organizations are organized on the basis of functional activity. This involves special expertise located in a number of discrete operating units, such as engine companies and truck companies. This generally good practice allows the organization to allocate work and achieve results that are measurable and can be accounted to a specific unit of the organization. However, often the organizational units can tend to be so focused on their tasks and responsibilities that the overall function of the organization is negated to a degree. One of the ways in which these disadvantages can be overcome is by the introduction of cross-functional teams that are tasked with a specific purpose, but the team members represent the unit capabilities of the organization. The secret to the success of these teams is that they focus on the organization as a whole and represent the sum of the interests, not just a particular unit. For example, the units within fire suppression may meet on a regular basis, but how often do they meet and work with the fire prevention section or apparatus maintenance? Most likely this occurs only when there is a problem. By having cross-functional meetings on a regular basis, problems can be prevented and the organization communicates and works in a more coordinated fashion.

Coordinating cross-functional teams is a time-consuming, often lengthy process, requiring some strife, realignment, and, above all, changed behavior on the part of every team member. From the extensive research into successful teams, a number of characteristics emerge that are present in high-performing cross-functional teams. It is useful to think of two sets of criteria that successful teams need to have in place: the preconditions and the characteristics of high-performing teams.

PRECONDITIONS OF SUCCESSFUL TEAMS

Preconditions must be provided outside of the team, usually by those who establish the team and set up the reporting relationships. Preconditions mostly have to do with the operating environment of the team and include the following.[10]

Purpose

Successful teams must have a clear sense of why they exist and the organizational strategic significance of why they have been established. Each team member needs to understand his or her role on the team and in the broader context of the entire organization.

Empowerment

Successful teams must have a strong sense of being in charge of their own destiny, within reasonable limits. They need to be responsible and accountable for what they do and recommend. Their recommendations must be taken seriously by those to whom they report. The success of the teams should be recognized, and the failures brought back for further analysis and review.

Support

Teams must be fully supported by the organization, especially by the person to whom they report. At times, others within the organization might be mistrustful of the team, because they were not included or the team was working on a high-profile and important issue. The team needs to be protected from this potential. However, the team activities must be explained, properly presented, and discussed at large within the organization at the proper time.

Objectives

Successful teams must translate their purpose into a series of measurable objectives that those responsible for the team can understand and accept. These objectives may not emerge initially, but when they do, they need to be explicit and be reviewed on a regular basis.

CHARACTERISTICS OF SUCCESSFUL TEAMS

Team characteristics describe those things that teams can do on their own behalf to make themselves successful. Getting these characteristics right implies that team members choose to conduct themselves in certain ways, monitor their behavior, and live up to the expectations of the other team members as well as to those of the organization. The training of the team members is important, and it is critical on certain tasks to ensure that the team members have the requisite skill and knowledge to be on the team. The characteristics that successful teams have are described in the following sections.

Interpersonal Skills

Successful teams develop the ability to work together without unproductive conflict. They do argue and disagree, but they do not allow differences to disrupt the purpose of the team. Generally, this means that there is a good degree of respect and decorum that is afforded to every team member. Everyone is a valued member and part of the team.

Participation

Good interpersonal skills generate a high degree of participation among team members. The team is organized so that everyone has the opportunity and is expected to contribute to the final product. Diversity of thought and action is a valuable part of this process as team members contribute their experience and understanding of the issues. Participation also implies that when a team member commits to doing something, it is completed as required.

Decision Making

Decisions are reached with the proper evaluation of information, analysis, and research. Many options and alternatives, as well as the consequences of the recommendations, must be evaluated. Team members must have a realistic view of the organization and environment in order to develop recommendations that the organization will support and implement. Team members must be committed to the final recommendations, regardless of their personal views.

Creativity

Successful teams always actively seek out new ideas, perspectives, and different ways of accomplishing things. People build on the thoughts and ideas of others, and this synergy creates new perspectives and solutions to the issues at hand. The creative process must be followed in order to have recommendations represent a new way of business, as opposed to just extending the old methods.

Managing the External Environment

Good teams operate in a way that ensures interaction with the organization, so that people know what is occurring to the extent that they want to. This serves a purpose in reducing suspicion and enables a higher degree of cooperation with others in the organization outside of the team. Periodic progress reports and other such mechanisms may be employed to achieve this.[11]

In comparing the preconditions and characteristics of successful teams with the ten principles of Kaizen, there is a high degree of congruence. Kaizen promotes a code of behavior based on openness, support, and self-discipline that works well with the observed behavior of successful teams. These principles work equally as well in the fire service environment, especially because the team concept is well ingrained in every firefighter. Successful fire service personnel managers use this natural advantage and apply the concepts of Kaizen, TQM, or whatever similar system works. This enables the fire department to move toward the future with a new perspective and purpose, not bound by the unnecessary traditions of the past.

Motivation

An organization is, in its most basic form, a system of cooperative human activity. Individuals create organizations to accomplish specific purposes. The organization attracts individuals who wish to satisfy their needs for money, interesting work, status, and so forth. In a perfect world, the organization would accomplish all of its objectives and the employees would satisfy all their needs in this process. The ideal is seldom attained. Motivation is important in an organization, especially with respect to high levels of productivity and performance. Motivation is the individual's needs, desires, and concepts that cause him or her to act in a particular manner. The task is for organizations to direct individuals so they can satisfy their needs as much as possible as they accomplish the objectives of the organization.

EXPECTANCY THEORY

One theory among many regarding motivation is expectancy theory. This theory treats motivation as a function of a person's expectations about relationships among his or her efforts, the effectiveness of those efforts, and the rewards they obtain. According to this theory, an individual is motivated to produce at a high level if he or she perceives these efforts resulting in successful performance. There is a perceived link between effort and performance. In addition, the individual must perceive that successful performance results in desired outcomes or rewards, which is the second important linkage in this theory. The complication is that for the same work, two people may have different desired rewards: One may desire money, whereas the other views money as unimportant (see following chart).

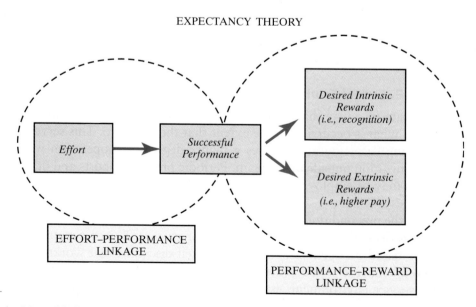

EXPECTANCY THEORY

Expectancy Theory

Desired outcomes can be intrinsic or extrinsic. **Intrinsic rewards** relate directly to the nature of the work itself and to whether it is interesting and challenging. **Extrinsic rewards** do not relate directly to the nature of the work itself, but may be linked to outcomes, such as salary increases and benefits. Research evidence supports the performance–reward linkage in expectancy theory; that is, individuals do continue to perform at a high level if they obtain the intrinsic and extrinsic rewards that they desire. Where the organization is unable or unwilling to provide for the proper linkage between effort–performance and performance–reward, frustration and a lack of motivation may result.[12]

intrinsic rewards
- relate directly to the nature of the work itself and to whether it is interesting and challenging.

extrinsic rewards
- do not relate directly to the nature of the work itself, but may be linked to outcomes, such as salary increases and benefits.

EQUITY THEORY

The aspects of the expectancy theory are also important with respect to motivating employees toward high levels of performance. In expectancy theory, individuals compare the ratio of their inputs and outcomes to the input–outcome ratio of another individual whom they believe to be comparable to them. If the individual does not perceive a balance in this formula, he or she attempts to restore it by working more or less efficiently or by trying to obtain greater rewards. For example, if a firefighter receives recognition and merit pay and another firefighter perceives that he or she works at the same level but does not receive merit pay, the second firefighter may adjust performance accordingly.

TWO-FACTOR THEORY

Frequently, the terms *satisfaction* and *motivation* are mistakenly interchanged. Frederick Herzberg developed what he called the "two-factor theory" concerning motivation. In this theory, **hygiene factors** refer to the context of the job as opposed to the content of the job. These factors include competence of supervisors, salary, company policies, and others. **Motivators,** however, make a person feel good about a job, so he or she strives to produce more. These factors include praise, a sense of responsibility, and successful completion of tasks.

hygiene factors
- refer to the context of the job as opposed to the content of the job. These factors include competence of supervisors, salary, company policies, and others.

motivators
- circumstances that make a person feel good about a job, so he or she strives to produce more. These factors include praise, sense of responsibility, and successful completion of tasks.

Hygiene Factors	Motivators
Supervision	Recognition
Interpersonal relations	Work
Working conditions	Responsibility
Salary	Advancement
Job security	Achievement

According to Herzberg, hygiene factors are independent from motivators. Hygiene factors can make an employee dissatisfied, but they do not contribute significantly to productivity. Motivators affect the individual's sense of intrinsic satisfaction and provide a positive feeling toward work. A firefighter who becomes dissatisfied with hygiene factors, such as pay and benefits or relations with other employees, may try to escape the work environment by increased absenteeism and other negative behaviors. However, even if the employee is satisfied with the hygiene factors, these factors do not significantly contribute to his or her level of motivation. Motivators act directly on the individual's level of productivity and create a positive environment and effort by the employee. The message is that organizations that concentrate only on hygiene factors such as pay and benefits may have satisfied employees, but not highly productive ones.

Fire department managers do not have significant control over many of the hygiene factors such as pay, benefits, and other factors in the context of the work environment. Dissatisfaction created by these factors is difficult for the manager to deal with, especially at the company or battalion level. In some cases, the manager must recognize the problem and still motivate employees as best as he or she can. When dealing with individuals, the same circumstances affect people differently. Fire department managers do have significant control over many of the motivational factors, such as recognition, stimulating work assignments, and the like. They should attempt to maximize these factors and take full advantage of the motivators that they can provide to the work environment.

In its most basic sense, job satisfaction is a positive motivational state that results from evaluating one's own work experiences. Job dissatisfaction occurs when one's expectations are not met. No single, simple formula can predict an individual employee's job satisfaction. Furthermore, the relationship between productivity and job satisfaction is not entirely clear. The critical factor seems to be what employees expect from their jobs and what they perceive as rewards from their jobs. Organizational commitment is the degree to which employees believe in and accept organizational goals and desire to remain with the organization. Various research studies have revealed that people who are relatively satisfied with their jobs are somewhat more committed to the organization. Commitment to the organization can be reflected in terms of lower absenteeism and turnover of employees. In my experience, it is clear that the satisfaction and commitment of employees to the fire service are very high. In most fire departments turnover is very low; in fact many firefighters work well beyond their minimum retirement time and essentially must be pushed out the door at some point. In comparison to other government agencies, the absenteeism is very low in most fire departments. This can be upset at times if firefighters feel that they are not appreciated by government or elected officials and the public. For the most part, it is unquestionable that firefighters are very committed to their work and go to extraordinary limits virtually every day of their careers. Nationally, the public esteem and trust in firefighters are very high, especially after the events of September 11, 2001. Firefighters appreciate and draw satisfaction from this, which is not found in most other professions.

MANAGERIAL GRID

The Managerial Grid Model is a behavioral leadership model developed by Robert Blake and Jane Mouton. The Blake Mouton Managerial Grid illustrates the concern for people and the concern for production found in successful managers and implies that the most successful managers emphasize both concerns strongly. This means that workers want leadership that provides good working conditions, job security, recognition, respect, and praise. It also means that leaders should use management skills to push for production goals as well as team harmony. The model proposes that when both people and production concerns are high, employee engagement and productivity increase accordingly. The Managerial Grid recognizes the possibility that some managers may lean more toward one of the two concerns than the other, or that the leader's concerns may shift depending on the particular situation. An experienced leader can read the conditions within an organization and understand the current motivational experiences in order to utilize the proper management for the current situation.

The Managerial Grid is based on two behavioral dimensions:

- *Concern for people*—This is the degree to which a leader considers the needs of team members, their interests, and areas of personal development when deciding how best to accomplish a task
- *Concern for production*—This is the degree to which a leader emphasizes concrete objectives, organizational efficiency, and high productivity when deciding how best to accomplish a task.

This model identifies five different leadership styles based on the concern for people and the concern for production.

Concern for People	High	Country Club management		Team management
	Medium		middle of the road management	
	Low	Impoverished management		Authority-compliance
		Low	Medium	High
		Concern for Production (Task)		

Authoritarian Leader (High Task, Low Relationship)

Leaders who get this rating are very much task oriented and are hard on their workers (autocratic). There is little or no allowance for cooperation or collaboration. Heavily task-oriented people display these characteristics: They are very strong on schedules; they expect people to do what they are told without question or debate; when something goes wrong they tend to focus on who is to blame rather than concentrating on exactly what is wrong and how to prevent it; and they are intolerant of what they see as dissent. This leadership style is often found in paramilitary organizations such as fire departments.

Team Leader (High Task, High Relationship)

This type of person leads by positive example and endeavors to foster a team environment in which all team members can reach their highest potential, both as team members and as people. These leaders encourage the team to reach team goals as effectively as possible, while also working tirelessly to strengthen the bonds among the various members. They normally form and lead some of the most productive teams.

Country Club Leader (Low Task, High Relationship)

This person predominantly uses reward power to maintain discipline and to encourage the team to accomplish its goals. Conversely, these leaders are almost incapable of employing the more punitive coercive and legitimate powers. This inability results from fear that using such powers could jeopardize relationships with the other team members.

Impoverished Leader (Low Task, Low Relationship)

These leaders use a "delegate and disappear" management style. Because they are not committed to either task accomplishment or maintenance, they essentially allow their team to do whatever it wishes and prefer to detach themselves from the team process by allowing the team to suffer from a series of power struggles.

Middle-of-the-Road Leadership (Medium Production/Medium People)

This style seems to be a balance of the two competing concerns. It may at first appear to be an ideal compromise. Therein lies the problem, though: When a leader compromises, he or she necessarily gives away a bit of each concern so that neither production nor people needs are fully met. Leaders who use this style settle for average performance and often believe that that is the most anyone can expect.[13]

It is important to recognize that the team leadership style is not always the most effective approach in every situation. Although the benefits of democratic and participative management are universally accepted, certain times call for more attention in one area than another. When organizations are faced with an economic hardship, people concerns may be placed on the back burner, for the short term at least, to achieve high productivity and efficiency. In fire departments, which are emergency oriented organizations, there are times, especially during emergencies, that the authoritative style is appropriate. Decisions have to be made and orders followed in very compressed time frames. What is important to understand is that this style does not have to carry over into the everyday nonemergency business of the department.

The Blake Mouton Managerial Grid is a practical and useful framework that helps managers critique their leadership styles. By plotting "concern for production" against "concern for people," the grid highlights how placing too much emphasis in one area at the expense of the other generally leads to low overall productivity. The model proposes that when both people and production concerns are high, employee engagement and productivity increase accordingly. Although the grid does not entirely address the complexity of "Which leadership style is best?" it does provide an excellent starting place to analyze one's skills critically and improve general leadership skills. This is really the main value of the grid—to think about leadership and the implications of one's particular style.

It is also important to understand that leaders may use the different styles of the Managerial Grid depending upon the circumstances they are faced with at any given time. While a leader may tend to be authoritative in handling discipline issues, he or she may be more of a team leader when working with the crew on a difficult assignment involving a fire prevention project. The main issue here is consistency, in that leaders use a style consistently to handle similar circumstances. A fire department officer cannot use an authoritative style to address a firefighter coming to work late one day and an impoverished style to handle the same case later in the week. Doing so would confuse everyone. Use the style that is most effective given the circumstances and use it in a consistent manner.

The theories reviewed in this section are basic and general motivational theories, explained briefly to make fire service personnel managers aware of them and their implications. However, people are different, and numerous factors affect an individual's personality and motivational level at any one time. Many of these theories may have nothing to do with the job. Therefore, it can be difficult at times to apply any of these theories to all individuals in the work environment. These theories are helpful in understanding what develops good motivation and productivity at both individual and organizational levels.

Performance Incentives

The most important factors in the successful performance and achievement of organizational goals are the ability and motivation of the people within the organization. To attract and retain such people, the organization provides incentives and rewards designed to focus the attention of employees to the specific behaviors that the organization considers necessary to achieve its desired objectives and goals. If performance incentives are to be useful in stimulating desired behaviors, they must reflect the demands of the employees whose behaviors they are intended to influence. Much of the work that firefighters perform is due to their desire to serve the public and to the intrinsic rewards inherent in this profession. Most of the fire departments in the United States are not driven by economic incentives; in fact, a good number are volunteer departments. This real and positive aspect of performance within the fire and rescue service is quite different from that of the the private sector. However, at some level an incentive program is necessary to motivate and properly recognize people for their contributions. This section focuses on several of the extrinsic **incentive systems** that are available to organizations.

incentive systems
■ designed to bridge the gap between organizational goals and objectives and employee expectations. They should provide a sufficient level of incentives to fulfill the basic needs of employees and direct their behavior toward the accomplishment of the stated mission.

The value of any kind of performance incentive system depends directly on the employee's perception of its worth. Individual perception relates to a range of factors, such as the employee's length of service, level in the organization, education, as well as to the physical and emotional state of the employee. Therefore, any organization must have a range of options available to provide performance incentives. Fire departments in general have a limited range of performance incentives available to them; however, many are evolving toward a greater number of options they can employ.

Incentive systems are designed to bridge the gap between organizational goals and objectives and individual employee expectations. To be effective, performance incentive systems should provide a sufficient level of incentives to fulfill the basic needs of employees and direct their behavior toward the accomplishment of the stated mission. In addition, the incentive system must be perceived as fair and possessing a high degree of equity. This essential equity can be demonstrated in at least three dimensions:

1. Internal equity—Are incentives fair in terms of relative worth of individual jobs within the organization?
2. External equity—Are the incentives paid for by an organization fair in terms of competitive market rates outside the organization?

3. Individual equity—Is each individual's incentive pay fair relative to similar performance within the same organization?[14]

The following reviews several of the more popular performance incentive programs available in both public and private sectors. Included are programs designed for short-term and long-term incentives.

MERIT PAY

The most traditional performance-based pay plan has been merit pay. The concept of merit pay has been implemented in a variety of ways, but most all plans share two basic components. First, some portion of the employee's pay is based on his or her rated performance in a given time period. Second, the merit increase awarded in one evaluation period is extended into the base pay in the future. Some merit pay incentive plans have a standard rate of percentage increase, and others include a range of pay increases from which the supervisor may select. Fundamental to a credible merit pay plan is a comprehensive performance appraisal system, as discussed in Chapter 8.

Critics of merit-pay-incentive plans indicate that the systems emphasize individual rather than group or team goals and that the actual merit pay award is not sufficient to inspire performance. In the early 1990s, the typical merit-pay difference between satisfactory and excellent performance at the $40,000 level averaged $17 a week after taxes.[15] The result is that the employees who are the most dissatisfied with this system are the ones the organization least wants dissatisfied—the top performers.

One of the more noted critics of merit-pay plans was W. Edwards Deming, the developer of total quality management programs. He stated that it is unfair to rate individual performance because "apparent differences between people arise almost entirely from the system that they work in, not from the people themselves." Deming also noted that the individual focus of merit pay discourages teamwork.[16] Perhaps the solution is to have an appropriate balance between individual and group or team objectives and performance in the merit-pay process.

PREMIUMS AND DIFFERENTIALS

Work premiums or differentials provide extra compensation for effort that is normally considered burdensome, distasteful, hazardous, or inconvenient. These premiums cover areas such as pay for overtime work, shift work, weekend or holiday work, and other situations that are unusual to the work environment. In some cases the organization has no choice, in that it must comply with federal wage and hour requirements, such as with overtime pay.

These types of incentives get the work done in less than ideal situations, but do not usually promote excellence or increased productivity. Many of the premium and differential items are a part of the collective bargaining agreement.

PAY FOR UNITS PRODUCED

Basing an employee's pay on some kind of measurable output is probably the oldest of all incentive programs. Frederick W. Taylor, the founder of scientific management, recognized that a slight addition to the employee's earnings could significantly enhance the individual's output. Three components provide the foundation for pay-for-units-produced plans: establishing the time required for the production of one unit, determining the standard level of performance in a standard week, and establishing an acceptable level of pay. If employees are able to produce more than the standard number of units, they are paid an additional rate for their increased productivity.

Pay-for-units-produced programs can be at the individual, team, or company level. They are difficult to use in service industries, especially those involved in public or emergency service.

LENGTH-OF-SERVICE AND SENIORITY AWARDS

Although seniority rewards are normally a direct part of the compensation system, many organizations recognize long service and loyalty with some form of award. It is an extra effort to ensure that the employees with long service are appreciated and are usually recognized with a plaque, lapel pin, or some other visible reward. It sets the example in the organization that long service and dedication are valued.

Many organizations have monetary awards and incentives to entice employees toward staying with the organization for an extended time. This practice reduces turnover and the filling of vacant positions within the organization. Several jurisdictions have length-of-service award programs that provide minimal retirement payments to volunteer firefighters. Most of these programs require active service of at least 25 years and a minimum age requirement in order to be eligible.

SUGGESTION PROGRAMS

Ever since the 1880s when the suggestion box was introduced, it has been a major factor in improving operations, productivity, products, and services. As an example, since 1898 when George Eastman presented a $2 award for a suggestion, the Eastman Kodak Company has accepted 740,000 ideas from a pool of more than 2.1 million.[17] Practically every incentive plan recognizes the importance of the suggestion and provides awards for stimulating employee creativity and innovation. To generate more suggestions and improve suggestion programs, organizations should always be specific when informing an employee about the reason an idea was not adopted. The rewards for suggestions should be in relationship to their value to the organization and its products or services.

This method taps into and takes full advantage of the talent within the organization. It is especially helpful to those who may have a valuable idea, but cannot get it up through the bureaucracy. The quality teams and quality circles of TQM have their roots in the suggestion box.

SPECIAL ACHIEVEMENT AWARDS

Organizations often recognize outstanding employee contributions through special awards. These awards are usually granted to a small number of employees on an annual basis. They are reserved for truly extraordinary performance well beyond the normal course of the employment activity. Some of these awards include a significant monetary bonus in addition to the recognition. These types of awards are publicized both within the company and to the media and professional journals. Many fire departments have an annual valor awards program to recognize the outstanding and heroic work of their members.

PROFIT SHARING

Profit sharing is any procedure under which an employer pays or makes available to regular employees, subject to reasonable eligibility rules and in addition to prevailing rates of pay, special current or deferred sums based on the profits of the business. Two major types of profit sharing are

Cash or Current Payment Plan. This plan provides for the distribution of profits by either cash or company stock, or by both, within a short period following the earning of the profit and the determination of the appropriate shares.

Deferred Plan. This plan provides for the placement of earned funds into an escrow account for distribution at a future date.

Numerous federal and state requirements apply to profit sharing plans. The profit sharing plans tie the compensation of the employee to that of the organization in a more direct way. As the company expands and profits, so do the employees via the sharing of profits. If a profit sharing plan is to have the proper motivational impact, the rewards offered to the employee must vary with the success of the business.

GAIN SHARING

Gain sharing plans have a rather extended history and have existed in a variety of forms. Both the well-known Scanlon plan and the Rucker plan were gain sharing plans. Gain sharing plans operate on the assumption that it is possible to reduce costs by eliminating wasted materials and labor, by developing new or better products or services, or by working smarter. The gain sharing plans operate on a high degree of cooperation and involvement between the employees and the employer. The employees involved develop and implement ideas related to productivity. The financial bonus is determined by a calculation that measures the difference between expected and actual costs during a bonus period.

Gain sharing programs are based on a measure of increased productivity, whereas profit sharing plans center on the degree of profit during a certain time period.

EMPLOYEE STOCK OWNERSHIP PLAN

An employee stock ownership plan (ESOP) is a defined contribution plan that operates like a qualified stock bonus plan. In an ESOP, the employer contributes stock to a stock bonus trust that qualifies as a tax-exempt employee trust. In most ESOPs, the stock allocation to a participant's account is in proportion to the relative base pay plus other short-term incentives from the organization for that year. Surveys indicate that companies with employee stock ownership plans are more profitable when compared to conventionally owned companies, and the more stock owned by employees, the greater the ratio becomes.

Successful performance incentive plans have several common characteristics. One of the most important is that they are based on a clear vision of the organization's goals and objectives. They operate by a compatible relationship between the individual or team and the organization, with all working toward and sharing in the success of the organization. Most incentive programs tend to allow for a long period of implementation and adjustment. The measurements used in the plans need to be valid, realistic, and stand the test of time during different business periods. Most of the performance incentive plans include a high degree of flexibility and discretion for the employee. The plans also include mechanisms so that they stay current and are not viewed as entitlement programs or as a part of the standard compensation package.

Fire departments have many opportunities to use the numerous performance incentive plans effectively. The only ones discussed that may not be available are the stock option plans, because most fire departments are a part of local government. However, some private sector fire departments and EMS companies use these programs with success. A fire service personnel manager should be aware of the various plan options and endeavor to incorporate them into the fire department whenever possible.

Summary

Measuring and managing performance are challenging enterprises and two of the keys to productivity and organizational effectiveness. Organizations can gain a competitive advantage by their wise and innovative use of human resources. The strong relationship between productivity and the systematic integration of effective personnel management with a flexible production system has been well established. Productivity concerns two specific assessments of performance: efficiency and effectiveness.

Performance management includes the activities of an organization that seek a competitive advantage through the efforts of its employees and the functions of the organization. Performance management ensures that the work of the individual is in concert with the goals of the organization and that total organizational performance is maintained at a high level. To be effective, performance incentive systems should provide a sufficient level of incentives to fulfill the basic needs of employees and direct their behavior toward the accomplishment of the stated mission. Motivation is the individual's needs, desires, and concepts that cause him or her to act in a particular manner. The task is for organizations to direct individuals so they can satisfy their needs as much as possible as they accomplish the objectives of the organization.

Review Questions

Use these questions to review the material in this chapter and for discussion purposes.

1. Identify and discuss three methods in which a fire department uses data in order to justify programs and budget requests.
2. What is the difference between *efficiency* and *effectiveness*?
3. Discuss two ways to improve the performance management of your fire department.
4. Identify and review the operation and appropriateness of four different performance incentive techniques.
5. What is *expectancy theory*, and how does it relate to the motivation of employees?

Case Study

The city manager comes to you as fire chief and says that a recent citywide study indicates that firefighters respond to emergencies only at a rate of approximately 15 percent of their on-duty time. The city is facing major budget reductions due to the general economy and a rapidly decreasing tax base within the city limits. The city manager expects that the fire department will have to make major cuts due to the inactivity of the fire department.

How do you respond to this? Draft a response back to the city manager indicating the efficiency and effectiveness of the fire department. What criteria will you use to support this?

Fire Service Professional Qualifications

This chapter contains information that will assist fire officers in meeting the job performance requirements (JPRs) of NFPA 1021, Standard for Fire Officer Professional Qualifications:

4.2 Human Resource Management— Fire Officer I

4.2.5 Apply human resource policies and procedures given an administrative situation requiring action, so the policies and procedures are followed.

4.4.1 Recommend changes to existing policies and/or implement a new departmental policy at the unit level, given a new departmental policy, so that the policy is communicated to and understood by unit members.

5.2 Human Resource Management— Fire Officer II

5.2.1 Initiate actions to maximize member performance and/or to correct unacceptable performance, given human resource policies and procedures, so that member and/or unit performance improves or the issue is referred to the next level of supervision.

6.2 Human Resource Management— Fire Officer III

6.4.4 Direct the development, maintenance, and evaluation of a department record keeping system, given policies and procedures, so that completeness and accuracy are achieved.

6.4.5 Analyze and interpret records and data, given a fire department records system, so that validity is determined and improvements are recommended.

6.4.6 Develop a model plan, given resources for an area to be protected, so that resource utilization is maximized.

7.2 Human Resource Management— Fire Officer IV

7.2.5 Evaluate an incentive program, given data, so that a determination is made regarding achievement of the desired results.

Endnotes

1. Jack F. Welch, A matter of exchange rates, *Wall Street Journal*, 1994.
2. U.S. Department of Labor, *High performance work practices and firm performance* (Washington, DC: U.S. Government Printing Office, 1994).
3. Accel Team, *Employee motivation: Theory and practice*, retrieved November 13, 2007, from http://www.accel-team.com
4. National Fire Protection Association and International City/County Managers Association, *Effective measures for fire prevention and suppression* (Quincy, MA: National Fire Protection Association, 1977).
5. National Fire Academy, *Personnel management for the fire and emergency services course* (Emmitsburg, MD: National Fire Academy/Fire and Emergency Services Higher Education, 2005).
6. Robert L. Mathis and John H. Jackson, *Human resource management*, 10th ed. (Mason, OH: Thomson Learning, 2003).
7. Wayne F. Cascio, *Managing human resources*, 7th ed. (New York: McGraw-Hill, 2006).
8. Donald E. Klinger and John Nalbandian, *Public personnel management: context and strategies*, 5th ed. (Upper Saddle River, NJ: Prentice Hall, 2002).
9. Accel Team, *Employee motivation*.
10. Associates of the Europe Japan Centre, *Kaizen strategies for improving team performance* (London: Pearson Education Limited, 2000).
11. Ibid.
12. R. E. Wood, A. J. Mento, and E. A. Locke, Task complexity as a moderator of goal effects, *Journal of Applied Psychology*, 72, 1987.
13. David Straker, *The Managerial Grid*, retrieved October 5, 2008, from http://www.ChangingMinds.org
14. Cascio, *Managing human resources*.
15. Randall S. Schuler and Susan E. Jackson, *Human resource management*, 6th ed. (New York: West Publishing, 1996).

16. W. Edwards Deming, *Out of the crisis* (Cambridge, MA: Center for Advanced Engineering Study, 1986).

17. Kodak forms board to consider ideas from its employees, *Wall Street Journal*, February 1984.

Suggested Readings

Associates of the Europe Japan Centre. 2000. *Kaizen strategies for improving team performance*. London: Pearson Education Limited.

Cascio, Wayne F. 2006. *Managing human resources,* 7th ed. New York: McGraw-Hill.

Clawson, James G. 2003. *Level three leadership,* 2nd ed. Upper Saddle River, NJ: Pearson Education.

Deming, W. Edwards. 1986. *Out of the crisis.* Cambridge, MA: Center for Advanced Engineering Study.

Klinger, Donald E., and John Nalbandian. 2002. *Public personnel management: context and strategies,* 5th ed. Upper Saddle River, NJ: Prentice Hall.

Kodak forms board to consider ideas from its employees. 1984. *Wall Street Journal,* February.

Mathis, Robert L., and John H. Jackson. 2003. *Human resource management,* 10th ed. Mason, OH: Thomson Learning.

Schuler, Randall S., and Susan E. Jackson. 1996. *Human resource management,* 6th ed. New York: West Publishing.

U.S. Department of Labor. 1994. *High performance work practices and firm performance.* Washington, DC: U.S. Government Printing Office.

Wood, R. E., A. J. Mento, and E. A. Locke. 1987. Task complexity as a moderator of goal effects. *Journal of Applied Psychology,* 72.

MFRI Archives.

13

Fire Service Professional Development

KEY TERMS

chief fire officer (CFO), *p. 362*

coaching, *p. 355*

job rotation, *p. 355*

mentoring, *p. 353*

professional development, *p. 350*

succession planning, *p. 354*

training activities, *p. 350*

OBJECTIVES

After reading this chapter, you should be able to:

- Describe professional development.
- Describe mentoring.
- Understand developmental opportunities.
- Describe fire and emergency services higher education.
- Describe fire service professional development.

Many key individuals must work together if an organization is to have an effective program for the professional development of its personnel. Professional development is directed at improving the performance of personnel in their current roles and preparing them to accept higher levels of responsibility within the organization. Management must provide this leadership and make the resources available to support individuals seeking professional development. Supervisors need to be flexible and encourage their personnel to take full advantage of the opportunities available. Finally, the individual employee must make the effort, for each one is ultimately responsible for his or her professional development. The fire and rescue service presents many opportunities for the professional development of its personnel.

Professional development includes much more than being proficient and knowledgeable about fire service–related subjects and materials. Fire officers need to have a good formal education that is broad in a number of subject areas, which means a college degree at the bachelor's level, and a master's degree is preferred for many upper-level positions. In the future, more progressive fire departments will require degrees before a firefighter is eligible to compete in the promotional process.

Professional Development

professional development
■ refers to acquiring the training and education required to enter a profession and maintaining the knowledge, skills, and abilities through continuing education.

Development of the organization's most important resource, its people, deserves a high priority in any organization. **Professional development** generally refers to acquiring the training and education required to enter a profession and maintaining the knowledge, skills, and abilities through continuing education. It is a lifelong process of learning to keep current with changing technology and practices in a profession, and it represents a journey, not a destination. Developing and implementing a program of professional development is often a function of the human resources or organizational development department. In the fire department setting, this function would most likely be coordinated from the fire training academy, or many departments now have separate career development sections or bureaus.

Professional development is different from training in many aspects. Training generally deals with "how" to accomplish something and is likely to be technical in nature. Professional development is directed more at "knowledge" that will enhance the organization or the individual's ability to deal with change and organizational or personal growth issues. Professional development, like education, increases one's knowledge and ability to make judgments, provide analysis, take responsibility, and to be strategic in one's thought process. It is of benefit to the member of the organization and to the organization as a whole.[1]

training activities
■ deals with "how" to accomplish something and is likely to be technical in nature.

The following chart summarizes professional development as opposed to **training activities**.

	Training	Development
Focus	Learn specific behaviors and actions	Understand information and concepts, develop judgment, and expand capability
Time	Short term	Long term
Measures	Performance appraisals, passing tests, and certifications	Competitive advantage, qualified people to promote, and satisfaction

Professional development opportunities in fire departments are very important. They assist in preparing firefighters for promotion to higher ranks and positions of greater responsibility. It tends to increase job satisfaction as well as retention. Fire departments must assist employees to focus on their careers to benefit both the individual and the department. Otherwise, the department will find it does not have people capable or willing to step up to higher-level responsibilities. Failure to achieve a level of success as well as the pride and accomplishment that come with this may cause firefighters to be dissatisfied or eventually leave the department. Creating career opportunities for professional development is important in maintaining a workforce focused on achievement. Although this is does not always mean a promotion, it may mean special projects, new assignments, and other challenging work projects. Professional development is forward looking; it should involve all levels of the department and may include academic pursuits, continuing education, specialized training and certification, and active participation in professional organizations, among other things.

One of the primary and easiest ways to take advantage of professional development programs is to be aware of what is currently available within one's department. The local fire training academy most likely has programs on a regular basis aimed at the professional development of the department membership. If it does not, then those in the department should encourage greater availability. Often a department can obtain speakers from the local college or university at no cost. In addition, some major corporations and businesses in the community may have interesting speakers available. In most cases, all one has to do is ask.

In the future, one of the major functions of the chief of training will be that of "professional development officer" who will deliver the training and manage the continuum of professional development. More precisely, this person will account for the achievement of competencies as fire departments move toward increased professional development. A number of factors will be required to achieve this, such as the following:

- Improve coordination with local colleges/universities and transcript collection for personnel files.
- Focus professional development planning to demonstrate how competency (through training, education, or correlating JPRs) is achieved and what college courses are approved or not given the remaining unmet competencies of the professional development plan.
- Ensure that succession planning is directly related to the professional development plan.
- Identify the opportunities for course work having a multitude of benefits in that it meets fire service certification requirements, as well as educational requirements, and training necessary for either state or federal mandates, such as hazardous materials or respiratory protection, for example.

At the state level, many excellent state fire training programs operate throughout the United States. Many states hold annual fire schools that offer a host of interesting speakers and programs on a wide variety of subjects. Missouri, Iowa, Kansas, and other states are well known for their annual fire schools. Because of their size, these larger programs can support the costs for professional presentations. Many state fire training programs present seminars on a regular basis, often at no cost to in-state students. For example, the Maryland Fire and

Rescue Institute hosts a program known as Executive Development Seminars. These seminars are specifically designed for the professional development of the career and volunteer officers of the state. They offer educational presentations covering a variety of managerial and supervisory subjects.

At the federal level, the National Fire Academy (NFA) has excellent cost-effective programs. Once selected, the student or his or her department pays only for meals while on campus; the round-trip travel, housing, and training are provided at no cost to the department by the U.S. government. The NFA programs cover a broad spectrum, are educationally sound, and are well presented by its faculty.

Beginning in October 2008, the National Fire Academy began issuing continuing education units (CEUs) for resident courses. Students who need CEUs to maintain certifications for employment purposes or professional associations will be able to take advantage of this opportunity. The number of CEUs is based on the length of the course and generally ranges from 5 to 8 CEUs. This new program extends the NFA's desire to promote continued education as a life-long commitment.

The Executive Fire Officer (EFO) program of the NFA has been particularly noted for excellence and is highly recommended for aspiring fire officers. The Emergency Management Institute (EMI), located with the NFA in Emmitsburg, Maryland, also offers a wide range of interesting programs aimed at the emergency management community. As an opportunity for professional development, both the NFA and the EMI provide excellent choices.

A number of national fire service conferences are also presented each year. Beyond the apparatus and equipment displays and the business meetings of the organizations, these conferences provide many educational opportunities. Conferences such as the Fire Department Instructors Conference (FDIC), IAFC Fire-Rescue International, IAFF Redmond Symposium, JEMS Conference, and others are well attended and recognized for their outstanding programs. Attendance in most cases is supported by the fire department. These seminars offer excellent networking possibilities of which those attending should take full advantage. Any of these conferences and many more are well worth the investment of time and money that it takes to attend. In addition, remember the value of attending conferences and professional development that are not fire service related to broaden one's horizons and provide a completely different perspective on issues. Many of the national conferences on management or human resources management would be good choices.

The National Fire Academy, in conjunction with seven universities, has developed the Degrees at a Distance Program to prepare graduates for leadership positions. Many regional colleges provide students with the opportunity to obtain a college degree through independent study and online learning. In these programs the student receives full upper-level college credit, but the emphasis is on collaborative learning with other students in a virtual classroom environment. These programs are specifically designed for students who work shift work and cannot attend a regular schedule of classes. Education is often the key to advancement, and this program offers degrees in fire service management and in many other concentration areas. Several states and many fire departments have programs to reimburse fire service students for the expense of attending college as well as educational incentives that include augmented pay and points for promotion.

Source: MFRI Archives.

Mentoring Programs

In addition to educational seminars, conferences, and obtaining a college degree, mentoring programs have been very successful in providing professional development in a personal way. **Mentoring** is a relationship between a junior and a senior colleague that is viewed by the junior colleague as a positive contribution to his or her development and future potential.[2] It involves a person who acts as a trusted counselor or teacher for someone else in the organization. That person provides coaching, visibility, role modeling, challenging work assignments, and direction to a subordinate in the mentor relationship. Such a relationship provides the opportunity to convey technical, interpersonal, and organizational skills from a more experienced person to the individual being mentored. Many fire departments have formal mentor programs to match potential mentors with those desiring such a developmental relationship. Others participate in a more informal manner; but one needs to be cautious in these arrangements to avoid the appearance of favoritism.

A number of guidelines can be useful in making the mentor relationship as productive as possible. The department needs to ensure that the program is voluntary on both sides. If someone is forced or coerced into a mentor relationship, it will not work and may in fact be damaging. Individuals who agree to serve as mentors will need training in this role, but they should be encouraged to develop their own style. The expectations of the mentor relationship should be realistic, and what can and cannot be accomplished needs to be made clear. Mentoring should be a positive experience, including the recognition of mentors for their contributions. The mentor should not be the direct supervisor of the person to be mentored; however, a good flow of information and communication to the supervisor can help to avoid conflicts and misunderstanding.

Mentors need to have a number of qualities that will enhance their ability to perform this role. In addition, they should have the support of the administration of the department in order to be successful. Some of the traits of successful mentors can be summarized as follows:

mentoring
- a relationship between a junior and a senior colleague that is viewed by the junior colleague as a positive contribution to his or her development and future potential.

Respect Within the Organization.
The mentor needs to be a person who is respected and seen as competent to perform in this role. A person with a diverse career who clearly understands the history of the organization and its future is a good choice.

Positive Attitude.
A good outlook on both life and the workplace is infectious. If the mentor is going to challenge and motivate the person to be mentored, then he or she needs to have a positive outlook and display this.

Communication Skills.
The mentor must be able to demonstrate, teach, inspire, and communicate this in the mentor relationship. Mentors also need to know when to listen and when not to provide advice, but allow the individual to seek his or her own way at times.

Good Values.
People of good character are able to set an example of personal and professional ethics that is important to the person being mentored. These lessons are imparted to the person being mentored more by action and example than by words.

Mentor programs can assist with the diversity of the fire department and provide people with limited access to those in power the opportunity to be recognized for their talents. Studies have produced evidence that underrepresented groups have less access to career-enhancing experiences than do majority members.[3] Women and minorities have benefited from mentor programs, which should be encouraged as a way to provide upward mobility for their careers. Mentor programs offer a unique method to provide for this, and the program should be diverse to best accomplish this goal.

Developmental Opportunities

Once the need for professional development is understood and accepted, then the organization must establish the opportunities and programs necessary for implementation. As with training programs, the first step is to conduct a needs analysis of both the organization and individual professional development needs. The fire department management has to establish priorities as to what type of professional development will be offered as well as determine who is eligible for which programs. The developmental opportunities need to be those that best promote the mission and success of the fire department.

succession planning
- the process of identifying the long-term needs of an organization in terms of human capability.

Succession planning is the process of identifying the long-term needs of the department in terms of human capability. Fire departments, unlike many other organizations, have an orderly chain of command, and the roles within the department are very structured. The issue is then "who" is ready to fulfill these roles and whether they possess the required knowledge and ability to move forward. Succession planning should be a component of the needs assessment that is performed.

Many different types of approaches and programs can be applied within an organization to improve the professional development of its workforce. The development opportunities can be accomplished by any organization without going outside of the organization to acquire program or resources. Technical training and educational opportunities are discussed later in this chapter. Some of the more popular methods of professional development are described in this section.[4]

COACHING

Coaching is the feedback and training that are provided to employees by their immediate supervisors. This really should occur on a continuing basis and be directed at the aspirations of the employee in line with organizational objectives. For coaching to be effective, an open and good line of communications needs to be established between supervisor and employee. Some organizations implement training programs to teach supervisors how to be better prepared and perform their role as a coach. As an example, a fire officer may have a firefighter that desires to be promoted to lieutenant. Conducting training sessions that are directed at the promotional exam as well as taking the time to explain supervisory practices would be seen as positive coaching experience.

coaching
- the feedback and training that are provided to employees by their immediate supervisor. This occurs on a continuing basis and is directed at the aspirations of the employee in line with organizational objectives.

JOB ROTATION

Job rotation is the process of presenting new experiences for the employee by changing the work environment. Job rotation is especially important in organizations like fire departments that have single-point entry systems and long-tenure employees. As firefighters move up in the organization, they need to benefit from different experiences and perspectives. Otherwise, when they eventually are promoted to a level that requires a broad-based level of experience, they will be less effective. Those fire officers who work only in fire operations during their career in many cases do not make well-rounded and experienced chief officers. As a firefighter and future fire officer, it is vital to learn from every assignment that one has in the department.

job rotation
- the process of presenting new experiences for the employee by changing the work environment.

Many progressive fire departments require that chief officers rotate assignments on a periodic basis to ensure the overall departmental perspective that is necessary for such positions. In fact, this experience is good for everyone, including firefighters who should rotate between different services and bureaus of the department. In my career I had served as the deputy chief of fire operations for several years and resisted a move to management services. However, when I later became fire chief, the time I spent in management services was by far the most beneficial to me. So even if firefighters do not agree with and appreciate a move at the time it occurs, they may later in their careers.

STAFF POSITIONS

Many fire departments and organizations have staff or assistant positions that work directly for a ranking member of the organization. These provide a great way to learn and gain experience for a number of reasons. One is that staff will be included in meetings and strategy sessions that offer great insight into how decisions are made and how the department is managed. In addition, staff or assistants will be able to observe senior managers and officers as they make decisions and direct the operations under their responsibility. A position such as this does require a good deal of confidentiality and one has to earn the trust of one's superior officers.

While in a staff position, it is a good time to learn by observing both the good and the bad that may occur within the department. Staff gains opportunities to understand how business is conducted and obtains what may take many years of hard earned experience to do otherwise. Staff also is introduced to people in positions of organizational power and begins to make helpful business contacts. This can be very valuable learning experience that will pay dividends to one's career for years to come.

COMMITTEE ASSIGNMENTS

Much of the work in organizations is accomplished through committees. Being assigned to important committee work provides not only the opportunity to broaden experiences but also a better understanding of the issues facing the organization. Committees provide insight into the personalities and group dynamics that often clash within an organization. This type of assignment will increase one's ability to work with people in groups to accomplish a certain objective.

If one is fortunate enough to be the committee chair, it can be a unique opportunity to demonstrate one's willingness and ability to work with people in a group setting. Superior officers will be watching and making judgments based on how well the committee functioned under one's leadership. Make every effort to be productive and urge the committee toward recommendations that are realistic and implementable. The experience gained in these settings will be important in future endeavors.

INTERNSHIPS

Some fire departments have internship programs, whereby lower-ranking officers are detailed to other departments of government or a private company to gain experience and a different perspective. This might involve a fire officer and a police officer switching positions for a few weeks or could be a very detailed internship lasting for a year or more. It could also mean working for a different agency at the state or federal level for a period of time.

Internships are very valuable experiences if managed properly by the providing and the receiving agency. They provide unique understanding and perspectives that are not likely to be obtained within a single department of government. Internships are routinely utilized in university-based educational programs, and their usefulness has been proven on numerous occasions. In smaller departments it may be difficult to have the staffing to do this, but if at all possible it is a worthwhile program.

Fire and Emergency Services Higher Education

The Fire and Emergency Services Higher Education (FESHE) program is a component of the National Fire Academy. The goal of FESHE is to standardize fire science education by bringing together people and educational systems in a cohesive manner to provide educational opportunity relevant to fire and emergency service providers. The concept of "seamless education" is a core component of FESHE whereby employment-related training is integrated with educational pursuits to provide a system that recognizes and values the accomplishments of the individual. In addition, the seamless education model allows for transfer credit from lower levels (associate's) through upper levels (bachelor's/master's) of the degree program with no loss of credit between the institutions. FESHE provides the forum and is the catalyst for the fire service and academia to work together in a coordinated fashion in the best interests of the student and the national fire service.

The Fire and Emergency Services Higher Education program has developed a collaborative system in which the education and developmental community pursue the following goals:

- National model for an integrated, competency-based system of fire and emergency services professional development
- National model for an integrated system of higher education from associate's to doctoral degrees
- Well-trained and academically educated fire and emergency services preparing the nation for all hazards[5]

Many firefighters and officers have earned college credits and training certificates obtained from various courses. However, this professional development is sometimes uncoordinated and fragmented, resulting in duplication of effort and inefficiencies for the student. Lack of coordination between fire-related training, higher education, and certification contributes to this problem. Collaboration and coordination are needed among all service providers responsible for fire and emergency services' professional development. Each has a major role to play in this new strategic approach to professional development. The goal is to transition the fire and emergency services from a technical occupation to a full-fledged profession similar to physicians, lawyers, and architects, who, unlike fire service personnel, have common course requirements within their respective degree programs.

The question of whether firefighting is a vocation or a profession generates a fair amount of debate. A "profession" is built on several major tenets, including an educational system and standards, certification by examination, reciprocity for practicing in different states, universally accepted standards of practice, and a professional development model, among others. The work accomplished by FESHE attempts to address the areas of a coordinated model curriculum for institutions of higher education and professional development. Dr. Denis Onieal, the superintendent of the National Fire Academy, authored a series of articles on the "professionalism" of the fire service; an excerpt from that series appears here:

> Professional status is a term that has been bandied about in the Fire and Emergency Services for years. What constitutes "professional" status is in the eye of the beholder. Were we to look at a "professional" independent of the fire service, to some, it means the performance of a series of skills in a manner that is far above average. To others, a professional is associated with performing skills "full-time," that is to say, for a living. Many feel that the distinguishing characteristics of a profession are years of formal education, approval of an accrediting board and continuing education requirements. More than likely, it is the last statement with which most would agree.
>
> Definitions aside, it is the walk down the main street in any city or town in America that demonstrates who in the community is professional. The physicians and nurses, the architects and engineers, the attorneys and the accountants are among the top professions in any community. What makes them so?
>
> Each has a unique set of knowledge and skills that are independent of a particular organization or place; they are "portable" skills and held in equal regard no matter where the person practices. In the process of becoming a professional, there is an accredited and independent testing process that assures competency to the public. Professionals are associated with others in their profession through some formal organization; they typically put service to others as more important than profit; and they assume responsibility for their professional acts. Typically, their profession has some continuing education requirements, and the work is client centered.[6]

The U.S. Fire Administration (USFA) hosts the annual FESHE conference on its campus in Emmitsburg, Maryland, each year. At the 2000 conference, two panel discussions were conducted. The first panel included fire service leaders representing national fire service organizations, and the second was comprised of state directors of fire service training. Both panels raised issues reflecting a national need for the following:

- Degree programs that teach critical thinking skills by requiring significant numbers of general education credits in addition to fire science courses
- Appropriate recognition of certification for academic credit and vice versa
- Associate's degree programs that are transferable to baccalaureate programs
- A model fire science curriculum at the associate's/bachelor's degree level that universally standardizes what students learn and facilitates the application of these courses toward certification goals
- Collaboration between fire service certification and training agencies as well as academic fire programs

NATIONAL PROFESSIONAL DEVELOPMENT MODEL

At the annual Fire and Emergency Services Higher Education (FESHE) conferences, efforts to address the challenges of a stovepiped system of professional development have been made by creating the National Professional Development Model that integrates training, education, and certification. This model depicts the coordination of education and training and establishes how they are both relevant and important to fire officers. This experience-based model encourages an efficient path for fire service professional development supported by collaboration between fire-related training, higher education, and certification providers. The model prescribes what these providers' respective roles should be and how they should coordinate their programs.

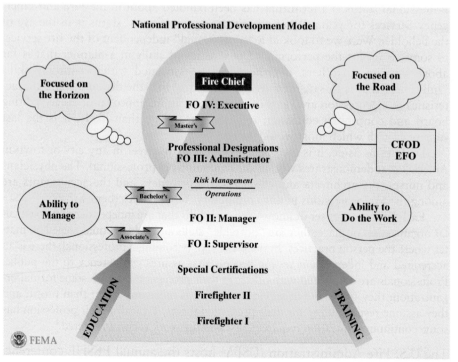

Source: NFA/FESHE, 2008.

The National Fire Academy (NFA) has facilitated the development of the National Professional Development Matrix (NPDM) that moves these models from concept to reality. The NPDM is designed for training and certification agencies and academic fire programs to assist emergency services personnel in their professional development planning. In addition, the NFA has produced a template that crosswalks NFPA 1021, Standard for Fire Officer Professional Qualifications, Fire Officer I through IV competencies with "national" level courses that included NFA training courses, FESHE model associate's and bachelor's courses, and general education courses recommended by the IAFC in its *Officer Development Handbook*. This has been very helpful to firefighters and officers who are pursuing training and academic opportunities to further their careers. It has been a very positive factor in creating the desired seamless education system for firefighters on a national basis. NFA urges states and fire departments to customize this template with their own standards, training, and college courses by convening professional development summits composed of fire departments, academic fire programs, state associations, and other key stakeholders.

DEGREES AT A DISTANCE PROGRAM

Increasingly, fire departments around the country are requiring fire officers to hold appropriate bachelor's and master's level degrees. Probably the most effective method that fire departments can employ to increase the educational accomplishment of their leaders is to require appropriate college degrees as a condition for promotion. The degree requirement must be structured in an appropriate manner, and time needs to be allotted for individuals to complete the degrees. Many departments make the educational requirements effective at a future date or stagger them in over a period of time. In addition, many fire departments and states have educational incentive or tuition reimbursement programs to assist fire service students.

Locally, numerous colleges and universities provide two- and four-year degree programs in fire science and/or administration. Over the past several years, a few master's degree programs in fire service management have emerged. As of the year 2000, there were approximately 222 two-year and 26 four-year fire science/ management programs in the United States. These programs typically serve both people who are employed on a full-time basis and those attending college classes.[7] Since this survey was conducted, undoubtedly many more programs have been created as the FESHE initiatives have been implemented.

The Degrees at a Distance Program (DDP), an online degree program sponsored by the NFA, has agreements with seven accredited colleges and universities throughout the country to offer bachelor's degrees with concentrations in fire administration/management and fire prevention technology. The Degrees at a Distance Program is an opportunity to take college courses that can be used toward a bachelor's degree with either of these concentrations. This program provides the fire service student with the opportunity to get a college education through independent study.

The Degrees at a Distance Program provides an alternative means for fire service personnel to earn a bachelor's degree or to pursue college-level learning in a fire-related course concentration without the requirement of having to attend on-campus classes. While independent study and distance learning have appealed to working adults nationally in growing numbers for the past few years, DDP is particularly attractive to fire service personnel whose fire departments' shift work normally makes classroom attendance difficult. With all the instruction delivered completely over the Internet, access to the courses is easy and learning is enjoyable. Students engage in online learning activities in which they collaborate with one another, engage in debate and discussion through frequent postings, research related Web sites, reference sources, and complete projects that are relevant to their own careers and fire departments.

Fire Service Professional Development

Numerous opportunities exist within the fire service for professional development at the local, state, and national levels. All one really needs is the time and financial resources to be in a position to select from a large array of programs and speakers. The goal, of course, is to find the professional development opportunities with the highest quality and relevance to one's position or future goals within the department. Students should not just attend conferences and seminars because they are available, but should have a plan as to what they want to achieve. This means that students will have thoroughly reviewed the agenda prior to the conference or seminar and carefully select the programs that they wish to attend. Having this level of focus will make their attendance more valuable and productive. Although in some cases it is good to just wander about and see what is new, and there is no harm in this, do not make this the normal pattern.

I have selected what I consider to be several unique fire service professional development opportunities to discuss in this section. These are not conferences or seminars, but are special and highly regarded opportunities that offer a very rewarding professional experience for the student. They are extended programs with highly qualified faculty that present on the latest and most important issues facing the fire service manager. Additionally, they are programs that are at the national level with student body that is diverse and represents a broad cross section of the country. Many of these programs are selective, and it will take some effort and level of accomplishment to attend. I would highly recommend attendance at any one of these opportunities as an excellent professional development experience.

HARVARD UNIVERSITY PROGRAM FOR SENIOR EXECUTIVES

The highly regarded *Senior Executives in State and Local Government* program at Harvard's John F. Kennedy School of Government provides experiences for participants both inside and outside the classroom to ensure that public officials are equipped on a daily basis to manage and lead results-driven government agencies and nonprofit organizations. In particular, this program provides an opportunity to

- Develop new conceptual frameworks for addressing program and policy issues
- Explore the relationship between citizens and their government
- Examine the ethical and professional responsibilities of leadership

- Exchange ideas with experienced faculty as well as an extremely diverse group of participants[8]

The 3-week intensive program is designed specifically to challenge assumptions about how leaders exercise leadership in the public sector. During the course of this program, students learn strategies for establishing meaningful, attainable organizational objectives. Attendees also investigate the process of developing and evaluating policy alternatives and consider the options for organizing and deploying resources to achieve these objectives. The case study method of analysis is utilized extensively in this course, and a high degree of interaction is expected from students.

Prospective students compete for the available fellowships, and a defined selection process determines who will attend. The program is directed at those in senior leadership positions; in the case of fire departments, this would be fire chief or deputy chief. One of the best features of the program is that it exposes other state and local leaders to the fire service, and they will learn from our experiences as well as others. Past attendees of this program rate it as one of the absolutely best professional development opportunities available.

The program is generously sponsored through a partnership among the International Association of Fire Chiefs (IAFC), the International Fire Service Training Association (IFSTA), the National Fire Protection Association (NFPA), and the Department of Homeland Security—United States Fire Administration (USFA). Each organization funds the tuition costs of two individuals for a total of eight fellowships.

NFA EXECUTIVE FIRE OFFICER

The Executive Fire Officer Program (EFOP), an initiative of the National Fire Academy, is probably the most prestigious fire-specific professional development opportunity in our profession. The program now requires a bachelor's degree for entry and takes 4 years to complete. It is a very selective program with a solid reputation for excellence. The EFOP spans a 4-year period with four core courses; each course is conducted at the NFA and is 2 weeks in length. Each EFOP participant must then complete an applied research project that relates to his or her organization within 6 months after the completion of each of the four courses. A certificate of completion for the entire EFOP is awarded only after the successful completion of the final research project.

The Executive Fire Officer Program is designed to provide senior officers and others in key leadership roles with the following:

- An understanding of
  1. the need to transform fire and emergency services organizations from being reactive to proactive; with an emphasis on leadership development, prevention, and risk-reduction;
  2. transforming fire and emergency services organizations to reflect the diversity of America's communities;
  3. the value of research and its application to the profession; and
  4. the value of lifelong learning.
- Enhanced executive-level knowledge, skills, and abilities necessary to lead these transformations, conduct research, and engage in lifelong learning.[9]

The fire officers who attend this course enhance their professional development through a unique series of four graduate and upper-division baccalaureate-equivalent courses conducted at the NFA as resident programs. The course requirements for the EFOP program follow:

Executive Development (R123)
Leading Community Risk Reduction (R280)
Executive Analysis of Fire Service Operations in Emergency Management (R306)
Executive Leadership (R125)

In addition, each year the USFA sponsors the EFOP Graduate Symposium. Attendance and participation in the symposium are limited to graduates of the EFOP. The symposium informs participants of outstanding applied research completed by EFOP participants during the previous year. It creates a forum for the exchange of cutting-edge trends and information. Recognized presenters from the private sector, education, and government provide an informational update and reinforce executive skill areas discussed within the EFOP experience. National Fire Academy faculty also provide presentations that reinforce and extend the EFOP experience. The EFOP symposium is always of high quality and a worthwhile education opportunity. Many jurisdictions are now either requiring or providing preference in the appointment of fire chiefs to those who are EFOP graduates.

CHIEF FIRE OFFICER DESIGNATION

chief fire officer (CFO)
■ an individual who has demonstrated requisite skills and abilities for the fire and EMS professions through structured education and proven leadership and management skills that display industry knowledge. Fire chiefs meeting the qualifications of the CFOD program can be so designated by the Center for Public Safety Excellence.

A **chief fire officer (CFO)** is an individual who has demonstrated requisite skills and abilities for the fire and EMS professions through structured education and proven leadership and management skills that display industry knowledge. Achieving professional designation as a CFO acknowledges that an individual has attained the status recognized by his or her superiors, subordinates, and colleagues. Many organizations within commercial, industrial, and governmental agencies also recognize this mark of distinction. Only those fire chiefs meeting the strict qualifications of the CFOD program can be so designated by the Center for Public Safety Excellence.

The Commission on Professional Credentialing (CPC), an entity of the Center for Public Safety Excellence (CPSE), administers the CFO Designation Program. The CPC includes nine members appointed by CPSE's board of directors. The Commission on Professional Credentialing consists of individuals from academia, federal and local government, and the fire and EMS profession. The CPC has the responsibility to set and maintain the professional designation standards for the program and makes decisions concerning individual designation, organization operation, and activities. The CPC manages the peer review process. The peer reviewers hold the CFO designation, recommend candidates for designation to the CPC, and determine whether individuals seeking designation meet program criteria and requirements. There are several benefits of the professional designation program for senior executive fire officers, which include the following:

■ Promoting excellence within the fire and EMS profession
■ Improving professionalism levels among senior fire officers
■ Providing a career path for new or prospective chief officers

- Ability to demonstrate superior characteristics attained by CFOs
- Assisting jurisdictional authorities in identifying individuals who possess superior skills, knowledge, and abilities prior to the hiring or promotion process
- Establishing benchmarks and core competencies for personnel serving in senior fire officer positions
- Developing an understanding of need for continuing education, training, and skill proficiency[10]

The CFOD program offers professional development opportunities, but mainly is a system to recognize the accomplishment and professional ability of fire chiefs. Applicants have to submit a very thorough portfolio of their work and accomplishments in a number of subject areas. The CFOD program consists of several components encompassing the professional and personal characteristic profile desired of a CFO, which include the following:

- Education
- Experience
- Professional development
- Technical competencies
- Contributions/achievements to the profession
- Association memberships
- Community involvement

After an analysis and approval of a panel of experts who determine the validity and level of accomplishment of the applicant, one can hold the chief fire officer designation and be recognized for this achievement.

NATIONAL FIRE SERVICE STAFF AND COMMAND COURSE

The National Fire Service Staff and Command Course is a high-intensity, concentrated program designed to involve the student in various peer group activities. The program has been presented since 1962 by the Maryland Fire and Rescue Institute of the University of Maryland at College Park and in cooperation with the International Association of Fire Chiefs (IAFC). The program is presented in Maryland and every other year in a different state.

The course is instructed in a seminar format and its goal is to expose fire service leaders to the latest issues and topics within the fire service environment. Experienced and noted guest lecturers present on relevant topics, and the program has been viewed as very successful in terms of exposure to new techniques and approaches to the successful resolution of organizational challenges. Many future and current leaders within the fire service have attended this course.

The program features subject matter experts presenting topics of significance to fire service leaders. Participation in the interactive exercises affords the students the opportunity to network, sharing insight through that process. This weeklong intense program is designed to expose emergency service officers to the modern concepts of emergency management by developing professional knowledge and technical skills. This program is open to all emergency service officers above the company officer level.[11]

INTERNATIONAL FIRE SERVICE JOURNAL OF LEADERSHIP AND MANAGEMENT

The purpose of the *International Fire Service Journal of Leadership and Management* is to advance the international fire service by publishing carefully selected, peer-reviewed articles that enhance the theory and practice of fire leadership and management. One of the keystones of actually being a professional is to have scholarly and research-based articles judged by a peer review process for publication. The *International Fire Service Journal of Leadership and Management* is sponsored by Oklahoma State University, Fire Protection Publications (FPP), and the International Fire Service Training Association (IFSTA).

The *International Fire Service Journal of Leadership and Management (IFSJLM)* is a refereed journal that publishes articles of interest to an international audience of academics and in-service fire and emergency management professionals. Although engineering and technology-based articles are welcomed, the primary impetus for creation of *IFSJLM* is to present the international fire service articles focusing on leadership and management. In addition, the journal offers a book review section of fire service textbooks.

All articles, unless otherwise noted upon publication, submitted to *IFSJLM* are peer reviewed. *IFSJLM* uses a double-blind review process: The author does not know who reviewed the article and the reviewers do not know who wrote the article. Thus, it is important that the author be identified only on the cover page. As they review an article, peer reviewers are trying to answer the following questions: Is the material in the article accurate and relevant? Is the article grounded in theory? Are the methods used in the study appropriate and appropriately used? Does the article significantly add to the understanding of fire leadership and management? If not, the article may merit publication, but not in *IFSJLM*.[12]

Each year prior to the beginning of the annual International Fire Service Training Association (IFSTA) Validation Conference, the *International Fire Service Journal of Leadership and Management (IFSJLM)* holds it annual Research Symposium. This symposium provides the opportunity for research papers to be presented in an academic format and allows for further discussion of the content.

Summary

Initial training and certification programs for fire service personnel are very important to one's safety and success. In addition, the skills and knowledge of the firefighter must continue to expand and increase as time moves forward. Professional development is directed at improving the performance of personnel in their current roles and preparing them to accept higher levels of responsibility within the organization. Adequate and sustained professional development provides the emergency services provider with the knowledge he or she needs for a career that continues to advance. Professional development, like education, increases one's knowledge and ability to make judgments, provide analysis, take on responsibility, and to be strategic in one's thought process. It benefits both the member of the organization and the organization as a whole.

Review Questions

Use these questions to review the material in this chapter and for discussion purposes.

1. Describe the difference between training and development.
2. A mentor needs to possess many qualities in order to be of value to the individual who is being mentored. Identify and describe at least four of these desired qualities.
3. Name four fire service conferences that would be a good choice for professional development, and detail your personal plan about how to gain the most from attending these conferences.
4. Describe the National Professional Development Model and explain what it means to you as a fire officer.
5. The Degrees at a Distance Program offers what advantages to the fire service student as compared to a more traditional classroom-based fire service degree program?

Case Study

You are a lieutenant in a small career fire department located in the southern United States. Your department's total staffing is 52 personnel working out of two fire stations. There is no fire training academy or training staff available on a full-time basis. The battalion chief attempts to provide training on an unscheduled basis, but other duties normally interrupt this effort. The state fire training academy is over 100 miles from your department's location.

You have a desire to acquire additional knowledge involving how to be a better manager and also how to maintain firefighter safety at a higher level. How would you go about accomplishing this and developing a plan that would make progress toward these objectives? Describe the professional development opportunities that you would recommend.

Fire Service Professional Qualifications

This chapter contains information that will assist fire officers in meeting the job performance requirements (JPRs) of NFPA 1021, Standard for Fire Officer Professional Qualifications:

4.2 **Human Resource Management—Fire Officer I**

    4.2.5 Apply human resource policies and procedures given an administrative

situation requiring action, so the policies and procedures are followed.

6.2 Human Resource Management—Fire Officer III

6.2.4 Describe methods to facilitate and encourage members to participate in professional development to achieve their full potential.

7.2 Human Resource Management—Fire Officer IV

7.2.3 Establish and evaluate a list of education and in-service training goals, given a summary of the job requirements for all positions within the department, so that all members can achieve and maintain required proficiencies.

Endnotes

1. Robert L. Mathis and John H. Jackson, *Human resource management*, 10th ed. (Mason, OH: Thomson Learning, 2003).
2. K. E. Kram, *Mentoring at work* (Glenview, IL: Scott Foresman, 1984).
3. F. S. Hall and M. H. Albrecht, *The management of affirmative action* (Santa Monica, CA: Goodyear, 1979).
4. Mathis and Jackson, *Human resource management*.
5. National Fire Academy, *Fire and Emergency Services Higher Education*, retrieved February 1, 2008, from http://www.usfa.dhs.gov/nfa/higher_ed/feshe/feshe_strategic.shtm
6. Denis Onieal, *Professional status: The future of fire service training and education* (Emmitsburg, MD: National Fire Academy, 2006).
7. Thomas B. Sturtevant, A study of undergraduate fire service degree programs in the United States—Fall 2000 (Doctoral dissertation, University of Tennessee, Knoxville, May 2001).
8. Harvard University, *Program for senior executives in state and local government*, retrieved February 1, 2008, from http://www.ksgexecprogram.harvard.edu/program/SL
9. National Fire Academy, *Executive Fire Officer Program*, retrieved February 1, 2008, from http://www.usfa.dhs.gov/nfa/efop
10. Center for Public Safety Excellence, *Chief fire officer designation*, retrieved November 15, 2008, from http://www.publicsafetyexcellence.org
11. Maryland Fire and Rescue Institute, *MFRI National Staff and Command Course*, retrieved November 15, 2008, from http://www.mfri.org
12. *International Fire Service Journal of Leadership and Management*, available from http://www.ifsjlm.org

Suggested Readings

Cascio, Wayne F. 2006. *Managing human resources*, 7th ed. New York: McGraw-Hill.

Dessler, Gary. 2001. *Management: Leading people and organizations in the 21st century*, 2nd ed. Upper Saddle River, NJ: Prentice Hall.

Hall, F. S., and M. H. Albrecht. 1979. *The management of affirmative action*. Santa Monica, CA: Goodyear.

International Fire Service Training Association. 2004. *Chief officer*, 2nd ed. Stillwater, OK: Fire Protection Publication, Oklahoma State University.

Kram, K. E. 1984. *Mentoring at work*. Glenview, IL: Scott Foresman.

Mathis, Robert L., and John H. Jackson. 2003. *Human resource management*, 10th ed. Mason, OH: Thomson Learning.

Mondy, Wayne R., and Robert M. Noe. 2005. *Human resource management*, 9th ed. Upper Saddle River, NJ: Pearson Education.

U.S. Department of Education. 2006. *Evidence of quality in distance education programs drawn from interviews with the accreditation community*. Washington, DC: Department of Education.

GLOSSARY

A

Accreditation Given to entities that have been recognized or approved by a higher institution for meeting a set of standards. Organizations that issue credentials or certify third parties against official standards are themselves formally accredited by accreditation bodies.

Active learning A process whereby one learns by doing, not by observing. This includes a higher degree of realism with active participants that will enhance retention.

Adverse impact Occurs when the selection rate for any race, sex, or ethnic group is less than four-fifths of the rate for the group with the highest selection rate.

Affective Includes those objectives that emphasize feeling and emotion, such as interests, attitudes, appreciation, and methods of adjustment.

Affirmative action Refers to the policies intended to promote access to education or employment aimed at an historically non-dominant group, typically minorities or women.

Arbitration The process in which union and management involved in a dispute agree in advance of a hearing to abide by the decision of an independent arbitrator.

Assessment The process of documenting, usually in measurable terms, knowledge, skills, attitudes, and beliefs. An assessment of the need for training will include a specific plan as to how it will be delivered and evaluated.

Attributes An approach to performance appraisal that concentrates on the extent to which individuals demonstrate certain desirable characteristics and traits within the job being evaluated.

Attrition A reduction in membership or personnel, normally through retirement, resignation, or other factors.

B

BARS Behaviorally anchored rating scales; various performance levels shown along a scale and described in terms of an employee's specific job behavior.

Behaviors An approach to performance appraisal that attempts to define the actual behaviors an employee must exhibit to be successful on the job.

C

Certification The confirmation of certain characteristics of an object, person, or organization. The certification is usually issued by an organization to an individual after successfully testing to the standards approved by an accredited agency.

Chief fire officer (CFO) An individual who has demonstrated requisite skills and abilities for the fire and EMS professions through structured education and proven leadership and management skills that display industry knowledge. Fire chiefs meeting the qualifications of the CFOD program can be so designated by the Center for Public Safety Excellence.

Coaching The feedback and training that are provided to employees by their immediate supervisor. This occurs on a continuing basis and is directed at the aspirations of the employee in line with organizational objectives.

Cognitive Includes those objectives that emphasize intellectual outcomes, such as knowledge, understanding, and thinking skills.

Collective bargaining The process by which union representatives for employees in a bargaining unit negotiate employment conditions for the entire bargaining unit.

Common law A rule made by judges as they resolve disputes between parties that are recorded and subsequently referred back to as precedent for making future decisions.

Compliance based Ethics codes that emphasize preventing unlawful or unethical behavior by increasing control and penalizing violators.

Compliance programs An understanding of what is required by law and a plan that addresses the specific needs of the organization regarding equal opportunity.

Conflict of commitment When otherwise positive activities may compromise the employee's basic job responsibilities.

Conflict of interest Typically involves potential financial gain or personal benefit to an employee or related individual.

Control of results Requires that the employees understand the desired level of performance against which they will be measured.

D

Development Is intended to identify problems in employee performance and then develop a plan of improvement to provide the necessary skill training or professional development.

Discipline The state of employee self-control, orderly conduct, and the adherence to established standards within an organization. Also refers to the actions imposed by an organization on its employees for failure to follow the organization's rules, standards, or policies.

Discrimination Refers to taking specific actions toward or against a person based on the person's group.

Disparate impact Occurs when substantial underrepresentation of protected class members results from employment decisions that work to their disadvantage.

Disparate treatment Occurs when protected class members are treated differently from others. If disparate treatment has occurred, then most likely intentional discrimination exists.

Diversity A society consisting of people of different ethnic and cultural backgrounds who possess different attitudes, values, and norms.

Doctrine of *stare decisis* The use of precedents to decide cases, which helps litigators anticipate how the courts may rule in cases that are similar to the precedent-setting case.

Due process Means that employees have a procedural route that provides the opportunity to explain their actions and confront disciplinary charges.

Duty of fair representation Union's duty to represent the interests of the union membership in a fair and impartial manner in consideration of the interests of all of the members.

E

EEOC U.S. Equal Employment Opportunity Commission; an independent federal agency created by Congress in 1964 to eliminate discrimination in employment; provides oversight and coordination of all federal equal employment opportunity regulations, practices, and policies.

Effectiveness The degree to which a goal has been achieved. It is the quality of the output measurement as compared to a standard.

Efficiency Is measured as a ratio of outputs produced to inputs consumed. This implies an ability to perform well while making good use of available resources.

Essential functions The fundamental job duties performed in a position; function may be essential because the position exists to perform that function, a limited number of employees perform the function, or the function is highly specialized.

Ethical behavior An individual's character traits, such as honesty, compassion, and courage, and as a standard of moral behavior that is accepted by society as being right.

Ethics From the Greek word *ethos* meaning character. It is a composite of good moral qualities, self-discipline, and judgment.

Evaluation Is intended to inform employees of their collective performance over a determined period of time. The evaluations are used to reward high performance and contributions to the success of the organization.

Extrinsic rewards Do not relate directly to the nature of the work itself, but may be linked to outcomes, such as salary increases and benefits.

F

Fact finding A neutral party, either an individual or panel, appointed to determine the facts of the dispute. A recommended settlement is proposed but is not mandated on either the union or management.

G

General duty Under OSHA regulations the duty an employer has to provide a place of employment that is free from recognized hazards.

Good faith bargaining An open, honest, and reasonable attempt put forth by both parties in the negotiation process toward reaching an agreement. Failure to bargain in good faith can result in an unfair labor practice.

H

Hostile environment Concept established by the U.S. Supreme Court in the *Meritor Savings Bank v. Vinson* case; refers to a situation in the workplace regarding a pattern of exposure to unwanted sexual behavior from persons and the employer takes no steps to discourage or discontinue such behavior.

Hygiene factors Refer to the context of the job as opposed to the content of the job. These factors include competence of supervisors, salary, company policies, and others.

I

Illegal acts Criminal offenses committed by an employee either on or off duty, as opposed to a rules violation.

Impasse Occurs when the two sides negotiating an agreement are unable to reach an agreement and become deadlocked. Alternative strategies are then employed to move the negotiations toward a successful conclusion.

Incentive systems Designed to bridge the gap between organizational goals and objectives and employee expectations. They should provide a sufficient level of incentives to fulfill the basic needs of employees and direct their behavior toward the accomplishment of the stated mission.

Instructor liability A situation in which an instructor is financially and legally responsible, such as when an act is committed that causes injury to another; therefore, the instructor may be responsible for any damage incurred. Liability may be civil or criminal.

Integrity based Ethics codes that define the organization's guiding values, create an environment that encourages and supports ethical behavior, as well as stress a shared accountability among employees.

Intrinsic rewards Relate directly to the nature of the work itself and to whether it is interesting and challenging.

J

Job analysis The systematic process of determining the knowledge, skills, and abilities required for performing jobs in an organization. It provides a summary of a job's responsibilities, its relationship to other jobs, and the working conditions under which the job is performed.

Job classification The process of categorizing positions according to the type of work performed, the type of skill required, or any other job-related factor.

Job description A document that provides information regarding the tasks, duties, and responsibilities of the job. These tasks, duties, and responsibilities are observable actions.

Job relatedness Selection standards relevant to the actual performance on the job. All selection and testing methods must be directly related to the actual job as identified in the job analysis.

Job rotation The process of presenting new experiences for the employee by changing the work environment.

Job specification A list of the knowledge, skills, abilities, and other characteristics that an individual must have in order to perform the job successfully. The KSAs are characteristics that are not directly observable; they are observable only when the person is doing the job.

K

KSAs Stands for knowledge, skills, and abilities. The knowledge refers to factual information necessary to perform a task, a skill is an individual's level of proficiency, and ability refers to a more general enduring capability that an individual possesses.

M

Management by objectives A system that features an agreement between the supervisor and the employee on the objectives for a particular period and a review of how well the objectives were achieved.

Mediation When an experienced third party assists labor and management to reach an agreement. The mediator takes an active role in bringing the sides together, but does not have the authority to make a final decision or settlement.

Mentoring A relationship between a junior and a senior colleague that is viewed by the junior colleague as a positive contribution to his or her development and future potential.

Mosaic society As organizations move toward a more diverse workforce, the diversity not only in terms of race, gender, and ethnicity but also in terms of age, sexual orientation, and other demographic variables.

Motivators Circumstances that make a person feel good about a job, so he or she strives to produce more. These factors include praise, sense of responsibility, and successful completion of tasks.

N

Negligent hiring A situation in which an employer is liable to a third party for injury because the employer knew or should have known of the employee's dangerous proclivities.

NFPA 1500 Standard on Fire Department Occupational Safety and Health Program, the purpose of which is to specify the minimum requirements for occupational safety and health as well as safety guidelines for those exposed.

NFPA 1582 Standard on Comprehensive Occupational Medical Program for Fire Departments, the purpose of which is to identify the minimum medical requirements to be a firefighter at the NFPA 1001 level.

O

Organizational values Define the acceptable standards that govern the behavior of people within organizations.

P

Performance appraisal The formal, systematic assessment of how well employees are performing their jobs in relation to established standards and the communication of that assessment to the employee and the organization.

Performance management The activities of an organization that seek a competitive advantage through the efforts of its employees and the functions of the organization. The goal is total organizational performance to be maintained at a high level.

Personnel control Emphasizes a reliance on the personnel involved to do what is best for the organization and provides them the assistance and resources to do so.

Personnel department An agency of government responsible for personnel management and includes at a minimum the activities of staffing, retention, development, adjustment, and managing change.

Political patronage A process whereby government jobs are decided on the basis of political loyalty.

Post-traumatic stress disorder A debilitating condition that affects people who have been exposed to a major traumatic event.

Prejudice A bias that results from prejudging someone on the basis of some trait.

Prima facie Means "on its first appearance," or "by first instance." In general, that on first examination, a matter appears to be self-evident from the facts or one of the determining factors.

Process An identifiable flow of interrelated events progressing toward some goal or consequence.

Productivity The ratio between inputs and outputs that indicates the value added by an organization. Productivity has to do with goods produced or services rendered in relation to cost in terms of time, finances, and other resources.

Professional development Refers to acquiring the training and education required to enter a profession and maintaining the knowledge, skills, and abilities through continuing education.

Protected class Individuals within a group identified for protection under equal employment laws and regulations.

Psychomotor Includes those objectives that emphasize motor skills, such as handwriting, operating machinery, and speaking.

R

Reasonable accommodation Requirement of the ADA of 1990 for an employer to provide reasonable accommodation to qualified individuals with disabilities who are employees or applicants for employment, unless doing so would create an undue hardship.

Recruiting image The general image of the department and the respect it has within the community in the recruiting process.

Recruitment The set of activities designed to obtain a pool of qualified job applicants for a certain position. It involves attracting individuals on a timely basis, in sufficient numbers, and with the appropriate qualifications.

Reliability Regarding a test or selection device, the consistency of the results or measurements that it produces. It is the degree to which a measure is free from random error.

Results An approach to performance appraisal that focuses on the attainment of objective and measurable results of the individual or group.

Risk control measures Their proper application reduces risks to an acceptable level or fosters recognition that the department cannot handle the event safely and will revert to a defensive strategy. The major categories are administrative controls, engineering controls, and personal protection controls.

S

Safety culture The elements necessary to establish a climate of consistently safe behaviors, which will result in a culture of safety by the human behavior and attitudes conforming to a respect for increased safety.

Selective recruitment In order to attract more women and minorities, recruiting techniques designed to reach a specific population of applicants.

Sexual harassment Unwelcome sexual advances, requests for sexual favors, and other verbal or physical conduct of a sexual nature when submission to or rejection of this conduct affects employment, unreasonably interferes with work performance, or creates an intimidating, hostile, or offensive work environment.

Special duty Under OSHA regulations the duty an employer has to comply with and enforce all OSHA standards and regulations.

Specific action control Attempts to ensure that employees perform certain actions that are desirable and avoid or not perform actions that are not desirable.

Stereotyping A process in which specific behavioral traits are ascribed to individuals on the basis of their apparent membership in a group.

Subject matter experts Job incumbents that have direct knowledge of the requirements of the jobs that they perform.

Succession planning The process of identifying the long-term needs of an organization in terms of human capability.

System A particular set of procedures designed to control a process in a predictable way.

T

Taxonomy A method for the classification of a system; with regard to training, a system of educational objectives.

Title VII Part of the Civil Rights Act of 1964 that prohibits employment discrimination based on race, color, religion, sex, or national origin.

Total quality management Recommends that the major focus of performance appraisal should be to provide employees with feedback in the areas that they can improve. The focus of TQM is on teams and groups of people achieving results as opposed to on individuals.

Training activities Deals with "how" to accomplish something and is likely to be technical in nature.

U

Undue hardship The significant difficulty or expense in relationship to the difficulty or cost of providing a specific accommodation.

Uniform Guidelines Guidelines developed by the EEOC that identify legally acceptable selection standards. They assist employers in determining the impact of their selection procedures based on the entire selection process.

Union certification An election of the union the NLRB or appropriate state agency conducts by secret ballot of the eligible employees that is then certified to represent the employees of that bargaining unit.

V

Validity The extent and accuracy to which the analysis measures what it was designed to measure. The validity of a job analysis is of critical importance to its use and justification of the results of the analysis.

Values Serve as criteria to judge the actions of individuals and organizations.

W

Weingarten rights The rights that union employees have to demand that a union representative be present at the interview, if the employee reasonably believes that the meeting may result in disciplinary action.

Workers' compensation A system that establishes an insurance program that compensates employees in the case of work-related injuries or their survivors in the case of their death. Benefits are provided regardless of fault or financial condition of the employer.

Workforce analysis An analysis conducted to determine where the department may be under represented in regard to gender and ethnic group representation as compared to the larger community.

Writ of certiorari Petition for review to the U.S. Supreme Court by an appellate court. If the U.S. Supreme Court refuses to review, then the decision of the appellate court becomes law in the circuit where it was decided.

INDEX